Methodological Problems in Reliability Study of Large Energy Systems (RSES 2017)

E3S Web of Conferences Volume 25 (2017)

Bishkek, Kyrgyzstan
11 - 15 September 2017

Editors:

Nicolai Voropai
Sergey Senderov
Alexander Michalevich

ISBN: 978-1-5108-5346-1

Printed from e-media with permission by:

Curran Associates, Inc.
57 Morehouse Lane
Red Hook, NY 12571

proceedings.com

Some format issues inherent in the e-media version may also appear in this print version.

Printed by Curran Associates, Inc. (2018)

For additional information, please contact EDP Sciences – Web of Conferences at the address below.

EDP Sciences – Web of Conferences
17, Avenue du Hoggar
Parc d'Activité de Courtabœuf
BP 112
F-91944 Les Ulis Cedex A
France

Phone: +33 (0) 1 69 18 75 75
Fax: +33 (0) 1 69 28 84 91

contact-edps@webofconferences.org

Additional copies of this publication are available from:

Curran Associates, Inc.
57 Morehouse Lane
Red Hook, NY 12571 USA
Phone: 845-758-0400
Fax: 845-758-2633
Email: curran@proceedings.com
Web: www.proceedings.com

TABLE OF CONTENTS

Statement of Peer review

In submitting conference proceedings to *Web of Conferences*, the editors of the proceedings certify to the Publisher that

1. They adhere to its **Policy on Publishing Integrity** in order to safeguard good scientific practice in publishing.
2. All articles have been subjected to peer review administered by the proceedings editors.
3. Reviews have been conducted by expert referees, who have been requested to provide unbiased and constructive comments aimed, whenever possible, at improving the work.
4. Proceedings editors have taken all reasonable steps to ensure the quality of the materials they publish and their decision to accept or reject a paper for publication has been based only on the merits of the work and the relevance to the journal.

Title, date and place of the conference

Methodological Problems in Reliability Study of Large Energy Systems (RSES 2017)
11-15 September 2017
Bishkek, Kyrgyzstan

Proceedings editor(s):

Prof. Nicolai Voropai, Melentiev Energy Systems Institute, Russia.
Prof. Sergey Senderov, Melentiev Energy Systems Institute, Russia.
Prof. Alexander Michalevich, Energy Institute of Belarus NAS, Belarus.
Prof. Husein Guliev, Azerbaijan Research and Design and Exploration Institute of Power Engineering, Azerbaijan.

Date and editor's signature

15.09.2017

17, avenue du Hoggar - PA de Courtabœuf – BP 112 - 91944 Les Ulis Cedex A (France)
Tél. : 33 (0)1 69 18 75 75 - Fax : 33(0)1 69 07 45 17 – www.edpsciences.org

 DOI: 10.1051/e3sconf/20172501001

Monitoring of energy security of the Republic of Belarus

Alexander Mikhalevich[1*], *Ivan* Filiutsich[1]

[1]Institute of Power Engineering, Laboratory of Energy Security, 15/2 Akademicheskaya Str., 220072, Minsk, Belarus

Abstract. Energy security is considered as one of the key elements of the national energy policy of the Republic of Belarus. Diversification of national energy mix, development of nuclear and renewable energy, improving energy efficiency are the most important energy goals in the country. To cope with energy security threats a system of quantitative indicators was adopted. Such an approach allows to implement effective monitoring of energy security state, helps energy sector decision makers to adopt right measures to eliminate potential risks as well as provides comprehensive tool for long-term energy sector planning.

1 Introduction

The key aspects of energy security of the Republic of Belarus were laid in several strategic documents adopted on the national level. These documents determine the main priorities for country's energy in short-term, middle-term and long-term perspective.

The Concept of Energy Security (here and below referred as the Concept) was approved by Decree of the Council of Ministers of the Republic of Belarus #1084 on 23.12.2015 and was considered the main document determining current and perspective state of energy security in the country as well as the main approaches for providing its reliable energy supply up to 2035 [1].
According to this Decree the National Academy of Sciences of Belarus was appointed the main national institution for implementation of monitoring of energy security. The Academy is responsible for annual reporting of state of energy security to the Council of Ministers.

2 Estimation of main energy security indicators

The method of indicative analysis was applied for quantitative assessment of energy security in Belarus [2].
According to this method assessment of state of the country's energy security is made on the basis of three possible values of indicators: normal, subcritical and critical.

The normal state of energy security implies normal values of almost all indicators.
Subcritical state of energy security is characterized by considerable threats to country's energy supply, its reliability, energy independence, energy efficiency investment activity. As a result, these threats require some correcting actions to prevent further worsening of energy security state.

According to the Concept energy security system of Belarus includes 11 indicators as well as its threshold and perspective values till 2035. It allows to implement an objective monitoring and control of energy security state.

Currently Belarus faces several considerable threats to sustainable and reliable energy supply. The main of them are:

- low geographical and structural diversification of energy import;
- possible restrictions of energy import from the dominant supplier because of economic and political reasons;
- development of energy transit routs by neighboring countries alternative to Belarus;
- high share of one energy source (natural gas) consumption for heat and electricity production;
- low per capita specific electricity consumption in Belarus in comparison to developed countries with similar climate conditions;
- expected increased imbalance of electric capacities in night hours after commissioning of Belarusian nuclear power plant.

Most of threats mentioned have a long-term character. As a result, Belarus needs a systematic approach to achieve main goals indicated in the Concept.
In 2016 both positive and negative tendencies of energy security in Belarus were indicated.
The main positive factors were :

- growth of share of domestic energy sources in total primary energy supply to 14.9% compared to 14.0% in 2015;
- domestic production of oil and natural gas in Belarus remained stable;
- in 2016 production of peat and wood fuel achieved 2.6 Mtce and increased to 0.3 Mtce compared to 2015;
- in 2016 Belarus continued to be an important transit country. Volume of oil supplied over the territory

* Corresponding author: amikhal@bas-net.by

of Belarus towards Adamovo hub (Poland) was 39.0 mio. t and to Brody hub (Ukraine) – 13.9 mio. t.

Table 1. Energy Balance of Belarus in 2015-2016

Indicator	Unit	2015	2016
Total primary energy supply	Mtce	36,5	35,6
Total primary energy production :	Mtce	5,1	5,3
oil	Mtce	2,4	2,3
natural gas	Mtce	0,4	0,4
peat	Mtce	0,3	0,5
biomass	Mtce	2,0	2,1
renewable electrycity	Mtce	0,02	0,02

- in 2016 average specific energy consumption of Belenergo power plants for electricity production was 230.4 g ce/kWh compared to 235.5 g ce/kWh in 2015;
- Belarus continued to increase its renewable generation capacity. It achieved 164.1 MW, or 1.8% of total installed capacity (as for 01.07.2016) compared to 141.1 MW in 2015;
- in 2016 decline of average price for imported natural gas took place in Belarus. For example, costs of imported natural gas for power generation sector reduced by 16.1% to 176.04 USD/thousand m. cub. compared to 2015 level. Average price of natural gas for gas distribution sector went down by 5% to 149.5 USD/thousand m. cub. compared to previous year.

In 2016 the main challenges for energy sector of Belarus were decline of electricity and heat consumption by national companies, increase of debts of energy consumers as well as investment cuts.

For example :

- total electricity consumption in Belarus was 36.1 billion kWh compared to 36.7 billion kWh in previous year. Decline of electricity demand in the country is considered a serious threat to national energy security taking into account start of operation of the nuclear power plant scheduled for 2020 and its expected annual generation of 18 billion kWh. While increase of electricity export is considered one of possible options to optimize load curve after commissioning of nuclear power plant only 160 mio. kWh were exported in 2016;
- decline of oil refineries' output took place in 2016 in Belarus. Two national refineries processed 18.6 tons of oil that was only 80.9% of 2015 level ;
- a considerable increase of electricity supply disruptions was fixed in 2016 in comparison with previous year (30416 and 21401 disruptions respectively). Most of electricity supply disruptions took place in grid 0,4-10 KV due to extreme climatic conditions;
- total debt for electricity purchased by consumers increased from 426.8 mio. BYN in 2015 to 764 mio BYN in 2016. Debt for heat supplied increased to 103.2 mio BYN compared to 64.7 mio. BYN in 2015.

Besides, there are some important issues in bilateral relations between Russia and Belarus which affect negatively energy security, including annually agreed volumes of oil supply to Belarusian refineries and tariff setting principles for natural gas market in both countries.

In 2016 the following tendencies of energy security were identified in Belarus:

- 4 indicators (domestic primary energy production to total primary energy consumption ratio; share of a dominant supplier in total energy import; energy sector capital expenditures to initial fixed assets value ratio; share of a dominant fuel in total energy consumption for electricity and heat production) had critical values;
- 4 indicators (total installed electrical capacity to maximal load ratio; energy sector accumulated depreciation to initial fixed assets value ratio; average number of daily electricity supply disruptions to total locality number ratio; energy import value to gross domestic product ratio) had normal values;
- 3 indicators (renewable energy production to total primary energy consumption ratio; share of a dominant fuel in total energy consumption; energy intensity of gross domestic product) had subcritical values.

In 2016 both indicators, characterizing energy independence of the country were improved. For example, total primary energy production to total energy consumption ratio amounted to 14.9% compared to 14% in 2015. Renewable energy production to total primary energy consumption ratio in 2016 achieved a subcritical value of 6%.

Decline of global prices for hydrocarbons was another factor which affected positively energy security of Belarus in 2016. Energy import value to gross domestic product ratio declined to 15.4% that corresponds to normal level of this indicator.

Diversification of national energy mix continued to be a challenge for the country in 2016. Share of a dominant fuel in total energy consumption amounted to 61.4% and increased by 1.4% in comparison with 2015. As a result this indicator had subcritical value. Belarus is still having a critical dependence on Russia in terms of energy imports: more than 99% of total primary energy were imported from that country in 2016 that implied critical value of the energy security indicator "Share of dominant energy supplies in total energy import".

Share of a dominant fuel (natural gas) in total energy consumption for electricity and heat production for national power company Belenergo was also extremly high and reached 96.7% in 2016. That means unsatisfactory deversification of energy supply by the national power sector and its high vulnerability to potential fuel supply disruptions.

Decline of electricity demand in the country due to recession of the national economy resulted in achieving value of energy security indicator "Total installed electrical capacity to maximal load ratio" of 175%.

Normal value was attributed to the indicator characterizing depreciation of fixed assets of energy sector of Belarus (amounted to 44% in 2016). However, investment activity of national power utility was below the critical threshold in 2016.

 DOI: 10.1051/e3sconf/20172501001

Table 2. Energy Security Indicators of Belarus

Indicator	Threshold Level		Value	
	Normal	Critical	2016	2035
Energy Independence				
Domestic primary energy production to total primary energy consumption ratio, %	30	16	14,9	20
Renewable energy production to total primary energy consumption ratio, %	14	5	6	9
Diversification of Energy Suppliers and Types of Energy				
Share of a dominant supplier in total energy import, %	65	85	>99	70
Share of a dominant fuel in total energy consumption, %	50	70	61,4	50
Reliability of Energy Supply, Reserve Capability and Distribution of Energy				
Total installed electrical capacity to maximal load ratio, %	140	95	175	145
Energy sector accumulated depreciation to initial fixed assets value ratio, %	45	75	44	<45
Energy sector capital expenditures to initial fixed assets value ratio, %	6	4	3,7	6,5
Share of a dominant fuel in total energy consumption for electricity and heat production, %	50	80	96,7	< 50
Average number of daily electricity supply disruptions to total locality number ratio, %	0,5	2	<0,5	0,4
Energy Efficiency and Economic Sustainability of Energy Sector				
Energy intensity of gross domestic product, kg ce/ MBYN (2005 prices)	160	485	378	268
Energy import value to gross domestic product ratio, %	15	30	15,4	15

To cope with these energy security challenges the following measures should be adopted in Belarus in short-term and mid-term perspective:

- to assure implementation of investment projects indicated in the State Program of Energy Saving for 2016-2020, the State Program of Energy Sector Development for 2016-202, the State Program of Comfortable Housing for 2016-2020 and other strategic and local documents and plans;
- strict implementation of construction schedule of the Belarusian nuclear power plant;
- to support activities aimed to increase electricity consumption in the country after commissioning of the nuclear power plant;
- to intensify bilateral negotiations with neighboring countries in terms of electricity export from Belarus;
- to support transition to a single market of oil, natural gas and electricity in the Eurasian Economic Union;
- to cooperate with Poland, Ukraine, Lithuania, Latvia in terms of diversification of oil and natural gas imports routs;
- to intensify dialog with the European Union in the field of energy cooperation to balance interests of Belarus as a transit country and Russia and EU – as a supplier and a buyer of energy resources respectively;
- to develop a national roadmap for Smart Grid technologies in Belarus taking into account increased disruptions of electricity supply;
- to assure implementation of obligatory expertise of large-scale investment projects in terms of their correspondence to the world's best practices in energy efficiency;
- to introduce a practice of development of regional sustainable energy plans for the period up to 2025-2030 which will include indicative targets of energy consumption, energy efficiency improvement, energy intensity and greenhouse emissions reduction;
- to introduce a practice of annual energy security monitoring not only on the national, but also on regional level.

3 Conclusion

Implementation of these measures will allow to achieve forecasted goals of the Concept as well as to reduce risks for energy security of Belarus and to promote sustainable development of the country.

References

1. *The concept of energy security of the Republic of Belarus: Resolution of the Council of Ministers of the Republic of Belarus, December 23, 2015, No. 1084.* http://minenergo.gov.by.
2. A. Mikhalevich. *Energy Bulletin*, **№ 20**, p. 61-72, (2015).

DOI: 10.1051/e3sconf/20172501002

Issues of an integral assessment of the state of Russia's energy security at the federal level

Sergey Senderov[1*], *Viktor* Rabchuk[1], and *Sergey* Vorobev[1]

[1]Melentiev Energy Systems Institute, 130 Lermontov str., Irkutsk, Russia

Abstract. The paper is devoted to the issue of an integral assessment of the state of Russia's energy security at the federal level. A methodology for the formation of such an assessment is proposed. Qualitative assessments of the state of energy security of the main objects of indicative analysis were formed on the basis of assessments of the qualitative state of the most important indicators of energy security. The paper presents a method of convolution of qualitative assessments of energy security for all objects of indicative analysis. As an example, integral qualitative assessments of Russia's energy security for 2015 and 2016 are formulated. The analysis of formulated qualitative assessments makes it possible to understand the main problems of ensuring Russia's energy security in the modern period.

1 Introduction

The main provisions of the Doctrine of Energy Security of the Russian Federation [1] contain various aspects of energy security (ES). These aspects should be assessed annually at the federal level. In [2, 3], approaches to the estimation of threshold and current values of the most important indicators of ES characterizing these aspects were described.

The problem of an integral assessment of the energy security of the country as a whole remains open to this day. What is this assessment and what are the directions and dynamics of this change from year to year?

Based on earlier studies [2, 3] and statements [1], 14 objects of indicative analysis (OIA) were identified. These objects include 47 indicators. How can we assess the ES of the country as a whole? The values of some indicators deteriorate over time in terms of ES, the values of others improve. How does the situation with the country's ES change in this case? To answer these questions, it is necessary to develop a methodical approach to the formation of an integrated assessment of the country's ES with any number of analysed indicators.

Current literature includes some works on energy security research. The most suitable ones are listed below. In Russia, this issue is most fully covered in [1-5]. Some researchers in the world have dealt with the identification of the most significant indicators. For example, this issue is covered in [6-9]. In [6], on the example of India, the authors describe the relationship of energy security with business, social processes and with the management of society. In [7] a look at the main problems of ensuring China's energy security based on the use of some major indicators. First of all, in this work, there are problems of importing energy resources to China. In [8] questions are raised about the problems of ensuring the energy security of Europe from the standpoint of the use of renewable energy resources. The paper analyzes the degree of participation in this process of the Desertec project (construction of the big solar power system in the Sahara). In [9] a general structure for the description and analysis of the energy security in an energy system is presented. The paper is devoted to the methodology of energy security research in the context of one system. Unlike energy independent Russia, it is important for most countries to monitor not only the structure of their energy balance. Particularly important is the share of the largest energy supplier in their total volume of energy resources import.

2 Methodological approach for the qualitative assessment of the OIA state from the standpoint of energy security

2.1 Qualitative assessment of the state of energy security indicators

The current values of the indicators are compared with their thresholds, justified in advance. For indicators of the federal level, this was done in [2].

The process of assigning the value of an indicator to a certain state can be represented as follows:

$$A_{ij} = f_{ij}(S_{ij}) = \begin{cases} N, S_{ij} < S_{ij}^{PC} \\ PC, S_{ij}^{PC} \le S_{ij} < S_{ij}^{C} \quad i = 1,n; j = 1,k \\ C, S_{ij} \ge S_{ij}^{C} \end{cases} \qquad (1)$$

where A_{ij} - qualitative assessment of the state of the j-th indicator of the i-th OIA; n - number of OIA; k - number of indicators in each OIA; S_{ij} - the actual value of the j-th

* Corresponding author: ssm@isem.irk.ru

 DOI: 10.1051/e3sconf/20172501002

indicator of the i-th OIA; S_{ij}^{PC}, S_{ij}^{C} - threshold values of the j-th indicator of the i-th OIA (pre-crisis and crisis, respectively); N, PC, C - possible state of energy in the aspect displayed by the j-th indicator of the i-th OIA (normal, pre-crisis and crisis, respectively).

2.2 Qualitative assessment of the OIA state from the standpoint of energy security

The next step should be to assess the qualitative state of each of the 14 OIA specified in the Energy Security Doctrine, [1]. All objects are characterized by corresponding indicators of the federal level in the number from 1 to 7. The process of forming such a qualitative assessment is presented below:

$$Q_i = F_i(A_{i1},...A_{ik}), i = 1, n \tag{2}$$

where Q_i - qualitative assessment of the state of the i-th OIA; n - the number of OIA; $A_{i1},...A_{ik}$ - qualitative assessments of the status of k indicators in the i-th OIA.

The rules F_i on which Q_i is determined can be different for each i-th OIA. The simplest way is to analyze the significance of each particular indicator in the summary significance of all indicators for the analyzed object. In other words, it is possible to determine the "specific weight" of a particular indicator in the sum of the weights of all indicators of the analyzed object. Further, we summarize all the specific weights of indicators whose values are in the same qualitative state. The sums of specific weights, grouped by qualitative conditions, will allow to form a qualitative assessment of the state of the whole object of analysis. At the first stage, considering the approximate equivalence of indicators within a single object, it can be assumed that they all have the same specific weight in the total sum of the weights. This assumption follows from the consideration of specific indicators for the relevant objects of analysis [2, 3].

After determining the specific weights of all indicators in the overall system of their significance (while we agreed that within the framework of one OIA they all have the same weight), it is possible to form a qualitative assessment of the state of each analysis object from the standpoint of energy security. This can be done as follows:

$$Q_i = \begin{cases} N, \sum_{j=1}^{k} V_{ij}^{N} \geq \delta_N \\ PC, \sum_{j=1}^{k} V_{ij}^{C} < \delta_C \; and \sum_{j=1}^{k} V_{ij}^{N} < \delta_N \\ C, \sum_{j=1}^{k} V_{ij}^{C} \geq \delta_C \end{cases} \tag{3}$$

where Q_i - qualitative assessment of the state of the i-th OIA; k - number of traced indicators in the i-th OIA; V_{ij}^{N}, V_{ij}^{C} - the specific weight of the j-th indicator in the i-th OIA, located in the area of normal and crisis values, respectively; δ_N, δ_C - values that characterize the level of achievement of a normal or crisis state, respectively.

Since we are talking about specific weights, it is clear that the sum of all the specific weights of indicators for one object is equal to one. In studies of the regional level of indicative analysis [4], the value of reaching the normal state δ_N was determined to be 0.7. A smaller value may mean that more than 30% of the sum of the specific weights of the indicators is in the zone of crisis values. The value δ_C was taken equal to 0.4 and means that the state of energy security can be characterized as a crisis if 40% or more of the sum of the specific weights of the indicators is in a state of crisis. All other cases characterized the pre-crisis state. When analyzing the federal level of ES it is proposed to use the same rules.

3 Formation of an integral qualitative assessment of the state of energy security for the country as a whole

3.1 Convolution of qualitative assessments of the state of energy security across all analysis objects

Carrying out the above procedures will allow to determine the qualitative assessment of the state of each OIA from the ES positions. Next, we will carry out the procedure for convolution of the qualitative states of ES of all OIA in order to obtain an integral qualitative assessment of the state of the country's ES. In this case, we are dealing with different objects of analysis (Table 1) in terms of the significance of the impact on the country's ES. To compare the significance of these objects with each other, we can use, as in [4], a method of pairwise comparison of their significance relative to each other. This can be done expertly using the method of interpolating the opinions of a group of experts in the field of ES. As a result of this comparison, it is possible to obtain a certain specific gravity of a specific analysis object in the total sum of "weights" of all the objects under study. This specific weight will show the significance of this object for ensuring the country's EE in the overall system of significance of the analyzed objects. We can determine this specific weight as follows:

$$V_i = \sum_{j=1}^{n} \upsilon_{ij} / \sum_{i=1}^{n}\sum_{j=1}^{n} \upsilon_{ij}, \tag{4}$$

where V_i - specific weight of the i-th object of analysis in the system of evaluated objects; n - number of OIA; υ_{ij} - the conditional significance of the i-th object of analysis in comparison with the j-th object.

The conditional significance υ_{ij} is the average value of the opinions of the experts on the relative importance of the i-th object of analysis before the j-th. The matrix of conditional comparative significance of objects of the analysis of ES is presented in Fig. 1.

To fill this matrix, the knowledge of experts in the field of ES is used. In determining the optimal number of experts, one can use the classical theory of tests, which shows the dependence of the increase in the reliability of the aggregate judgments of the subjects on the number of subjects [5-7].

	1	2	3	...	n
1	1	υ_{12}	υ_{13}	...	υ_{1n}
2	υ_{21}	1	υ_{23}	...	υ_{2n}
3	υ_{31}	υ_{32}	1	...	υ_{3n}
...	...	...	...	1	...
n	υ_{n1}	υ_{n2}	υ_{n3}	...	1

Fig. 1. The matrix of conditional comparative significance of OIA of ES

These studies show that the median value of the accuracy of the estimates obtained by the three experts is an order of magnitude and twice that of those obtained by one or two experts, then increases insignificantly, reaching a maximum in the range of five to nine experts. If the "importance weights" of experts are equal, then their optimal number can be within 3-5 people.

3.2 Integral qualitative assessment of the ES state at the country level

The steps above allow us to make a qualitative assessment of the state of the EE at the country level. The process of obtaining such an assessment is presented below:

$$G = \begin{cases} N, \sum_{i=1}^{n} V_i^N \geq \gamma_N \\ PC, \sum_{i=1}^{n} V_i^C < \gamma_C \, u \sum_{i=1}^{n} V_i^N < \gamma_N \\ C, \sum_{i=1}^{n} V_i^C \geq \gamma_C \end{cases} \quad (5)$$

where G - a qualitative assessment of the state of the country's ES; N - number of analyzed analysis objects; V_i^H, V_i^K - the specific weight of the *i*-th object of analysis, located in the area of normal and crisis values, respectively; γ_N, γ_C - coefficients that characterize the level of achievement of a normal or a crisis state, respectively

Values γ_H and γ_K can be adopted according to the same logic as for the values δ_H, δ_K in the description of (3).

Further, it is possible to form a qualitative assessment of the state of energy security of the country as a whole. At the same time, it is important to understand that this qualitative assessment is formed on the basis of the numerical values of the most important indicators of the federal level.

Any qualitative assessment must necessarily be accompanied by supporting materials containing information on the state of individual objects of analysis. This information allows us to understand which aspects of the provision of EB should first of all be paid attention.At the same time, each OIA is characterized by specific indicators of EB. An analysis of the current values of these indicators allows us to determine the directions in which it is necessary to initiate activities to improve the situation with the provision of ES.

4 An example of practical research

4.1 Assessment of the qualitative state of energy security for the objects of indicative analysis

We will apply the above methodological approach to assess the real situation with the provision of Russia's ES at the federal level. Information on the OIAs and specific indicators that characterize these objects is presented in the list below:

Object 1. The ratio of the annual increase in the balance reserves of primary fuel-energy resources (FER) to the volumes of their extraction.

Indicators 1.1-1.4 "Ratio of the increase in recoverable balance reserves of the type of primary FER for the accounting year to the total volume of production of this type of FER in the country for the same year" for oil, gas, coal and uranium, respectively (for uranium for 5 years).

Object 2. The share of natural gas in the structure of the balance of primary FER.

Indicator 2.1 "Share of natural gas in the structure of the balance of primary FER"

Object 3. Dynamics of the level of prices inside the country for the main types of FER.

Indicators 3.1-3.6 "The annual index of changes in prices (tariffs) in the country's average for the type of FER for the accounting year in relation to the previous year" for gas, coal, fuel oil, light oil products (in general), electricity, heat, respectively.

Object 4. The amount of non-payments in the retail and wholesale electricity markets.

Indicator 4.1 "Relation of the amount of non-payments to sales organizations in the country as a whole from retail electricity consumers for electricity supplied to them (consumers) in the accounting year to the total selling price of all electricity supplied to retail customers in the same year, attributed to the same indicator for previous year".

Indicator 4.2. The same for electricity generating companies from grid companies.

Indicator 4.3. The same for network companies from sales organizations.

Object 5. Implementation of investment programs by the fuel and energy complex (FEC).

Indicators 5.1-5.5 "The ratio of the actual annual volumes of investments in the FEC for the forthcoming

DOI: 10.1051/e3sconf/20172501002

analysis of the three-year period to the volume of investments planned by the investment programs of this sector over the same period" for production and transport of oil and gas condensate; Oil refining; Gas industry; Coal industry; Electric power industry, respectively.

Object 6. The relative change in the specific energy intensity of GDP.

Indicator 6.1 "Relative change in the specific energy intensity of Russia's GDP in the calculated year as compared to the previous year.

Object 7. The share of fossil primary FER used in the domestic market in the total volume of their extraction.

Indicators 7.1-7.3 "Ratio of the share of the fossil primary FER type used in the domestic market in the accounting year in the total volume of its production in the country in the same year to the same indicator in the previous year" by types of FER: oil (oil refining); gas; coal respectively.

Object 8. Terrorist acts on fuel and energy complex objects, main factors, measures taken.

Indicator 8.1. "The ratio of the number of officially recorded acts of terrorism at the FEC facilities in the calculated year to the number of such acts in the previous year".

Indicator 8.2 "The ratio of the total amount of economic damage from terrorist acts at FEC facilities in the calculated year to the same value in the previous year"

Object 9. Stability of providing consumers with various types of FER (including in the regional context).

Indicators 9.1-9.5 "Relative total shortage of FER type for the country as a whole in the calculated year" for gas, coal, fuel oil, light oil products in general, electricity, respectively.

Indicators 9.6-9.8 "The level of potential supply of demand for the type of FER in the territory of the federal district in the conditions of sharp cooling" for boiler-furnace fuel in general; Electricity; Thermal energy, respectively

Object 10. Strikes at FEC facilities, main factors.

Indicator 10.1 "The ratio of the number of strikes at FEC facilities in the whole country in the calculated year to the same indicator in the previous year".

Indicator 10.2 "The ratio of total damage across the country from strikes at FEC facilities in the calculated year to the same indicator in the previous year".

Object 11. Physical depreciation of fixed production assets by branches of the FEC and related industries.

Indicators 11.1-11.7 "Ratio of physical depreciation of fixed production assets in the FEC in the accounting year to the same indicator 5 years ago" for the FEC as a whole; Gas industry; Extraction and transport of oil; Oil refining; Coal industry; Electric power industry; Power machine building, respectively.

Object 12. Emergency situations at the FEC facilities.

Indicator 12.1 "Relation of the total number of emergencies in the FEC in the calculated year to the same value in the previous year".

Indicator 12.2 "The ratio of total economic damage from emergencies to the FEC as a whole in the calculated year to the same value in the previous year".

Object 13. The share of exports of Russian FER in total exports, including the share of the Asia-Pacific region, in total exports.

Indicator 13.1 "Ratio of the share of Russian export of FER in the calculated year in the total volume of the country's exports to the same amount in the previous year".

Indicator 13.2 "Share of the Asia-Pacific region in total exports of Russian FER in the calculated year".

Object 14. Share of oil refining and petrochemical products in total exports of Russian FER.

Indicator 14.1 "Ratio of the share of oil refining and petrochemical products in the total export of FER in the calculated year to the same indicator five years ago".

As mentioned above, we assume that the weights of all indicators in each OIA are equal. According to this, the specific weights of the indicators were calculated in Table 1. Also in Table 1, Qualitative assessments of the state of specific indicators are presented. They were formulated in accordance with (1).The process of forming these estimates was presented in [3]. In this paper, due to save space, this process is not given, but Table 1 shows the results for 2015-2016. [2, 3].

All indicators of the federal level in Table 1 are distributed according to the corresponding analysis objects. A qualitative assessment of the ES state for each object is formed according to (3). The data in Table 1 allow one to view the criticality of the state of an object of analysis from the ES positions. The evaluation of this state is performed according to (2) and (3). It can be seen from the table 2 that in 2015-2016 four OIAs (Nos. 5, 6, 11, 13 (list of OIAs)) were in the crisis zone from the ES positions, and two objects (Nos. 2 and 4) were in the pre-crisis zone. At the same time, the situation at OIA no. 2 "The share of natural gas in the structure of the balance of primary FER" improved in 2016 from crisis to pre-crisis, thanks to the reduction of this share below the threshold of 55%. Eight objects (No. 1, 3, 7, 8, 9, 10, 12 and 14) were in an acceptable state from the ES standpoint. At the same time, the situation at OIA no. 7 "The share of fossil primary FER used in the domestic market in the total volume of their extraction" has radically changed to normal by 2016. This was due to a relative decrease in gas use and increased use of coal. This led to an increase in the balance of the species structure of domestic consumption of boiler-furnace fuel.

4.2 Integral qualitative assessment of the state of Russia's energy security at the federal level

The next step is the formation of an integral qualitative assessment of the state of Russia's energy security at the federal level. It is possible to implement it after the formation of the specific weights of each OIA in the common system of weights of all objects. This is done according to (4) and as shown in Fig. 1 using the principles of pairwise comparison of the conditional significance of specific OIA. This process takes into account the independent opinions of four experts in the field of energy security research. The objects of analysis in pairs were correlated, so that the sum of their relative

DOI: 10.1051/e3sconf/20172501002

Table 1. The results of the qualitative assessment of the ES state for OIA.

OIA, year of analysis		The qualitative state of the ES indicators								ES state			Assessment
										C	PC	N	
										Transition boundary			
		The indicator number in OIA								**0,4**		**0,7**	
		1	2	3	4	5	6	7	8	Sum of specific weights			
1	Specific weight	0,25	0,25	0,25	0,25								
	2015	N	N	N	N					0	0	1	**N**
	2016	N	N	N	N					0	0	1	**N**
2	Specific weight	1											
	2015	C								1	0	0	**C**
	2016	PC								0	1	0	**PC**
3	Specific weight	0,17	0,17	0,17	0,17	0,17	0,17						
	2015	N	N	N	N	N	N			0	0	1	**N**
	2016	N	N	N	N	N	N			0	0	1	**N**
4	Specific weight	0,33	0,33	0,33									
	2015	PC	C	PC						0,33	0,67	0	**PC**
	2016	PC	C	PC						0,33	0,67	0	**PC**
5	Specific weight	0,2	0,2	0,2	0,2	0,2							
	2015	PC	N	C	C	C				0,6	0,2	0,2	**C**
	2016	C	N	C	C	C				0,8	0	0,2	**C**
6	Specific weight	1											
	2015	C								1			**C**
	2016	C								1			**C**
7	Specific weight	0,33	0,33	0,33									
	2015	PC	C	C						0,67	0,33	0	**C**
	2016	PC	N	N						0	0,33	0,67	**N**
8	Specific weight	0,5	0,5										
	2015	N	N							0	0	1	**N**
	2016	N	N							0	0	1	**N**
9	Specific weight	0,13	0,13	0,13	0,13	0,13	0,13	0,13	0,13				
	2015	N	N	N	N	N	N	PC	N	0	0,13	0,87	**N**
	2016	N	N	N	N	N	N	PC	N	0	0,13	0,87	**N**
10	Specific weight	0,5	0,5										
	2015	N	N							0	0	1	**N**
	2016	N	N							0	0	1	**N**
11	Specific weight	0,14	0,14	0,14	0,14	0,14	0,14	0,14					
	2015	C	C	C	PC	PC	PC	PC		0,43	0,57	0	**C**
	2016	C	C	C	PC	PC	PC	PC		0,43	0,57	0	**C**
12	Specific weight	0,5	0,5										
	2015	N	N							0	0	1	**N**
	2016	N	N							0	0	1	**N**
13	Specific weight	0,5	0,5										
	2015	C	PC							0,5	0,5	0	**C**
	2016	C	N							0,5		0,5	**C**
14	Specific weight	1											
	2015	N								0	0	1	**N**
	2016	N								0	0	1	**N**

weights in the pair was 1 (for example: 0.7 and 0.3 or 0.2 and 0.8, etc.). From the sum of the experts' opinions on the conditional significance of one OIA to another, we obtained averaged arithmetic values of comparative estimates in all possible pairs of objects [3]. This was done by interpolation. On this basis, the specific weights of each OIA in the system of evaluated objects were calculated. The result of this work is presented in the corresponding column of Table 2. Comparative analysis of OIA specific weights showed that objects nos. 11, 10, 5 that ranked 1, 2, and 3 places, respectively, can be recognized as the most significant for evaluating of energy security. Knowing the specific weights and qualitative assessments of the ES state for each OIA (Table 1), according to (5), we can to form an integral qualitative assessment of Russia's ES at the federal level. The process of obtaining this assessment is also presented in the Table 2.

The data in Table 2 show that the state of Russia's EB at the federal level improved qualitatively, moving from the crisis situation in 2015 to the pre-crisis in 2016.

This was facilitated by the following processes that were mentioned above in the comments to Table 1:

Table 2. Formation of an integral assessment of the Russia's ES in 2015, 2016.

OIA	Specific weight	Qualitative assessment					
		2015			2016		
1	0,073	N			N		
2	0,078	C			PC		
3	0,066	N			N		
4	0,070	PC			PC		
5	0,085	C			C		
6	0,078	C			C		
7	0,065	C			N		
8	0,064	N			N		
9	0,047	N			N		
10	0,088	N			N		
11	0,091	C			C		
12	0,079	N			N		
13	0,058	C			C		
14	0,059	N			N		
Integral assessment							
ES state		C	PC	N	C	PC	N
Transition boundary		0,4	-	0,7	0,4	-	0,7
Sum of specific weights		0,455	0,070	0,475	0,312	0,148	0,540
Integral assessment of Russia's ES		**C**			**PC**		

- transition from a crisis to a pre-crisis state of the OIA "The share of natural gas in the structure of the balance of primary FER" due to a decrease in this share below the crisis level of 55%;
- transition from the crisis to the pre-crisis state of the OIA "The share of fossil primary FER used in the domestic market in the total volume of their extraction". This was due to the relative reduction in the use of gas and increased use of coal, which contributed to a better structure of the domestic consumption of energy fuel.

5 Conclusion

The paper shows the principles of conducting a comprehensive indicative analysis of the state of Russia's energy security at the federal level. A methodical approach to the formation of such an integral qualitative assessment is presented. An example of the formation of this estimate is based on information on the current status of the indicators of energy security at the federal level. It is shown that there are positive tendencies forming a qualitative improvement in the state of ES from 2015 to 2016. Periodical carrying out of such an assessment is necessary. The analysis of the dynamic range of qualitative assessments of the state of OIAs will allow us to comprehensively assess the directions and scale of the implementation and development of trends in the state of the country's ES in general.

Obtaining such an assessment is not an end in itself, it must necessarily be accompanied by assessments of the situation for OIAs with access to the current values of the most important indicators of the federal level.

The application of such an integrated approach will make it possible to more reasonably formulate the necessary areas of activity and specific measures to increase the level of ES in the conditions of realization of threats to Russia's ES.

References

1. Bushuev V.V., Voropai N.I., Senderov S.M., Saenko V.V., *Economika regiona*[Economy of the region], **2**, 40-50 (2012).
2. Senderov S.M., Rabchuk V.I., *Izvestia RAN. Energetika*[Proc. of the Russ. Acad. of Sciences. Pow. Engin.], **5**, 3-15 (2015).
3. Senderov S.M., Rabchuk V.I., Pyatkova N.I., Vorobev S.V. *Ensuring Russia's energy security: choosing priorities* (Nauka Publ., Novosibirsk, 2017).
4. Pyatkova N.I., Rabchuk V.I., Senderov S.M., Slavin G.B., Cheltsov M.B. *Energy Security in Russia: Problems and ways of decision* (SO RAN Publ., Novosibirsk, 2011).
5. Senderov S.M., Pyatkova N.I., Rabchuk V.I., *2008 IEEE PES General Meeting, Pittsburgh, PA, USA*, 30-35 (2008).
6. Bambawale M.J., Sovacool B.K., *Energy Policy*, **39 (3)**, 1254-1264 (2011).
7. Leung G.C.K., *Energy Policy*. **39 (3)**, 1330-1337 (2011).
8. Lilliestam J., Ellenbeck S., *Energy Policy*, **39 (6)**, 3380-3391 (2011).
9. Hughes L., *Energy Policy*. **42 (C)**, 221-231 (2012).
10. Petrov E.G., Novozilova M.V., Grebennik I.V., Sokolova N.A. *Methods and means of decision-making in socio-economic and technical systems* (OLD-plus Publ., Herson, 2003).
11. Libby R., Blashfield R., *Organizational Behavior and Human Performance*, **21**, 121-129 (1978).
12. Einhorn R., Hogarth R., Klempner E., *Psychological Bulletin*, **84**, 158-172 (1977).

Information support of intelligent energy systems and its impact on ensuring adequacy

Yuri Chukreyev[1*], *Mikhail* Chukreyev[1]

[1]Komi Scientific Centre of the Ural Branch of the Russian Academy of Sciences, ISE and EPN, 167000, Russia

Abstract. The issues of the influence of the changed conditions for obtaining probably-determined information on the operation of generating equipment and power consumption regimes on the decisions taken to ensure the required level of reservation of the territorial zones of the UES of Russia in the current regulatory environment to the indicators of balance reliability are considered. Comparison of the present conditions for presenting the initial information for the estimation of the balance sheet reliability of electric power systems with the conditions for the development of methodological recommendations (MR) for the design of power systems in 2003 and not approved by the Ministry of Energy of Russia MR 2012.

1 Introduction

The globalization of the electric power industry has led to the creation of unified energy systems (hereinafter referred to as "the EEC"). They connect the individual parameters of the regime of individual countries or their groups. An example is the Unified Energy System of Russia (hereinafter - UES of Russia), the unification of the power systems of the European Union, the unification of the power systems of North America. The reliability of the power balance of large EPS should be analyzed both in the whole system and in the context of its separate geographically allocated zones. In our country, within the framework of the Unified Energy System of Russia, there are zones in the form of integrated energy systems (UPS) with their fragmentation into separate territories. At the same time, the reliability of ensuring the power balance of each zone can be provided in different ways. In the European Union, the reliability of the power balance is ensured within each national energy system. In North America, 8 zones are identified (the habitable regions of the USA, Canada and Mexico). In them, the reliability of the power balance is ensured by both own sources and the generation of adjacent zones, but according to the previously agreed power supply rules. The feature of the UES of Russia is to ensure reliability both in North America, but without restrictions on the supply of power. By the mid-70s of the last century, a fairly strict hierarchical system was built on the basis of methods implemented in programs that make it possible to manage the development of the UES of the country for a perspective of 5 to 20 years. The task of ensuring balance reliability in this system has been given quite a lot of attention. This is necessary to justify the normative values of capacity reserves.

Reforming the industry in our country, started in the mid-1990s, led to a decrease in interest in issues of ensuring balance reliability when considering issues related to the long-term planning of the UES of Russia. This was facilitated by a number of factors: a sharp decrease in electricity consumption (by one third), a change in ownership patterns, deterioration in the quality of management personnel in the electricity sector, the closure of industry institutions and much more.

In the Federal Law "On Electric Power Industry" [1] the necessity of solving this problem was partially restored. The issues of justifying the means of ensuring reliability were legislatively assigned to the leading organization in the electric power industry of JSC "SO UES". Starting from 2009, in accordance with the Regulations approved by the RF Government Resolution[*1], JSC "SO UES" together with JSC "FGC UES" annually performs the work "Scheme and program for the development of the UES of Russia for a seven-year period" (hereinafter - SaPD UES of Russia) . We can say that from that time a new stage in the task of justifying the means of ensuring reliability under the management of the development of the UES of Russia began.

The importance of the problem of justifying the capacity reserves at the current stage of the country's electric power industry development is characterized by the assignment of Deputy Prime Minister A. Dvorkovich to the Ministry of Energy of Russia and the Ministry of Economic Development of Russia[*2] on the improvement of the long-term planning system in the electric power industry, UES of Russia. Below, the issues of

[*1] Rules for the development and approval of schemes and programs for the future development of the electric power industry, approved by Government Decree No. 823 of October 17

[*2] The original 8928 order of the government decree of 26.11.2016 for 5 liters.

[*] Corresponding author: chukreev@iespn.komisc.ru

modernization of software complexes are considered in order to take into account the changed information content of the PBN assessment task and their impact on the decisions taken to justify the means of ensuring the reliability of the UES of Russia.

2 Software product justification reserve capacity and upgrade to modern conditions

The solution of the problem of the justification of normative power reserves and especially its compensation (operational) component required in the development of future balance of power. Their normalized values are justified in the preparation of methodical recommendations for the design of energy systems (MR) [2]. The rationale of capacity reserves is impossible without the development of software assessment indicators of adequacy. While developed in the post-Soviet period software is quite acceptable for modern conditions in terms of their methodological content. Currently, these include "Yantar" (ISEM SB RAS) [3] and "Orion-M" (ISE and EPN, Komi SC of the Ural branch of the RAS) [4]. The latter is more adapted to the changes. And more importantly he is more focused on the solution of the problem of the justification of power reserves. This software was used to design an MR 2003 and 2012, approved the Ministry of energy.

The software complex "Orion-M" is an upgrade of «Orion» [5]. It was widely used in the branches of the Institute "Energosetproekt" in the 80s of the last century to analyze the balance reliability of prospective options for the development of EPS for 10-20 years. And in the beginning of 90-ies was used to monitor the reliability of the UES of the former USSR for the accounting and for the next three years (JSC "NIIPT", now "STC UES"). In the period of reforming of the industry and the adoption of a Government resolution from 2009 to 2013 in "Orion-M" brought great changes [4]. They are to some extent influenced decisions on the justification of means of redundancy in the UES of Russia.

At the present stage the most appropriate indicators of adequacy from the standpoint of justification of the means of reliability, are:

– the probability of deficit-free operation of territorial zones of the EPS ($\rho = 1 - J_л$), where $J_л$ – the integral probability of power shortage;

– the average number of days of power shortage (loss of the anticipated load – *LOLE* [6]);

– the average number of hours per year of power shortages (loss of load hours – *LOLH*).

It should be noted that in the "Orion-M" provides for a wide range of indicators of adequacy, including all of the above.

3 Upgrading software "Orion-M"

The intellectualization of the power system in the aspect of obtaining reasonably reliable retrospective information about the generating and network equipment, and modes of consumption opens new opportunities for the methodological support of the objectives of adequacy. Statistical processing of information on the operation of generating equipment for a sufficiently long period of time allowed to identify certain trends in the change in the norms of their emergency planned repairs [4]. They are associated with the presence of huge untapped excess capacity [7]. Statistical processing of retrospective information on power consumption modes allowed to reveal the presence of correlation dependencies of random deviations from average hourly loads between individual territorial zones of the UES of Russia [8].

The above-mentioned factors have a significant impact on indicators of adequacy. Consequently, on their means of support and should be considered in the applied program complexes. All versions of "Orion" [4,5] and other available [3], the formation of a deterministic random States at each discrete load change is carried out by statistical modeling methods (SM) (Fig.1).

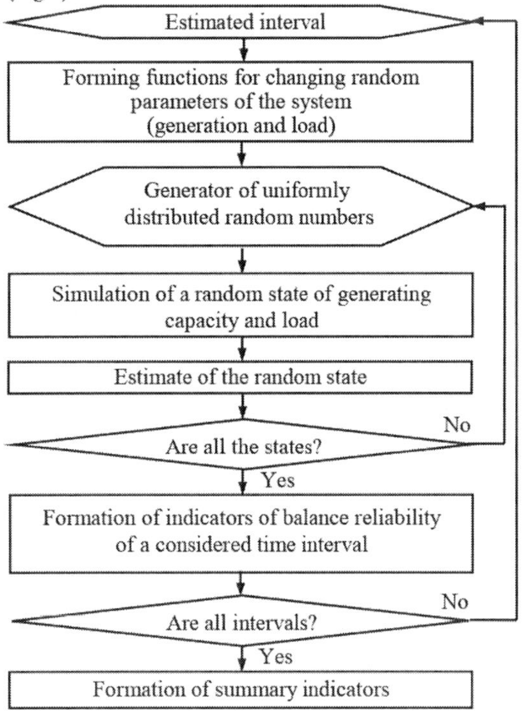

Fig. 1. Enlarged flowchart for estimating the balance reliability indicators of loading schedule intervals

Hitherto it was believed that random changes in the load in each territorial area variables are independent, both from the generation and loads of other areas. In this case, to increase the effectiveness of SM in a software was used, the composition of the distribution laws of random variables. In Fig. 2 shows the density ranges of probability decline in generating capacity ($f(P_г)$) and the random deviations of load ($f(P_н)$), and the composition of their distributions ($f(P_c)$). While SM is performed on the function tracks the distribution of generation and load ($F(P_c)$). Definition, subject only to a random value generation, the load remains constant and equal to the average degree of the graph ($\boldsymbol{P}_н$, Fig. 2).

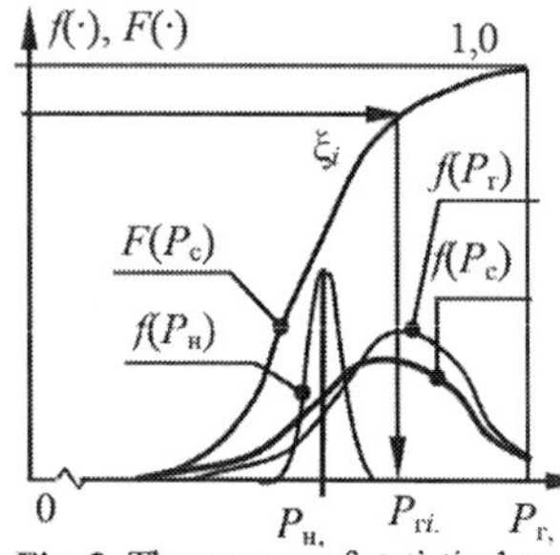

Fig. 2. The process of statistical modeling in "Orion-M"

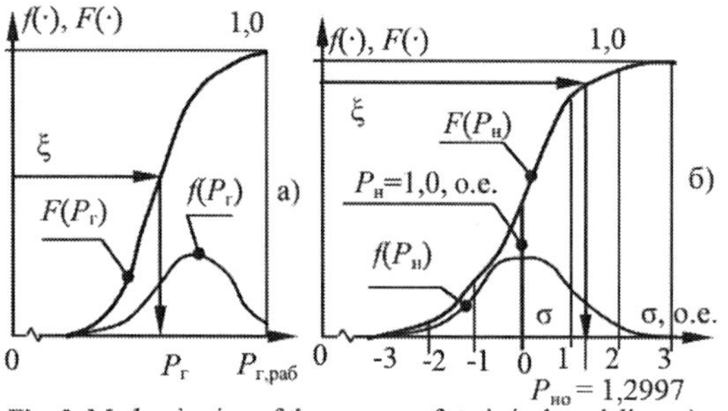

Fig. 3. Modernization of the process of statistical modeling: a) on generation, б) on the load

Account correlations of random fluctuations of the load at different territorial areas requires the implementation of the modeling process separately for the generating capacity and load (Fig. 3). Organization of the modeling process for existing generating capacity differs from the previous case. Instead of the function $F(P_c)$ is applied ($F(P_г)$) (Fig. 3, a). To account for correlation of random fluctuations of load in different zones it is necessary to form the functions of their changes (under hypothesis of normal probability distribution law). The function is built in relative units, normalized to the average

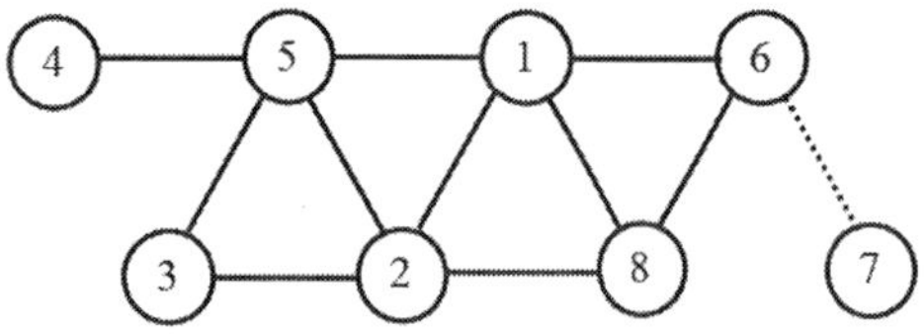

Fig. 4. Model of the settlement scheme of the UES of Russia. UPS: 1 - Ural; 2 - Average Volga; 3 - South; 4 - Northwest; 5 - Center; 6 - Siberia; 7 - The Far East; 8 - Kazakhstan.

quadratic deviation (the standard deviation σ, Fig. 3,б), Analysis of retrospective information, made in the [8] for the model design scheme of UES of Russia, represented in Fig. 4, allowed to form a correlation matrix $\boldsymbol{R}$ (1).

$$\boldsymbol{R} = \left\| \begin{matrix} 1 & 0{,}71 & 0{,}54 & 0{,}50 & 0{,}58 & 0{,}68 \\ 0{,}71 & 1 & 0{,}72 & 0{,}49 & 0{,}78 & 0{,}50 \\ 0{,}54 & 0{,}72 & 1 & 0{,}23 & 0{,}53 & 0{,}30 \\ 0{,}50 & 0{,}49 & 0{,}23 & 1 & 0{,}83 & 0{,}56 \\ 0{,}58 & 0{,}78 & 0{,}53 & 0{,}83 & 1 & 0{,}47 \\ 0{,}68 & 0{,}49 & 0{,}30 & 0{,}56 & 0{,}47 & 1 \end{matrix} \right\| \quad (1)$$

In the modeling of random fluctuations of load by statistical modeling methods used decomposition of this matrix into factors by the method of Cholesky [9]. Then, the random vector of deviations of the power consumption from the average value is determined by the expression-statement $\Delta P = \sigma \times Lr \times P_{н0}$ ($P_{н0}$ – simulated power load in p. u., in Fig. 3 equal 1,2997). At constant correlation matrix R matrix Lr takes the form presented in (2).

$$\boldsymbol{Lr} = \left\| \begin{matrix} 1 & 0 & 0 & 0 & 0 & 0 \\ 0{,}708 & 0{,}706 & 0 & 0 & 0 & 0 \\ 0{,}543 & 0{,}468 & 0{,}698 & 0 & 0 & 0 \\ 0{,}503 & 0{,}193 & -0{,}188 & 0{,}821 & 0 & 0 \\ 0{,}579 & 0{,}527 & -0{,}042 & 0{,}516 & 0{,}345 & 0 \\ 0{,}679 & 0{,}019 & -0{,}108 & 0{,}235 & -0{,}186 & 0{,}66 \end{matrix} \right\| \quad (2)$$

Let the methods of statistical simulation of the generated load vector with the values of power in relative terms: $P_{н0}$ = {1,3204; 0,500; -0,766; -1,394; 1,2997; -1,3998}. A new vector $P_{н\sigma} = Lr \times P_{н0}$ given (2) takes the form $P_{н\sigma}$ = {1,3204; 1,288; 0,4165; -0,24; 0,2102; -0,5059}. The transition to named units is performed according to the expression $P_{H_j} = \bar{P}_{H_j} + P_{H\sigma} Ч \sigma$ ($\bar{P}_{H_j}$ - the average load of *j*-th zone in MW, σ – the standard deviation load *j*-th territorial area in MW).

4 Analysis of the rationale of reserve revisions MR 2003 and 2012[*3] years

The authors involved in the calculations for the justification of compensatory reserve capacity in these MR. This was the basis to the study of the peculiarities of the methodical principles of assessment indicators of adequacy [7] is applied in the calculations. The results obtained for "Orion-M", made in 2011-12, agree rather well with the results obtained in accordance with the MR 2003 and Another was to be expected, since information components of generating equipment accidents and random deviations of load from its mean values show no significant change. Confirmation of what was said characterizes the table. 1. It shows the value of the full reserve capacity, obtained in accordance with the MR 2003 and "Orion-M" in 2011-2012, For obvious

[*3] MR on the design of power systems / JSC "Institute" Energosetproekt ", 2012 (approved by NP" NTS UES "in July 2012, but not approved by the Ministry of Energy of Russia).

temporal reasons, the results given for the balance of power already held in 2012 (SaPD 2011-2017).

Table 1. The total reserve capacity for 2012

	Irregular load maximum, MW	Total reserve capacity in accordance with MP 2003, MW / %	Estimated values of the total reserve capacity (MW /%) when preparing the 2012 MR		
			When optimizing the bandwidth of links [4]	With the existing communication bandwidth [4]	New MR 2012 (not approved by the Ministry of Energy of the Russian Federation)
UES of Russia without East	147 785	23738/ 16,06	24529/ 16,6	22974/ 15,55	30140/ 20,4
UES of Russia (European part)	116 945	19881/ 17	20241/ 17,31	18980/ 16,23	23340/ 19,93
Ural	350 05	6362/ 18,17	6944/ 19,84	6263/ 17,89	7010/ 20
Av. Volga	173 16	2187/ 12,63	2213/ 12,78	15,07/ 8,7	2860/ 16,5
South	140 19	1988/ 14,18	2222/ 15,85	2049/ 14,62	2750/ 19,5
Northwest	141 51	2982/ 21,07	2815/ 19,89	2933/ 20,73	2700/ 19
Centre	364 54	6362/ 17,45	6046/ 16,59	6229/ 17,09	8020/ 22
Siberia	308 40	3857/ 12,51	4291/ 13,91	3994/ 12,94	6800/ 22

In lines 3 and 4 of the table. 1 shows the value of the full reserve capacity, obtained by [10]. In the calculations of the norms for carrying out planned repairs of generating equipment and strategic reserve capacity have not changed significantly compared with the MR 2003 were considered two scenarios for the capacity of the links. One without optimization (line 4), the second with optimization for equal unit costs in the development of one kW of generating capacity and reserve capacity of the links (line 3).

In the 2012 edition of the MR magnitude of repair and the strategic power reserves adjusted by the Customers of the work of JSC "so UES" and JSC "Institute "Energosetproekt" (line 5 of the table. 1) [10]. Maintenance reserve main causes are the transition to a conclusion in repair of the condition as well as opened due to a reduction of power consumption, the possibility of medium and capital repairs in the winter time, including the month of December. These circumstances when processing the retrospective information has led to a significant increase in the proportion of fixation of reserve capacity. For information in the former Soviet Union reserves for capital repairs of virtually all of United energy systems was zero. Rules for the strategic reserve in all of United energy systems was increased from 1-2 % to 3 % [10].

5 The impact of information support of modern smart grids on the justification of power reserve

In this section the results justify the compensation of reserve capacity in relation to the forecasted 2022 in accordance with SaPD UES of Russia in 2016 to 2022. Since the development is not approved by the Ministry of energy MR 2012 has been quite a dramatic change in the content objectives of the evaluation indicators, the carrying of reliability of the UES of Russia. This requires an estimation of the impact of these changes on decision-making on justification of compensation (operational) power reserve. Under content we understand statistical information about the generating equipment (unplanned and planned conclusions in repair) and probabilistic specific information about the deviations of load from its time of change for the average daily schedule of the month of December [8].

Table 2. Average statistical indicators of the location of generating equipment in emergency and scheduled repairs as a percentage of the irregular maximum load

№		UES of Russia without East	Name UPS					
			Ural	Av. Volga	South	Northwest	Centre	Siberia
1	2022 MR 2012	4,42 5,99	5,88 6,67	2,84 5,36	3,96 5,26	4,62 7,22	3,03 6,91	5,48 4,38
2	2022 MR 2012	– –	4,09 2,59	3,27 3,82	4,68 3,85	4,82 4,06	3,89 2,52	4,07 2,88
3	2022 MR 2012 MR 2003	8,87 6,80 4,04	10,8 5,30 4,91	9,72 7,50 2,65	4,80 3,90 3,00	7,40 4,40 4,91	6,70 8,30 6,24	11,8 9,00 3,09
4	Irregular load maximum, MW	155860	37390	17096	16831	15151	39266	30126

Their larger characteristics as a percentage of the irregular maximum load is given in table. 2. Comparison subject information about the random parameters of the system, recent and relevant information used in the preparation of MR 2012 and 2003, as reflected in the literature. For these sections the content of tasks in the rows of the table. 2 shows:

– first – mathematical expectation (m.e.) output in unscheduled emergency repairs of the generating equipment;
– second – random characteristics the medium quadratic deviation load percentage;
– the third – value of capacity reserves for planned repairs;
– in the fourth – magnitude irregular highs loads 2022 in accordance with SaPD UES of Russia in 2016-22.

The table shows clearly significant changes in the random parameters of the generating equipment operation and the deviations in the power consumption regime that have occurred since the development of the MR 2012, as a result of the statistical processing of retrospective information in [8] (lines 2 and 3). A significant increase in the share of generating equipment that is in scheduled repair (line 4), as well as practically unused due to the presence of excess capacity, makes it necessary to conduct studies on their accounting when simulating emergency outputs of equipment for repairs. In Table. 3 presents the results of the influence of the noted changes in the information content of modern intellectualized EPS on the amount of the compensation power reserve.

Table 3. The values of the compensation capacity reserve for 2022 (MW /%)

	Options	UES of Russia without East	Number and name UPS					
			1. Ural	2. Av. Volga	3. South	4. Northwest	5. Centre	6. Siberia
Calculated on the "Orion-M" (information on the accidental characteristics of the output for the repair of generating equipment in accordance with the adopted in the 2012 MR)								
1.	The standard deviation is as in the MR 2012 and 2003	14053 9,02	3584 9,59	628 3,67	1437 8,54	1879 12,40	4465 11,37	2060 6,84
2.	Standard deviation based on historical information	15228 9,77	4034 10,79	628 3,67	1562 9,28	2004 13,23	4540 11,56	2460 8,17
Calculated for "Orion-M" (information on the random characteristics of the operation of the generating equipment and the standard deviation of the load in accordance with the processing of historical data)								
3.	The s.d. of independent interconnected UPS	11920 7,65	3834 10,25	78 0,46	1194 7,09	1504 9,93	2855 7,27	2455 8,15
4.	The s.d. is correlated in accordance with (1) and (2)	15610 10,02	4894 13,09	528 3,09	1544 9,17	1629 10,75	3990 10,16	3025 10,04
5.	The s.d. - correlated with independent repairs	14100 9,05	4411 11,80	420 2,46	1441 8,56	15,27 10,08	3506 8,93	2794 9,28

The first two positions of the table. 3 show the performance effect of energy consumption on the means of ensuring the reliability of the assumption of independence of random fluctuations of load in the form of standard deviations. The information about the accident generating equipment for cleanliness of experiment adopted in accordance with the requirements of MR 2012 (reference books). It was expected that in accordance with row 3 of the table. 2, the effect of random fluctuations of the load of the means of ensuring the reliability should be more important. However, in practice it is seen from table 3 this is not entirely confirmed.

A more significant discrepancy in the results justify the means of ensuring the reliability observed on the two following items in the table (3 and 4). Them for the same information about the random generation of hardware (from retrospective information) considered two approaches to the modeling of random fluctuations of load. First, as in the formation of the MR 2003 and 2012 (row 3). The second (line 4), taking into account the dependence of the random deviations between the UPS (see section 2). It can be seen that taking into account the correlation leads to a significant increase in the share of the compensation capacity reserve for the UES of Russia as a whole from 7.65 to 10.02%.

Note that the compensatory reserve capacity for UES of Russia as a whole (line 3), using data on the state of generating equipment based on the processing of retrospective decreased by 2.22 % compared to its value when using the same information from the reference data (line 2). This is consistent with the first row of the table. 2, where the mathematical expectation of output in unscheduled repairs of generating equipment for the UES as a whole differ for these calculations, more than half of the item (4,42 and 5.99 %). More significant decrease of compensatory reserve capacity (2.22 %) is easily explained by the reduction not only of the expectation of unscheduled repairs of the equipment, but also the dispersion of the distribution density $f(P_r)$ (Fig. 2 and 3). For UES of Russia these changes is shown in Fig. 5. For each Union has consistently presented the mathematical expectation of unplanned insights generating equipment in repair and the necessary amount of the compensation allowance when using retrospective information. And these same settings, but using data on the generating equipment in accordance with reference books, as in MR 2012. In all member States except the Middle Volga, there is a strong correlation – the expectation is always less compensatory reserve. In the Middle Volga, the picture is somewhat different and this can be explained by the presence of this significant energy system reserves bandwidth connections with a sufficiently large capacity of the UPS Centre and the Urals (Fig. 4).

Apart in the study presented results (line 5), obtained taking into account the actual reduction in generating capacity by the magnitude of planned repairs of equipment and strategic reserve (terminal equipment). This required major modifications to "Orion-M". Modified the principles of simulation random variable generation and loads, and methods of forming the distribution function $F(P_r)$ (Fig. 2 and 3). This has affected the composition of equipment is really involved in the coating loading (without components in planned repairs and the needs of the strategic reserve). The magnitude of compensatory reserve capacity for UES of Russia as a whole fell by almost one point (from 10.02 to 9.05 %). In separate energy systems of UES of Russia the decrease is also observed that fluctuation from 0.61 % (South) to 1.29 % (Urals).

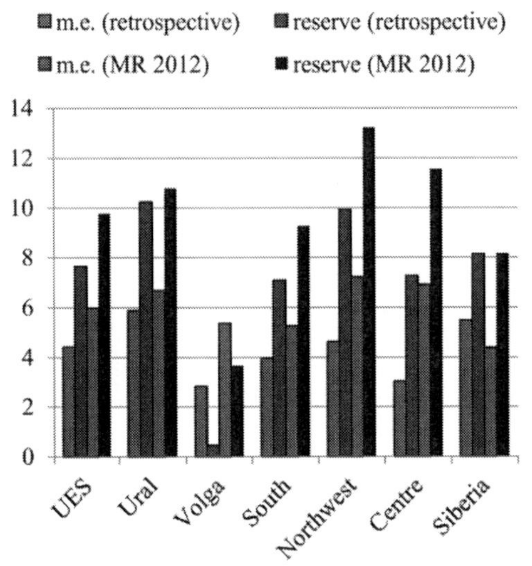

Fig. 5. Interrelation of changes in the random parameters of the power systems of the unit and calculated parameters of the compensation power reserve

In the last two rows (4 and 5) I would like to mention another important fact. Accounting the influence of random deviations of load between separate power systems of the UES of Russia dramatically (by 5-7 times) reduces the probability of over-bandwidth links in comparison with the assumption of independence of these deviations (row 3). This is further evidence of the unreliability of calculations on a substantiation of reserves bandwidth connections. They must first be determined by ensuring static and dynamic stability of power systems and transfer of large amounts of power due to the presence of different timing of the passage of the maximum load a separate grid UES of Russia.

6 Conclusion

Intellectualization UES of Russia, basing on analysis of retrospective information to identify the dependence of the random deviations of load between separate power systems within UES of Russia. Generate new performance standards for unplanned and planned repairs of generating equipment. This has led to the need to modernize the software and carry out comparative calculations on a different informational basis. Studies have shown the versatility of existing software ("Orion-M"), the possibility of its adaptation to changing conditions, as a methodological and information aspect.

References

1. Federal Law "On Electric Power Industry" of March 26, 2003 No. 35-03.

2. Methodical recommendations on the design of power systems. Approved by the order of the Ministry of Energy of Russia of June 30, 2003, No. 281 (Ministry of Energy of Russia, 2003)

3. G.F. Kovalev, L.M. Lebedeva *Reliability of power systems* (Novosibirsk: Science, 2015)

4. Yu.Ya. Chukreev, M.Yu. Chukreev *Models for estimating the balance reliability indicators for managing the development of electric power systems* (Syktyvkar, Komi Scientific Center of the Ural Branch of the Russian Academy of Sciences, 2014)

5. Yu.Ya. Chukreev *Models for ensuring the reliability of electric power systems* (Syktyvkar, Komi scientist. Center of UrO RAN, 1995)

6. R. Billinton , R.N. Allan *Reliability Evaluation of Power Systems. Second Edition.* (New York and London, Plenum Press, 1996)

7. Yu.Ya. Chukreev, M.Yu. Chukreev *Energy Policy.* **Issue 2.** - P. 39-50 (2017).

8. Report on R & D "Substantiation of the normative values of the components of the total reserve capacity in the context of the UPS and UES of Russia as a whole in the planning of their development" (Syktyvkar, Komi Scientific Center of the Ural Branch of the Russian Academy of Sciences, 2016)

9. V.M. Verzhbitsky *Fundamentals of numerical methods* (Moscow, High School, 2009)

10. Report on research "Correction of methodological recommendations for the design of the development of power systems" (Moscow, JSC "Institute" ENERGOSETPROEKT", 2011)

DOI: 10.1051/e3sconf/20172501004

Search of critically important objects of the gas industry with the method of determining critical elements in networks of technical infrastructures

Sergey Vorobev[1*], *Alexey* Edelev[1] , *Elena* Smirnova[1]

[1]Melentiev Energy Systems Institute, 130 Lermontov str., Irkutsk, Russia

Abstract. The paper shows the application of the method of determining critical elements in the networks of technical infrastructures in the search and analysis of the importance of critical gas industry facilities. The purpose of this method is to search for critical elements and sets of elements in network infrastructures and their ranking by importance. The paper presents an adaptation of the method for determining critical elements in networks of technical infrastructures for the model of gas supply system functioning, which consists in describing the response of the gas supply system to the failure of a single element or a group of them. Conclusions are made about the expediency of using this method in research into the search for critical gas industry objects and determining their significance.

1 Introduction

One of the main aspects of ensuring energy security is to create the conditions for the greatest possible degree of satisfaction of consumers with energy resources in emergency situations. Investigation of this aspect of energy security requires the definition of critical objects for the fuel and energy complex in general and for individual energy systems. It means the definition of those facilities, partial or complete failure of which can cause significant damage to the country from the fuel and energy sector.

The need to define and analyse the critical objects in the fuel and energy sector is also dictated by the fact that the functioning of the fuel and energy complex and its constituent cells is the most important factor in ensuring the normal vital activity of the economy of the whole country and its individual regions.

Abroad, there are many studies devoted to the study of the vulnerability of certain critical infrastructures in the event of an accidental disturbance, a deliberate attack or a natural disaster. In the case of insufficient knowledge of the infrastructure, probabilistic methods of risk analysis are used to assess its vulnerability [1, 2]. If historical data are stored, then the theory of statistics is used to analyze and predict the impact of natural disasters on the performance of infrastructure [3]. To take into account the topology of the infrastructure, when analyzing its structural vulnerability, network approaches are used, for example, the theory of complex networks [4]. Recently, in foreign studies, more attention has been paid to the consideration of interconnected infrastructures [5] and the impact of interaction between them on their vulnerability [6, 7].

In this paper we will talk about the gas industry. The matter is that today more than 90% of Russian natural gas is extracted in one gas producing region (Nadym-Pur-Tazovsky district of the Tyumen region). This region is located in 2-2.5 thousand km from the main gas consumption areas and in 4-5 thousand km from the countries that import this gas. Thus, all Russian gas is transported over long distances through the system of main gas pipelines, which have a large number of crossings and bridges, moreover, the threads of powerful gas mains are often laid at a small distance from each other. At present, in the gas transportation system of Russia, it is possible to note more than 20 crosses of main gas pipelines that potentially dangerous for system. Previously, a number of studies devoted to the identification of critical objects in the gas industry were carried out [8]. The list of crossings of main gas pipelines in the Unified Gas Supply System of Russia has been determined, the violation of which will lead to a relative deficit of gas supplies for the system as a whole of 5% or more.

Also, conducted studies have shown that along with main gas pipelines crossings, a significant shortage of gas supplies through the system (5% or more) may result in a disruption of the functioning of both a separate section of the main gas pipelines located between the node compressor stations and the disruption of several such sites. The determination of such sections of main gas pipelines and their combinations is complicated by the fact that in the current configuration of the Unified Gas Supply System of Russia there are about 500 edges. Therefore, at this stage of the study of critical objects of the gas industry, the authors proposed the use of the method of determining critical elements in the networks

* Corresponding author: seregavorobev@isem.irk.ru

of technical infrastructures [9]. Its application is justified by the need to assess the importance of the gas sector's critical objects.

2 Definition and ranking of critical elements

The definition of critical elements, as a rule, is a simple task when considering only single failures. When considering several simultaneous failures, this task can become much more complicated.

It is especially difficult to determine the critical groups of elements with the synergistic effect. In this context, the synergistic effect means that the negative consequences of the group's failure are wholly higher than the total impact of individual failures of the elements within the group. In other words, the failure of a group of two elements with serious negative consequences can give a synergistic effect if the failure of each of the elements in itself is not the cause of any significant consequences.

The method for determining critical elements in technical infrastructure networks [9] facilitates the definition and ranking of such groups of elements (as well as groups of elements whose failure does not give a synergistic effect). The found critical elements or sets of elements can then be studied in more detail using probabilistic methods of risk analysis [10].

The essence of the method is to investigate the set of failures, each of which represents a set of faulty elements, has only one negative effect on the system is characterized by a size that specifies the number of elements whose failure occurs simultaneously.

The size of the set of failures n is chosen by the researcher depending on the total number of elements of the system t. However, based on practical considerations, n should not exceed 3 or 4, since the number of possible failure sets, equal to $t!/((t-n)!*n!)$, Grows rapidly with increasing n. The ranking of failure sets is carried out in accordance with the magnitude of their synergistic effects.

If the set of failures F consists of n elements $c_1, \ldots, c_n$, $n>1$, then $F=\{ c_1, \ldots, c_n \}$. The elements of the set of failures can be divided into subsets of S. This partition can be performed in several ways. Let V_i denote the set containing the subsets of S for the particular partition method F and P denote the number of partition methods. The concrete subset belonging to V_i is denoted by S_{ij}. If we denote the number of such subsets of m, then V_i will consist of the series $S_{i1}, \ldots, S_{im}$. Since the subsets are constructed on the basis of the breakdown of S, then all the elements contained in the subsets also belong to the set of failures and each element can be contained in only one subset for each partitioning method.

A lot of failures have a synergistic effect only if the negative consequence of failures $C(F)$ is greater than the sum of the consequences for subsets of the set F for all possible partitions $V_1, \ldots, V_p$.

$$C(F) > \sum_{j=1}^{m} C(S_j^i) \forall V_i : \tag{1}$$

$$F = \{c_1, \ldots, c_n\}, n > 1,$$

$$S_j^i \subset F, S_1^i \cap \ldots \cap S_m^i = \varnothing, S_1^i \cup \ldots \cup S_m^i = F, j = 1, \ldots, m$$

$$V_i = \{S_1^i, \ldots, S_m^i\}, i = 1, \ldots, p.$$

The synergistic effect of the set of failures $C_{syn}(F)$ is defined as the difference between the consequences due to the set of failures and the largest sum of consequences from subsets for all possible partitions of V.

$$C_{syn}(F) = C(F) - \max_{V_i}\left(\sum_{j=1}^{m} C(S_j^i)\right) \tag{2}$$

The synergistic effect cannot be calculated on the basis of the consequences of the individual considered failures, since it manifests only when all elements of the set are simultaneously rejected, that is, it is a consequence additional to the effect of subsets of failures. For example, the synergistic effect of a set of failures of size 3 sets cannot be calculated by summing the consequences of its subsets of size 2 and 1. Thus, such critical sets of failures cannot be determined only by considering combinations of elements that are critical in themselves.

When considering the issue of two simultaneous failures, the criticality of a particular element is considered as a vulnerability of the system to failures of this element and one of the other elements. There are a large number of sets of failures of size 2, which include this element, and every set of failures is associated with a certain consequence. Thus, the vulnerability of a system can be described with a set of failure sets, including a description of the consequences of failure of each set of failures. Vulnerability indicators that help to compare the criticality of different elements can be obtained by analysing a set of failure sets. One such method is the averaging of the consequences of all sets of failures that contain a certain element. The obtained value can be interpreted as the average value of the consequences due to failures of a particular element and another element selected randomly (for failure sets of size 2).

The indicator by which the elements making the main contribution to the synergistic effect of a certain set of failures can be determined in [2] is calculated as follows

$$Con_{size=n}(c_i) = \frac{\sum C_{syn}(F \mid c_i \in F, n)}{\sum C_{syn}(F \mid n)} \tag{3}$$

$\sum C_{syn}(F \mid c_i \in F, n)$ then the sum of the synergistic effects of all sets of failures of size n that contain the element c_i.

$\sum C_{syn}(F \mid n)$ is the sum of the synergistic effects of all sets of failures of size n. The indicator (3) expresses the contribution of synergetic consequences from a particular element to the aggregate synergistic effect of a set of failures of a certain size. Thus, the exponent (3) will be high for such an element c_i, which is contained in a large number of sets of failures with a significant synergistic effect.

3 Determination of critical elements in the energy system

Essential aid in the research of the critical elements in the gas industry can be provided by geoinformation systems (GIS) as a means of visualizing information describing complex spatially-distributed systems, such as energy systems. The basic principles of mapping energy system schemes and searching for weak places were developed in the process of creating and using the problem-oriented GIS "Oil and Gas of Russia" for studying the survivability of individual energy systems, in particular, gas supply systems, oil and oil products systems [11-13]. The scheme of the energy system in the GIS is presented in the form of a map consisting of two layers with vector data. The first layer, represented by point objects, characterizes the condition of the producers, and also shows whether the needs of energy consumers are completely satisfied. The second layer, represented by lines, gives an idea of the degree of congestion of the transport subsystem. Attributive information on the production and consumption of energy resources is associated with the point objects of the first layer, and information on the sections of the main pipelines (throughput, number of pipes, their diameter, etc.) connects with the lines of the second layer.

To assess the importance of the critical elements in the Unified Gas Supply System of Russia using the method of determining critical elements in technical infrastructure networks, it is necessary to adapt the latter to the functioning model of the energy system. This adaptation consists in describing the reaction of the energy system to the failure of a single element or a group of them. The adaptation of the method for assessing the significance of the critical elements in the Unified Gas Supply System of Russia is shown below.

The flow distribution model in the Unified Gas Supply System of Russia in the "Oil and Gas of Russia" is designed to assess the production capabilities of the Unified Gas Supply System of Russia in conditions of various kinds of disturbances. The purpose of such studies is to minimize gas deficits at the consumption sites. The Unified Gas Supply System of Russia in the model is represented as a set of three subsystems: gas sources, main gas transport network and consumers.

When solving the problem of estimating the state of a system after a perturbation, the criterion of the optimality of the distribution of flows is the minimum gas deficit in the consumer with minimum costs for delivering gas to consumers. This problem can be solved by finding the maximum flow over the network. The problem of the maximal flow was investigated in [14] and is formulated as follows:

$$f \to max, \tag{4}$$

with follow conditions

$$\sum_{i \in N_j^+} x_{ij} - \sum_{i \in N_j^-} x_{ji} = \begin{cases} -f, j=O \\ 0, j \neq O, S \\ f, j=S \end{cases} \tag{5}$$

$$0 \le x_{ij} \le d_{ij}, \text{ for all of (i, j).} \tag{6}$$

In this formulation, f is the maximized variable corresponding to the maximum flow.

With complex schemes, which are the Russian and European gas transmission networks, there may be several solutions, that is, several possible maximum flows. Then it is advisable to talk about minimizing the cost of delivering gas to consumers and using the Basaker-Gowen algorithm [14]:

$$\sum_{(i,j)} C_{ij} x_{ij} \to min, \tag{7}$$

$$x \in X^*. \tag{8}$$

Its application allows us to determine the maximum flow of the resource at its minimum cost or the optimal volumes of daily gas from the underground gas storages, which maximally provide the given volumes of gas supply to consumers at the minimum costs for extraction, transportation of gas and its withdrawal from underground gas storages.

O - the node number of the common source; S - is the number of the common drain node; N^+_j – is a subset of edges entering into the node j; N_j – is the subset of outgoing edges from the node j; f – is the value of the total flow over the network; x_{ij} – is the flow along the edge (i, j); d_{ij} – restrictions on the flow along the edge (i, j); X^* - is the set of solutions of problem (4)-(6); C_{ij} – specific costs for gas transportation.

The node O is connected by fictitious edges with all real sources of energy resources, and node S - with all consumers.

Similarly to [1], the total shortage of energy resources to consumers is used to estimate the impact of multiple failures F:

$$C(F) = \sum_{i \in N_S^+} d_{iS} - f \tag{9}$$

N^+_s - is a subset of the edges that "enter" the common sink S,

d_{iS} – gas demand in the country and deliveries for export. The synergistic effect of the failure set F is calculated according to (2).

4 Assessment of importance of critical elements of the gas industry

The scheme of the Unified Gas Supply System of Russia used for the calculations in this paper on the flow model takes into account all the main features of the operation of the Unified Gas Supply System of Russia and contains:

- 382 nodes, including:
- 28 gas sources;
- 64 gas consumers (subjects of the Russian Federation);

- 24 underground gas storage facilities;
- 266 nodal compressor stations;

- 486 edges representing main gas pipelines and outlets to distribution gas networks.

On this scheme of the Unified Gas Supply System of Russia several calculations were made to determine the criticality of individual edges in the gas transmission network:

1 - the operation of each edge of the scheme was interrupted in turn, and the total daily gas deficit was determined by the system (million m3 per day) as a result of such a violation;

2 - the functioning of a system with multiple failures n = 2;

3 - the functioning of a system with multiple failures n = 3.

Baseline data for calculations, such as daily gas production, consumption, export and import volumes, throughput capacity of existing gas pipelines, are taken in accordance with official statistical information [15-17] for 2016.

Table 1. Results of determining the criticality of individual edges in the gas transmission network

Segment of main gas pipeline	Gas deficit, mln. m^3/day.	Total gas deficit, %	The value of the criterion of the object when	
			n=2	n=3
A	412,70	21	0,0739	0,0756
B	412,70	21	0,0752	0,0784
C	412,70	21	0,0763	0,0807
D	295,01	15	0,0573	0,0614
E	295,01	15	0,0573	0,0614
F	175,54	9	0,0352	0,0424
G	174,75	9	0,0351	0,0423
H	160,68	8	0,0327	0,0401
I	156,40	8	0,0354	0,0418
J	156,40	8	0,0369	0,0444
K	152,54	8	0,0313	0,0386
L	112,24	6	0,0218	0,0263
M	92,70	5	0,0237	0,0318
N	82,70	4	0,0255	0,0348
O	79,20	4	0,0303	0,0411
P	78,72	4	0,0302	0,0410
Q	74,71	4	0,0161	0,0211
R	74,71	4	0,0161	0,0211
S	51,60	3	0,0226	0,0325
T	51,60	3	0,0227	0,0324

Table 1 clearly demonstrates that the first 5 segments of main gas pipelines "A" - "E" are critical both in terms of the maximum gas deficit in consumers and in their contribution to the synergistic effect that is possible with simultaneous disruption of the functioning of several segments of main gas pipelines. Violation of the functioning of these segments will lead to the emergence of a significant gas deficit in consumers, from 15 to 21% in total across the gas transmission network. All these factors make it possible to classify the segments of main gas pipelines "A" - "E" to critical objects of gas industry.

Violation of the functioning of each of the following three segments of main gas pipelines "F" - "H" will lead to a significant gas deficit at consumers, up to 9% in total throughout the system, but the contribution to the synergetic effect from them will be lower than the disruption of the functioning of any segment of "I "-" J ". The failure of each of these five segments of main gas pipelines will lead to a significant gas deficit in consumers, 8-9% in total throughout the system. Therefore, segments of main gas pipelines "F" - "J" can also be attributed to critical objects of gas industry.

A similar situation can be traced with segments of main gas pipelines "O" - "R" and "S" - "T", the synergetic effect in case of a violation of any of these segments will be higher than violation of segments "K" - "N" and "Q" - "R" respectively. It should be noted here that segments of main gas pipelines "N" - "T" are not critical objects, since the total gas deficit in the system in case of a violation of their functioning is <5%.

Violation of the functioning of several unconnected segments of main gas pipelines as well as the disruption of the functioning of the intersection of main gas pipelines are more likely to cause more harm to the system than the disruption of the functioning of one segment of main gas pipelines. Taking this fact into account, we can speak about the greater importance of segments of main gas pipelines as critical objects of gas industry with a large value of the criticality index.

5 Conclusion

The article shows the application of the method of determining critical elements in the networks of technical infrastructures when analysing the importance of the gas industry's critical objects. This method involves a systematic assessment of the consequences of failures of infrastructure elements to determine their criticality and is characterized by a sufficiently high computational resource capacity. It was originally used for the analysis of the electric power system, however, after appropriate adaptation, which consists in describing the response of the system to element failure, is suitable for studies of the vulnerability of gas supply systems. A detailed analysis of the results of the assessment of the importance of segments of main gas pipelines on two indicators is presented: the total gas deficit and the element criticality index. This indicator can be used to establish the order of priority of measures to increase the survivability of the Unified Gas Supply System of Russia.

Also, the suitability of the presented method in the studies of the gas sector in the part of determining segments of main gas pipelines with the greatest contribution to the possible synergetic effect in the case of disruption of the functioning of several segments was noted in the article.

In the future, it is possible to apply the method of determining the critical elements in the networks of technical infrastructures to analyse the possible synergistic effect of the gas industry's critical objects.

References

1. E. Zio, *Reliability Engineering & System Safety*, **94(2)**, pp.125-141. (2009).
2. G.E. Apostolakis, *Risk analysis*, **24(3)**, pp.515-520. (2004).
3. H. Liu, R.A. Davidson, T.V. Apanasovich, *Reliability Engineering & System Safety*, **93(6)**, pp.897-912. (2008).
4. L. Cuadra, S. Salcedo-Sanz, J. Del Ser, S. Jimenez-Fernandez, Z.W. Geem, *Energies*, **8(9)**, pp.9211-9265. (2015).
5. M. Ouyang, *Reliability engineering & System safety*, **121**, pp.43-60. (2014).
6. S. Wang, L. Hong, X. Chen, *Physica A: Statistical Mechanics and its applications*, **391(11)**, pp.3323-3335. (2012).
7. J. Johansson, H. Hassel, *Risk and Interdependencies in Critical Infrastructures: A Guideline for Analysis*. London: Springer-Verlag, pp. 49–66. (2012).
8. S.M. Senderov, V.I. Rabchuk, A.V. Edelev, Izv. Ros. Akad. Nauk. *Energetika* – № **1**. - p. 70-78. (2016).
9. H. Jonsson, J. Johansson, H. Johansson, *Proceedings of the Institution of Mechanical Engineers, Part O: Journal of Risk and Reliability*. Vol. 222. **No. 2**. P. 235-243. (2008).
10. J. Johansson, H. Hassel, E. Zio, *Reliability Engineering & System Safety*. **No. 120.** P.27-38. (2013).
11. A.V. Edelev, S.M. Enikeeva, S.M. Senderov, *Computational technologies*. – Vol. 4, **№ 5**. – p. 30 – 35. (1999).
12. N.I. Voropai, S.M. Senderov, A.V. Edelev, *Energy* **42**, pp. 3-9, (2012).
13. S.V. Vorobev, A.V. Edelev, *Software products and systems*. – **№3**. – p. 174 – 177. (2014).
14. L.R. Ford, D.R. Fulkerson. *Flows in networks*. Princeton University Press. 276p. (1966).
15. *Export of the Russian Federation of the most important goods in 2011 - 2016 (according to the Federal Customs Service of Russia)* http://customs.ru/ index.php?option=com_ newsfts&view =category&id= 52&Itemid=1978&limitstart=60.
16. *InfoTEK Monthly oil and gas magazine*. **№1**, – P.154. (2017).
17. *Ministry of Energy of the Russian Federation. Statistics*. http://minenergo.gov.ru/activity/statistic.

Main trends and scale of changes in state of energy security of the regions of Russia

Elena Smirnova[1,*], *Sergey* Senderov[1]

[1]Melentiev Energy Systems Institute, 130 Lermontov str., Irkutsk, Russia

Abstract. The paper is devoted to the analysis of the main trends and the scale of the change in the state of energy security of Russian regions. The analysis of the problems was carried out in accordance with the methodology for assessing the state of energy security of the Russian Federation at the regional level. The regions of Russia with unsatisfactory state of energy security, as well as the dynamics of changes in the state of energy security in all regions of Russia in the period from 2011 to 2016 are determined. The factors and reasons for the formation of negative and positive trends are analysed.

1 Introduction

Energy security (ES) - the state of protection of citizens, society, the state, the economy from the threat of a deficit in providing their energy needs with economically accessible energy resources of acceptable quality, from threats of disrupting the continuity of energy supply [1, 2]. In fact, energy security is the balance of energy supply and demand or the lack of deficit of energy balance.

To assess the state of the ES regions of Russia and, to identify emerging trends in this regard, appropriate monitoring and analysis of the state of domestic energy is needed. Such an assessment should be carried out in accordance with the methodology for monitoring the state of Russia's ES at the regional level, that is, on the basis of monitoring the most important indicators of the functioning of energy at a given point in time.

This paper presents materials that characterize the state of ES in the regions of Russia across all federal districts. Such estimates were obtained with the direct use of the monitoring technique for the state of ES [3], provided with the necessary illustrative material. Individual regions are ranked according to the state of their ES. Studies conducted in different time frames (2011, 2015 and 2016), allow us to assess the direction of the change in the situation with the provision of ES in the regions of Russia and the dynamics of these changes over the past six years.

In all developed countries, the issues and problems of ensuring an acceptable level of energy security are receiving increasing attention. For example, in [1] the situation with ES indicators and the situation in Europe as a whole is reflected. In [5,6] the same information is presented about Turkey and the countries of Asia. In general, the literature on the assessment of energy security focuses on the problems of diversification of energy supply sources, diversification of suppliers of imported energy. The issues of long-term political stability in the regions under consideration and issues of improving the situation with the provision of ES by regions of the world are touched upon. For the purpose of carrying out the corresponding analysis, specialists develop and use various indicators and systems of indicative analysis, adapted for a particular region [2-4]. At the same time in Russia, there is a system of indicators that allows us to assess the situation with energy security in the approach to the realities of modern Russia and its specifics. This takes into account the peculiarities of each region. In general, the system of indicative analysis used provides concrete benchmarks aimed at improving the situation in individual regions and in the country as a whole. This paper is devoted to the description of the regional aspect of this problem.

2 Elements of the methodology for monitoring the state of Russia's ES at the regional level

Currently, the evaluation of the state of the ES regions of the country [3,10] is based on the use of indicators distributed across three interrelated blocks, Table. 1. The values of indicators characterize the conditions for securing ES in the region in aggregate. Separation of individual blocks is necessary to obtain an idea of the most important aspects of securing the ES regions. Each of the presented in Table 1 indicator has its own, expertly generated and justified, threshold values, that is, those values that determine the boundary of the transition of the actual value of the indicator from one area of qualitative states to another.

* Corresponding author: smirnova.e.m@isem.irk.ru

 DOI: 10.1051/e3sconf/20172501005

The methodology uses two such thresholds for each indicator. The pre-crisis threshold value of the indicator means the boundary value between the normal and pre-crisis state of energy in the aspect described by this indicator. The crisis threshold means, respectively, the boundary between the pre-crisis and crisis states.

As a result of comparing the actual values of specific indicators with their thresholds, the level of crisis indicators is estimated. But in this case it is not yet possible to judge the crisis of the state of the region's ES in general. Some indicators can have acceptable from the EB positions, the values of others may be in crisis or pre-crisis conditions. The picture can vary from region to region and from year to year. Accordingly, in order to form a final qualitative assessment of the state of ES in the region, it is necessary to convolve qualitative assessments of the status of individual indicators in a single integrated assessment of the ES of the analysed territory.

Table 1. Composition of the most important indicators of regional energy security.

1. Block of production and resource provision of the region's fuel and energy supply system
1.1. The ratio of the total available capacity of the region's power plants to the maximum electric load of consumers on its territory. 1.2. The ratio of the amount of available capacity of power plants and the throughput of interconnections between the region and neighboring consumers to the maximum electric load on its territory. 1.3. Opportunities to meet the needs of Boiler Heating Oil (BHO) from the region's own sources.
2. The block of reliability of fuel and energy supply of the region
2.1. The share of the dominant resource in the total consumption of BHO in the territory of the region. 2.2. The share of the largest power plant in the installed electric capacity of the region. 2.3. The level of potential supply of demand for fuel in the conditions of a sharp cooling (10% consumption of consumption) in the region.
3. Block of the state of Basic Production Assets (BPA) energy systems in the territory of the region
3. Block of the state of BPA energy systems in the territory of the region 3.1. Degree of depreciation of the BPA in the energy sector of the region. 3.2. The ratio of the average annual input of installed capacity and reconstruction of power plants in the region over the last 5-year period to the established capacity of the region.

The state of a particular indicator, depending on the location of its values on the state scale, is estimated as follows:

$$f(S_i)=\begin{cases} \mathrm{N},\ S_i < S_i^{PC} \\ PC,\ S_i^{PC} \leq S_i < S_i^{C} \quad i=1,n \\ C,\ S_i \geq S_i^{C} \end{cases} \tag{1}$$

where n – is the number of indicators being evaluated; Si - actual (expected) value of the i-th indicator; S_i^{PC}, S_i^{C} - the pre-crisis and crisis threshold values of the i-th indicator; N, PC, C - qualitative assessment of the state of energy in the aspect indicated by the i-th indicator: normal, pre-crisis and crisis, respectively.

Due to the fact that the indicators chosen for the assessment are not of equal importance in their importance, an integral evaluation takes into account the significance of each specific indicator in their common set. Or the "weights" of specific indicators in the overall system of their values. The specific weight of the indicator in the total sum of weights is determined by the formula:

$$V_i = \sum_{j=1}^{n} \upsilon_{ij} / \sum_{i=1}^{n} \sum_{j=1}^{n} \upsilon_{ij}, \tag{2}$$

where V_i - is the specific weight of the i-th indicator in the system of indicators being evaluated; V_{ij} - conditional significance of the i-th indicator in comparison with the j-th indicator.

When carrying out the above-described procedure for actions, a qualitative assessment of the state of individual regions of Russia is formed:

$$Q = \begin{cases} \mathrm{N},\ \sum_{i=1}^{n} V_i^{N} \geq \delta_N \\ PC,\ \sum_{i=1}^{n} V_i^{C} \prec \delta_C \ and \sum_{i=1}^{n} V_i^{N} \prec \delta_N,\quad i=1,n \\ C,\ \sum_{i=1}^{n} V_i^{C} \geq \delta_C \end{cases} \tag{3}$$

where Q - an integral assessment of the qualitative state of the region's energy security; V_i^N, V_i^C - the specific weight of the i-th indicator, located in the area of normal and crisis values, respectively; δ_N, δ_C - coefficients that characterize the level of achievement of a normal or a crisis state, respectively. Carrying out such an assessment allows on the basis of monitoring the dynamic range of the integral indicator values, to understand in which direction the level of ES in a particular region is changing and which region should, first of all, pay attention to the level of the federal district or country.

3 Analysis of the state of the ES regions of Russia

In order to demonstrate the efficiency of the methodological approach discussed above, to highlight and assess some trends in the changing of the most important factors for ensuring the safety of subjects located on the territory of the Russian Federation, the following was done. The subjects of the Russian

DOI: 10.1051/e3sconf/20172501005

Federation for analysis are located in the territories of all federal districts. According to these districts, an information base has been created, based on statistical information [11-13]. In the dynamics from 2011 to 2016. Values of all indicators listed above are estimated. At the same time, the values of the indicators were quantitatively correlated with their threshold values, and all regions were allocated to the respective groups of territories. With the help of this, it became possible to assess the qualitative state of each indicator in the relevant subject of analysis. As a result of the general analysis of data on 78 subjects of the Russian Federation, it is possible to characterize briefly the energy inherent in the territories of the federal districts, from the standpoint of energy security requirements.

Based on the results of the analysis conducted in the dynamics from 2011 to 2016. The share of regions where the situation from the ES's point of view has not fundamentally changed is 82%. Information on the state of these regions and their location in the area of a crisis, normal or border pre-crisis state is shown in the graph (Fig. 1).

In general, the unsatisfactory situation of ensuring the ES for the constituent entities of the Russian Federation is shown on the graph (Fig. 1), shows that most of the regions are located in the area of crisis and pre-crisis values of the indicator weights.

Let's dwell on regions where the situation with providing ES has changed qualitatively, these regions are presented in Table 2.

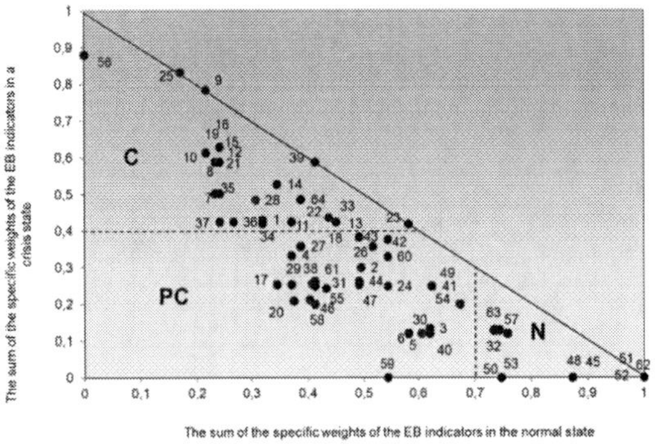

Fig. 1. The final state of the ES of the subjects of the Russian Federation, 2016

1 - The Republic of Karelia; 2 - the Republic of Komi; 3 - the Vologda region; 4 - the Kaliningrad region; 5 - the Leningrad Region; 6 - Murmansk region; 7 - Novgorod Region; 8 - Pskov Region; 9 - Belgorod Region; 10 - Bryansk region; 11 - Vladimir Region; 12 - Voronezh Region; 13 - Ivanovo Region; 14 - the Kaluga area; 15 - Kostroma Region; 16 - Kursk Region; 17 - Lipetsk Region; 18 - Moscow region; 19 - Orel region; 20 - Ryazan Region; 21 - Smolensk region; 22 - Tver Region; 23 - Yaroslavl Region; 24 - The Republic of Dagestan; 25 - Kabardino-Balkaria Republic; 26 - Republic of North Ossetia-Alania; 28 - the Chechen Republic; 29 - Stavropol Territory; 30 - Krasnodar Territory; 31 - Volgograd Region; 32 - Rostov Region; 33 - Republic of Kalmykia; 34 - Republic of Bashkortostan; 35 - Republic of Mordovia; 36 - The Republic of Udmurtia; 37 - Chuvash Republic; 38 - the Orenburg region; 39 - Penza region; 40 - Perm Territory; 41 - Samara Region; 42 - Saratov Region; 43 - Ulyanovsk Region; 44 - Sverdlovsk Region; 45 - Tyumen region; 46 - Chelyabinsk Region; 47 - Altai Territory; 48 - Kemerovo Region; 49 - Novosibirsk Region; 50 - Omsk Region; 51 - Tomsk Region; 52 - Krasnoyarsk Territory; 53-Irkutsk Region; 54 - Republic of Buryatia; 55 - The Republic of Tuva; 56 - The Altai Republic; 57 - the Republic of Sakha (Yakutia); 58 - Primorsky Territory; 59 - Khabarovsk Territory; 60 - the Amur Region; 61 - Kamchatka Territory; 62 - Sakhalin Oblast; 63 - Chukotka Autonomous District; 64 - Jewish Autonomous Region.

Table 2. Assessment of the qualitative state of certain regions of the Russian Federation

Year	The order numbers of the estimated ES indicators *								The sum of the specific weights by state			Quality condition ES
	1.1	1.2	1.3	2.1	2.2	2.3	3.1	3.2	Boundaries of states			
	Specific weights of indicators											
	0,104	0,138	0,133	0,120	0,079	0,170	0,127	0,129	C[a]	PC	N[b]	
1	2	3	4	5	6	7	8	9	10	11	12	13
Archangelsk region												
2011	N	N	C	PC	N	C	PC	N	0,303	0,247	0,450	**PC**
2016	N	N	C	PC	N	C	N	C	0,432	0,120	0,448	**C**
Tambov Region												
2011	N	N	C	C	PC K	PC	C	PC	0,380	0,378	0,242	**PC**
2016	N	N	C	C	PC	PC	C	C	0,509	0,249	0,242	**C**
Tula region												
2011	N	N	C	C	PC	PC	C	C	0,509	0,249	0,242	**C**
2016	N	N	C	C	N	PC	PC	N	0,253	0,297	0,450	**PC**
Karachay-Cherkess Republic												
2011	N	N	C	C	N	N	N	N	0,253	0	0,747	**N**
2016	N	N	C	C	N	N	N	PC	0,253	0,129	0,618	**PC**
Astrakhan region												
2011	N	N	N	N	PC	N	PC	C	0,129	0,206	0,665	**PC**
2016	N	N	N	N	PC	N	PC	N	0	0,206	0,794	**N**
Mari El Republic												
2011	N	N	PC	C	N	C	PC	C	0,419	0,26	0,321	**C**
2016	N	N	N	C	N	C	PC	N	0,29	0,127	0,583	**PC**
Republic of Tatarstan												
2011	N	N	PC	C	N	C	PC	C	0,419	0,26	0,321	**C**
2016	N	N	N	C	N	C	PC	PC	0,29	0,256	0,454	**PC**
Kirov region												
2011	N	N	C	C	N	N	C	C	0,509	0	0,491	**C**
2016	N	N	C	C	N	N	C	N	0,38	0	0,62	**PC**
Nizhny Novgorod Region												
2011	N	N	N	C	N	N	PC	C	0,249	0,127	0,624	**PC**
2016	N	N	N	C	N	N	PC	N	0,12	0	0,88	**N**
Kurgan region												
2011	N	N	C	C	C	N	C	C	0,588	0	0,412	**C**
2016	N	N	C	C	PC	N	PC	N	0,253	0,206	0,541	**PC**
The Republic of Khakassia												
2011	N	N	N	PC	C	N	PC	N	0,079	0,247	0,674	**PC**
2016	N	N	N	PC	C	N	N	N	0,079	0,12	0,801	**N**
Transbaikal region												
2011	PC	PC	N	PC	PC	N	PC	C	0,129	0,568	0,303	**PC**
2016	N	N	N	PC	N	N	PC	N	0	0,247	0,753	**N**
Magadan Region												
2011	N	N	C	C	C	N	PC	C	0,461	0,127	0,412	**C**
2016	N	N	PC	C	C	N	PC	C	0,328	0,26	0,412	**PC**
Republic of Crimea												
2011	C	N	C	C	N	N	PC	C	0,486	0,127	0,387	**C**
2016	C	N	C	C	N	N	PC	N	0,357	0,127	0,516	**PC**

[a] The state of ES in the region is recognized as a crisis if the sum of the specific weights of the indicators in the state "C" exceeds 0.4

[b] The state of ES in the region is recognized as normal if the sum of the specific weights of the indicators in the state "N" exceeds 0.7

In 2016, the crisis situation was noted in all regions of the Central Federal District, with the exception of the Lipetsk, Ryazan and Tula oblasts. In the Tula region (Table 2), the crisis situation shifted to the pre-crisis one due to commissioning of capacities in 2013-2015. (At Novomoskovskaya GRES 190 MW and the start-up of two hydroelectric generators at Cherepetskaya GRES with a capacity of 225 MW each).

For the analysed six-year period, the situation improved in such regions as the Republic of Mari El, the Republic of Tatarstan, Kirov, Kurgan, Magadan oblasts and the Republic of Crimea (Table 2). Improvement of the situation was due to the active policy of renewal of BPA the conduct of planned capital repairs in electricity and heat and power and made it possible to bring the situation in the regions into the region of pre-crisis values from the crisis. For example, in the Republic of Tatarstan, 590 MW of new generating capacities were commissioned in the last three years of the analysed period (2014-2016). Gas production in the republic increased almost threefold, which increased the ability to meet the needs of the BHO from the region's own sources and switched to the range of normal values by the corresponding indicator (1.3) (Table 2).

In the Republic of Crimea, the situation is improving every year, through the implementation of projects for the construction of generation facilities on the territory of the republic and the provision of reliable and uninterrupted power supply. The region has considerable potential for the development of alternative energy sources, such as solar and wind. In 2014, a wind farm with a capacity of 25 MW was commissioned, and in 2015-2016, Commissioned four lines of energy bridge to connect the grid of the Crimea to the UPS of Russia (IPS South), with a total capacity of 800 MW. However, at the same time, wear of power equipment on the peninsula is about 70%, which requires appropriate attention and taking measures to reduce it.

It is worth paying attention to those regions where the situation with ES has stabilized in six years, and by 2016 has moved into the region of acceptable values. There are only four such subjects (5% of the total number of analysed): the Astrakhan Region, the Nizhny Novgorod Region, the Republic of Khakassia and the Trans-Baikal Territory. The conditions for improving the situation were: modernization of electric power equipment, the annual commissioning of new generating capacities and, as a consequence, a decrease in the share of a large source in the region's available electric power (index 2.2), as well as an increase in the ability to meet the needs for (BHO) from the region's own sources.

Table 3 presents an assessment of the state of the regions of the Russian Federation in terms of indicators for 2016.

Table 3. Assessment of the state of the ES regions of the Russian Federation by indicators for 2016, %

Integral assessment of the state of ES by regions	1.1	1.2	1.3	2.1	2.2	2.3	3.1	3.2
N	79	92	36	13	53	56	10	46
C	17	4	55	69	27	27	20	41
PC	4	4	9	18	20	17	70	13

It can be seen from the table that in 69% of regions the crisis situation arises from the high share of the dominant resource in the total consumption of the region's BHO (indicator 2.1). These include all regions of the Central, North Caucasian, Volga Federal Districts, as well as the Kaliningrad, Leningrad, Murmansk, Novgorod and Pskov regions of the North-West Federal District, which do not have sufficient sources of their own for the production of BHO, and the dominant type of fuel is the gas. As shown in Table 3, for indicator 1.3 (the ability to meet the needs of the BHO from the region's own sources) - the share of regions in crisis is 55%.

4 Conclusion

Based on the results of the work done, the following conclusions can be drawn.

In general, the situation of providing ES by subjects of the Russian Federation is unsatisfactory. Most of the regions are located in the area of crisis and pre-crisis values of indicator weights.

It can be noted that a relatively high percentage of regions with an acceptable state from the EB's point of view on indicator 1.2 (The ratio of the sum of the available capacity of power plants and the capacity of interconnections of the region with neighbouring to the maximum electric load of consumers in its territory) is 92% and according to indicator 1.1 (The ratio of the total The available capacity of the region's power plants to the maximum electric load of consumers on its territory) - 79% of the regions.

The pre-crisis situation in 70% of the regions, according to Table 3, is affected by the unsatisfactory state of indicator 3.1 - the degree of depreciation of the basic production assets of the energy sector in the region. Since in most of the constituent entities of the Russian Federation the current values of the degree of wear and tear of equipment are 50-60%.

References

1. V.V. Bushuev, N.I. Voropay, A.M. Mastepanov, Yu.K. Shafranik *Energy Security of Russia* (Siberian Publishing Company RAS, Novosibirsk, 1998)

2. N.I. Pyatkova, S.M. Senderov, M.B. Cheltsov et al. *Izvestiya RAN. Energetika*, **6** (2000)

3. S.M. Senderov, N.I. Pyatkova, V.I. Rabchuk, G.B. Slavin, S.V. Vorobiev, E.M. Smirnova (ISEM SB RAS, Irkutsk, 2014)

4. Rocco C, Tarantola S, Costescu Badea A, Bolado Lavin R. ,ESREL, Prague, Czech Republic (2009)

5. G. Kocaslan *International Journal of Energy Economics*, **4**, 4 (2014)

6. H. Cabalu, C. Alfonso, A.Dorsman et al. (eds.), *Energy Economics and Financial Markets* (Berlin, Heidelberg 2013)

7. Lefèvre N., *Energy Policy* (2009)

8. B. Kruyt, D.P. van Vuuren, H.J.M. de Vries, H. Groenenberg, *Energy Policy*, **37**, 6 (2009)

9. J.C Bollen, MNP, Bilthoven (2008)

10. S.M. Senderov *Safety of objects of the fuel and energy complex*, **1**,1 (2012)

11. Statistical form of Rosstat *Information on balances, receipt and consumption of fuel and heat, collection and use of used oil products* (2011-2016)

12. Statistical form of Rosstat *Information on the use of fuel and energy resources* (2011-2016)

13. Statistical form of Rosstat *Technical and economic indicators of power plants, district boilers* (2011-2016)

Research of critical energy infrastructures taking into account energy security

Natalia Pyatkova[1,*], *Natalia* Beresneva[1], *Alexey* Edelev[1]

[1]Melentiev Energy Systems Institute, 130 Lermontov str., Irkutsk, Russia

Abstract. The article suggests an approach to the study of critical energy infrastructures. The existing definitions of the critical infrastructure are given, the feasibility of researching the fuel and energy complex combining the energy systems as a critical infrastructure is justified. The developed models for carrying out research to identify critical objects in the energy sector are presented. A developed package of applied programs for calculating and analyzing FEC states in dynamics as a critical infrastructure is described.

1 Introduction

Investigations of critical infrastructures have become one of the new directions for research of reliable energy supply to consumers in conditions at a high level of development of information technologies.

Abroad, studies of critical infrastructures began in the 90s of the last century [1-3], research centers have been established.

At the same time, in our country the terminology on critical infrastructures is currently not settled. In particular, in [4] critical is understood "infrastructure, the elements of which are of particular importance for the state and society, the failure or limitation of performance would lead to long disruptions in supply, serious violations of public security or other dramatic consequences" In accordance with this definition to critical infrastructures with complete certainty includes sectoral energy systems and the fuel and energy complex (FEC), which is an interconnected infrastructure, consisting of separate energy systems. This is confirmed by the importance of these structures for the economy of the state and society and is the main part of researching the problems of the country's energy security.

2 Research of critical energy infrastructures work from the positions of energy security

The energy security of the country and its regions is essentially a balance of energy supply and demand [5]. The main content of research tasks to ensure energy security is:

- forecasting the conditions of functioning and development of energy systems (ES, ie critical infrastructures) and the fuel and energy complex (a complex consisting of critical infrastructures), taking into account possible emergency situations (critical and emergency situations);
- assessment of the condition of the energy systems in these conditions, identification of potential critical objects (CO) in them;
- selection of possible directions and measures to prevent or reduce the negative impact of contingencies, taking into account specialized safety measures (reserves and holdings), as well as the capabilities of the energy systems for critical (CS) or emergency (ES) situations (partitioning systems for localization of accidents, regulation energy consumption, disconnection of individual consumers, etc.).

Studies of critical energy infrastructures from the ES standpoint are carried out in the framework of computational experiments based on the information base presented by the technical and economic characteristics of energy facilities, reporting data on the state of energy systems, and the results of research into the development of the fuel and energy complex. These studies are based on the territorial-production models of the fuel and energy complex, reflecting the technological cycle of energy resource transformations, imitating the work of key industry objects under given operating conditions. The main research tool is information support programs for modeling FEC states in abnormal situations, with subsequent calculation and evaluation of these states. An important component of the state assessment is the identification of potential CO, the determination of the degree of their criticality in the context of the situations studied. CO in this case are the most vulnerable elements of critical infrastructures, which can cause significant disruptions in the operation of reliable energy supply to consumers. Therefore, when allocating the CO in the fuel and energy complex, the conditions for the maximum possible satisfaction of consumers should be taken into account in case of

* Corresponding author: beresneva@isem.irk.ru

realization of contingencies. This task is directly related to:
- identification and neutralization of various threats to sustainable fuel and energy supply;
- analysis and assessment of the energy systems state in the CS and ES, which are the implementation of ES threats;
- advance preparation of FEC facilities and systems for operation during emergencies caused by the implementation of threats of various types.

In practice, the emergence of abnormal situations is associated with the implementation of natural (cooling) and technogenic (transport accidents, energy production and production facilities) threats. When realizing natural hazards, deviations of the maximum seasonal heating loads of regions can be observed depending on their climatic conditions and on the composition of consumers in them up to 20-30%. This may be accompanied by a significant increase in demand for energy resources for both the single climatic zone and several neighboring regions.

Technogenic threats are most critical for the gas industry and the power industry. In the gas supply system, one of the most dangerous factors is the possibility of damaging transcontinental gas pipelines coming from a large gas producing region through the Urals to the European part of Russia. In the electric power industry, significant shortages can be caused by disconnections between individual power systems or emergencies at specific power plants (for example, a major accident in the Moscow power system in 2005 at the Chagino substation, an accident at the Sayano-Shushenskaya HPP in 2009, a system accident at the power facilities of the SFO due to a failure at the Bratsk hydropower plant in June 2017).

It is possible to evaluate the functioning of systems under these conditions in the framework of model studies based on the apparatus of dual estimates [7]. Dual estimates in this case can serve as a measure of the scarcity of a resource, while showing how changing the restriction of this resource per unit can lead to a change in the value of the objective function by the value of the dual estimate [6]. In our case, such a resource may be the production capabilities of the technological methods under consideration (capacity for extraction, production, transportation of energy resources, stocks for different purposes). For objects whose production capacity is sufficient to perform the required technological operations, the value of the corresponding dual variable will be zero. For potential CO, the values of the dual variable will be less than zero. The priority of the CO in this case is determined by ranking the values of increments in production capacities. Assessment of the state of systems in critical and emergency situations is supposed to be carried out on the basis of economic-mathematical models and research tools described below. A generalized scheme of model studies of critical energy infrastructures from the standpoint of providing ES is shown in Fig. 1

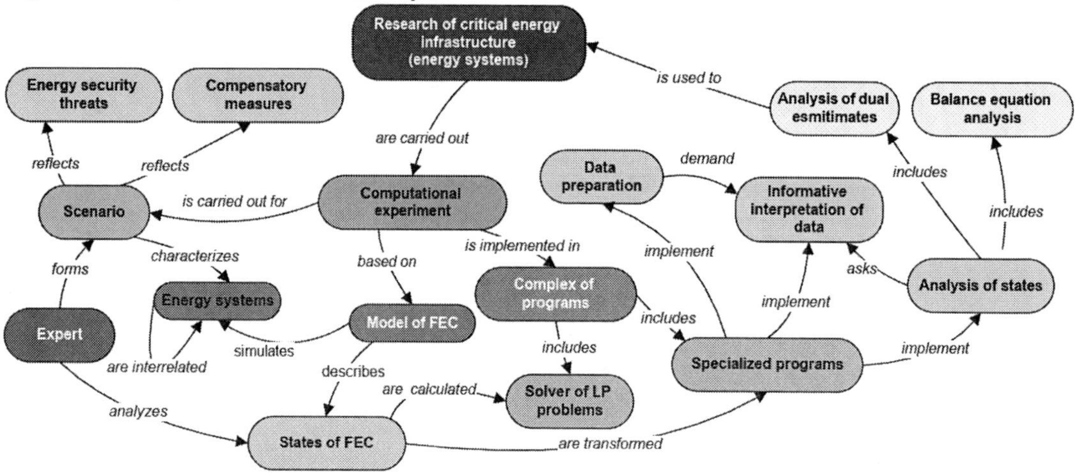

Fig. 1. Scheme for model studies of critical energy infrastructures

3 System of mathematical models for the study of critical infrastructures from the of energy security positions

A review of the work of foreign researchers on the modeling of critical infrastructures showed that these models have features that allow them to be applied only for these countries [8].

At present, a large number of models [9] have been developed in foreign countries for the study of ship systems, campuses, large electricity distribution networks, waterways and much more, and their main task has been to integrate them into a single hardware and software system that will provide managers with appropriate the ability to make an informed and correct decision, getting a reply to the question "what needs to be protected or hit and how much it will cost (spend ie the efforts and resources)".

Therefore, to study the behavior of Russia's critical energy infrastructures, original models for these studies were proposed.

Possible large-scale consequences from emergency situations in the energy sector require a special analysis of the behavioral features of critical infrastructures (energy systems) near their ultimate state. To carry out this analysis is possible only in the case of a hierarchy of studies based on FEC models. Only such a scheme of calculations will provide the conditions for obtaining and coordinating the results of research on the functional properties of individual energy systems and structural relationships in the fuel and energy sector, and will jointly take into account the physico-technical and technical-economic characteristics of the objects under consideration. At the same time, two tasks must be solved:
- the task of allocating potential CO in the operation of the fuel and energy sector in abnormal situations;
- the task of assessing the extent of their impact on the functioning of the fuel and energy complex in the context of the development of potential threats to ES.

These tasks are solved within the framework of the joint use of the models of the functioning and development of the fuel and energy complex, conceptually identical in the content aspect (Fig. 2), expressed in the allocation of fuel and energy resources, facilities (energy and economic-organizational) and multi-level territorial entities. Objects of models are described by technological and economic characteristics, processes in them (technologies) of resources conversion taking place. The nature of the latter determines the type of objects, identifies the energy (the objects of extraction and processing of resources, the objects of their storage and transportation) and organizational-economic (various categories of consumers, branch management organizations) objects. The interaction of objects, their association into groups and their territorial binding in models is regulated by managerial and organizational ties applicable, in particular, to territorial entities.

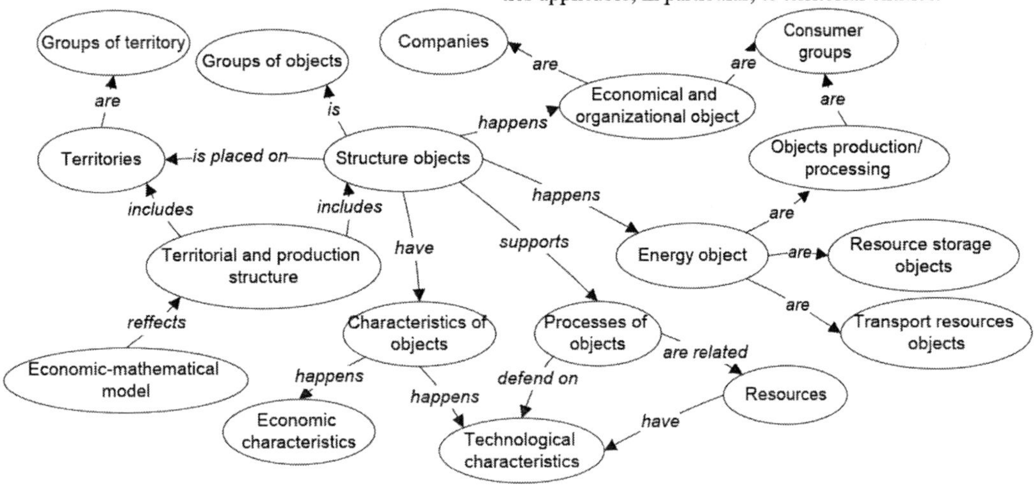

Fig. 2. The informative characteristic of the models for researches fuel energy complex

These models, different in the territorial, temporal and technological hierarchy, in the context of the interrelationship between the tasks of functioning and development of the fuel and energy complex, have the following features:

1. In the models of fuel and energy complex functioning there is a detailed representation of the objects, the working capacity of which is considered in three time intervals (from day to quarter). The need for such graduation is caused by the features of the operation of individual energy systems in emergency situations, with a maximum load of their production capacities, especially during the heating season (for example, taking into account the December maximum load in the electric power system and taking into account the daily gas withdrawals from the UGS at the end of the heating season). At the same time, administrative units of the country (as a rule, subjects of the Russian Federation) are considered territorially.

2. In the models of fuel and energy complex development, at the technological level, groups or types of fuel and energy complex are identified, tied to the federal districts of the country. At the same time, forecasting is carried out over time intervals in the annual section.

This model system is based on the model for estimating the territorial-production structure of the fuel and energy complex, taking into account the requirements of the ES, which was considered in detail in [5,10]. This model can be used in two modes:
- in the mode of determining the optimal development of energy technologies (taking into account structural redundancy in the form of reserves of capacities, fuel reserves, interchangeability of energy resources) and optimal distribution of energy resources consumed,
- in the mode of determining the shortage of energy resources (FER deficits) in the whole country and for individual regions.

This model is characterized by the hierarchy of levels shown in Fig. 3. Technologically, it consists of sectoral subsystems of the energy complex (gas, coal, oil refining (in terms of fuel oil) industries, electricity and heat power).

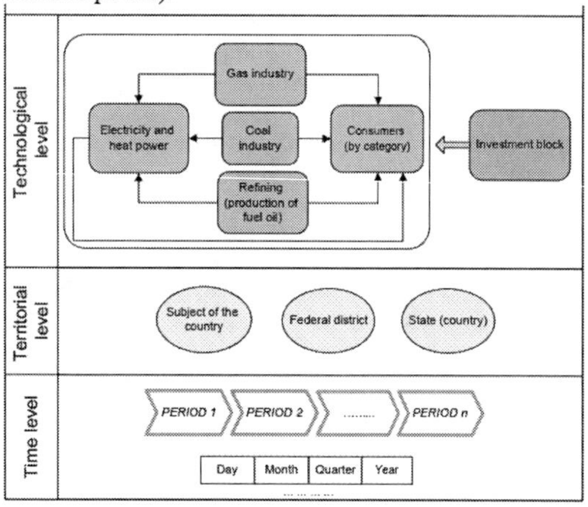

Fig. 3. Territorial, temporal and technological structure of the fuel and energy complex model

In the final implementation, it is supplemented by a financial unit describing the investment costs for the reconstruction, modernization of existing facilities, the withdrawal of obsolete equipment, the commissioning of new facilities at power industry facilities. It also takes into account the dynamics of the situation [11], which made it possible to trace such features of the multi-step process of the development of the fuel and energy sector, as:

- commissioning of new production facilities;
- dismantling and conservation of old facilities,
- reconstruction of facilities with changing technological scheme.

The dynamics in the model is taken into account in the form of T independent static blocks, each of which describes all the territorial and technological links of the fuel and energy complex as applied to stage t of the settlement period. Dynamic relationships between blocks are constructed using equations that formulate for all x_j objects of the FEC the condition of continuity of their productivity at different stages of the calculation period. For the first stage, this condition is written as

$$x_{i1}^{o} + x_{i1}^{c} + x_{i1}^{d} = P_{i0} \qquad (1)$$

And for subsequent stages in the form of equations

$$x_{it-1}^{o} + x_{it-1}^{c} + x_{it-1}^{n} = x_{it}^{o} + x_{it}^{c} + x_{it}^{d} \qquad (2)$$

where

P_{i0} - productivity (object) i to the beginning of the calculation period,

x_{it-1}^{n} - productivity of the new part of the object i on the stage t-1,

x_{it}^{o} - operating productivity (object) i in a step t,

x_{it}^{c} - preservation of an object i on the stage t,

x_{it}^{d} - liquidation of the object i on the stage t.

For the convenience of forming ties the equation (2) is divided into two parts

$$-x_{it-1}^{o} - x_{it-1}^{c} - x_{it-1}^{n} + Z_{it-1} = 0 \qquad (3)$$

$$-Z_{it-1} + x_{it}^{o} + x_{it}^{c} + x_{it}^{d} = 0 \qquad (4)$$

Z_{it-1} - intermediate variable characterizing the full performance of the object i in the early stages t. It takes into account the power outflow in phase t and the introduction of new power levels in the time step $t+1$.

In general, the model is used to determine the following characteristics (indicators):

- the size of shortages in certain types of energy resources for the categories of consumers considered, allocated to territorial associations and the country as a whole, as the magnitude of the discrepancy between the given demand and the possibility of producing this type of energy resource (taking into account reserves, the possibility of replacing this type of energy resource in others consumers, etc.);
- changes in the carrying capacity of inter-regional transport links, determined by comparing the corresponding indicators of the variant under consideration with the initial one;
- the recommended rational use of production capacities of energy facilities, as well as the distribution of certain types of energy resources by categories of consumers.

4 Information support for research

Information support for research on the dynamic model of the fuel and energy complex is realized in the modernized "universal" module of the distributed PPP "Corrective" [12]. Adaptation of this module required the organization of work with dynamic links of neighboring time states, in terms of the organization of calculations - the implementation of a new scheme for the formation of design files, conceptually expressed in the creation of a single file with descriptions of time states and communication equations between them. As a result of the adaptation, a software module with rigid data representation structures has been created that supports automatic data collection from sources (statistical reporting forms and subject databases), automatic generation of model calculation files, search for potential CO with subsequent identification (including, Cartographic) in the context of the results of the calculation. The procedure for preparing data is set out in the rules for assembling states that describe the basic technologies of the model used, with reference to information sources, in the rules for specifying dynamic relationships between neighboring states. The rules for assembling states are represented by a block of rules for transforming resources within territories, rules for transporting resources between them. The rules for specifying dynamic relationships between neighboring FEC states are represented by the characteristics of these states and the characteristics of the bonds themselves. The essence of the implementation of dynamic relationships is as follows:

1. Organization of separate modeling of static FEC states by reference years, separate modeling of dynamic relationships between formed static states.
2. The use of a unique unified encoding for modeling the static states of the fuel and energy complex, extended by additional identification of the temporal affiliation of states at the stage of calculation.
3. The use of a unique (different from the representation of the FEC states) a single encoding for modeling the dynamic relationships between states. Support for the typing of these links throughout the time interval, with reference to specific state technologies.
4. Support for the mechanism for defining dynamic relationships between neighboring time states in the form of "rules" describing the resources and technologies of the associated static states, resources and technologies of dynamic links of these states, including an auxiliary variable to reduce the coupling stiffness for debugging the model.

The research technology implemented in the module is represented by the following key stages:

1. The stage of the formation of initial data, including their automatic preparation on the basis of available sources (subject tables, subject databases), expert correction of the collected data, including a possible correction of the rules for specifying dynamic links.
2. The stage of preparation and execution of calculations, including the generation of autonomous states of the fuel and energy complex by reference years, the generation of communication equations between them, the formation of a single file in the solver format for subsequent calculations.
3. Representation and analysis of calculation results in tabular and cartographic form.

This technology completely reflects the iterative cycle of research, allows to obtain the existing solution taking into account the possible regulation of the rigidity of dynamic links.

5 Experimental studies for estimation of the condition in critical infrastructures of energy in conditions of non-state situations

With the use of developed models and software in the framework of research on the impact of energy security threats on reliable energy supply to consumers, a number of experimental calculations were carried out. The first cycle of calculations concerned the assessment of the state of the branch systems in the complex when the natural threat of "cooling" was realized in separate and in two neighboring regions. The next cycle of calculations was connected with the assessment of short supply of natural gas to regions from production centers in emergency situations complicated by distortions in the fuel and energy balance of European regions towards this resource.

The analysis of readiness to provide demand for boiler-furnace fuel (BFF), electric and thermal energy in the conditions of sharp cooling, was performed at the level of federal districts according to data on production and consumption for 2016 [13]. The calculations were carried out for the January days, corresponding to a 10% increase in the average daily heat and power demand in areas subject to peak temperatures. To such territories were assigned:

- alternately North-West, Ural, Siberian and Far-Eastern Federal Districts;
- the following combinations of federal districts: Northwest - Central, Northwest - Volga, Northwest - Urals, Privolzhsky - Central, Volga - Urals, Privolzhsky - Southern, Ural - Siberian.

The characteristics of the obtained results are exemplified by the example of cooling in the Urals and Volga FD, Urals FD, presented at the level of relative daily deficits in BFF and heat (Fig. 4).

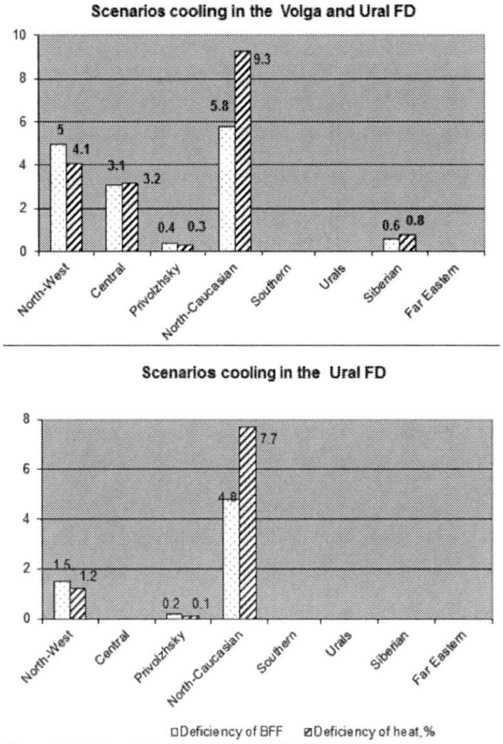

Fig. 4 Relative daily deficiency of BFF and heat in conditions of cold

The main regularity of these results is the emergence of a deficit among fuel consumers in the European regions of the country. And the main part is made up of natural gas, which is the main fuel in these regions. The increased demand for electricity, according to calculations, was repaid through diversification of electricity generation at various types of power plants and the use of reserves of generating capacity in the presence of fuel in thermal power plants. At the same time, the lack of gas at TPPs was compensated by the reserves of fuel oil.

DOI: 10.1051/e3sconf/20172501006

The second cycle of calculations was aimed at analyzing the consequences of accidents on non-trunk gas pipelines in the northern regions of the Urals Federal District, for which an average of more than six accidents per year is recorded. As calculations show, as a result of the accident in this region, characterized by a high density of gas transport flows, the potential shortage of gas to major consumers can be about 14% for the days of the accident. Most significantly, this shortfall can affect consumers in the North-West (short supply of gas - 37%) and Central (short supply - 21%) federal districts. The effect of the negative impact of an accident can be compensated by the interchangeability of fuels. The analysis of the results showed that, taking into account the possibilities of substituting gas with heating oil, taking into account the use of reserves and reserves in energy systems, the shortage of gas in the whole country in this situation will amount to 4.5% of the requirements of the period under review. The deficit of fuel oil in this case is 1.6%. n general, this emergency situation can cause a deficit in consumers of fuel and energy resources at a rate of 2.8% of the total quarterly demand for them. Based on the calculations carried out, it can be argued that, given the existing share of gas in the European countries' fuel and energy unit, more than 70%, negative consequences are possible due to the inability to compensate for missing gas volumes by other energy resources in a major accident on the main gas pipelines.

6 Conclusion

The model-tools presented in the article make it possible to carry out research to assess various contingencies in the fuel and energy complex and branch energy systems. The results obtained with their help confirm the working capacity of the created tools, the possibility of using them:

- to take into account inter-industry links in research on energy security;
- for the assessment of the state of energy and fuel and energy complex in emergency situations with different options for their development;
- to identify restrictions on energy facilities (transport, production, production, storage), which adversely affect the reliable fuel and energy supply to consumers.

The development of research involves the elaboration of a methodical tool for identifying potentially weak links in the fuel and energy complex that can significantly affect reliable energy supply to consumers and the behavior of computational experiments based on this apparatus.

References

1. Executive Order. 13010. Critical Infrastructure Protection. Federal Register. Vol. 61. **№. 138**. July 17, pp. 3747-3750 (1996)
2. S. Rinaldi, J. Peerenboom, and T. Kelly. IEEE Control Systems Magazine, IEEE, December, pp. 11-25 (2001).
3. C. Keating, R. Rogers, D. Dryer, A. Sousa-Poza and other. Engineering Management Journal.Vol. 15, **№ 3** (2003).
4. Protect critical infrastructure. The concept of basic measures of protection https://www.google.ru/url?sa=t&rct=j&q=&esrc=s&source=web&cd=1&ved=0ahUKEwigkZ6K_PrSAhUCZCwKHfWBAuIQFggcMAA&url=https%3A%2F%2Fwww.bmi.bund.de%2FSharedDocs%2FDownloads%2FDE%2FBroschueren%2FSprachvarianten%2FBasisschutzkonzept_kritische_Infrastrukturen_russisch.pdf%3F__blob%3DpublicationFile&usg=AFQjCNEjmCRfUtwKQGtk9uxHwqHBquPANA&sig2=fS5phhQcKNaE41QBH_qSuQ&cad=rjt
5. N.I. Pyatkova and others, *Energy security of Russia: problems and solutions* (SB RAS Publishing House, Novosibirsk, 2011)
6. V.I. Zorkaltsev, *Methods for forecasting and analyzing the efficiency of the fuel supply system* (Nauka, Moscow, 1988)
7. V.A.Sakovich, *Optimal solutions to economic problems*. (Vysh. School, Minsk, 1982).
8. M. Dunn, and I.Wigert. International CUP Handbook, *An Inventory and Analysis of Protection Policies in Fourteen Countries*. (Swiss Federal Institute of Technology, Zurich ,2004).
9. D. D. Dudenhoeffer, M. R. Permann, and M. Manic, Submitted to Proceedings of the 2006 Winter Simulation Conference, pp. 478-485 (2006)
10. N.I. Pyatkova, S.M. Senderov, E.V. Pyatkova, Izvestiya RAN. Power Engineering, **2,** 7 (2014)
11. A.A. Makarov, L.A. Melentiev, *Methods of research and optimization of the energy economy*.(Nauka, Novosibirsk, 1973)
12. A.V. Edelev, S.M. Senderov, N.I. Pyatkova, Problems of management. **2,** 7 (2015)
13. *Ministry of Energy of the Russian Federatio, Statistics* http://minenergo.gov.ru/activity/statistic.

Energy security problems and features of its research at the present stage

Natalia Pyatkova[1,*], *Victor* Rabchuk[1]

[1]Melentiev Energy Systems Institute, 130 Lermontov str., Irkutsk, Russia

Abstract. The paper offers a modern interpretation of the term "energy security". Strategic threats to energy security are identified for the modern conditions; their dynamics and transformation are shown at different time stages of the Russian energy sector development. Also the work presents methodological and modeling developments on the study of energy security problems. A scheme is proposed for researching the country's fuel and energy complex, taking into account the requirements of energy security, the methods and models used are analyzed. The paper describes the features and new challenges in the research of energy security at present and the need to develop modern methods, models and tools.

1 Introduction

In recent years, there has been a sharp increase in interest in the issue of energy security (ES). To date, the problems of ES have been outlined and numerous publications on this problem have been published.

For the first time, issues of energy security were recognized as a problem in industrialized countries during the oil crisis of 1973-74, when, following the next outbreak of the Middle East conflict, a multiple increase in oil prices followed. At the same time, the essence of ES was formulated by the World Energy Council: ES is the belief that energy will be available in the quantity and quality that are required under given economic conditions. To date, a collection of articles by foreign researchers on various aspects of energy security was published [1].

The problems of ES in Russia began to be actively discussed after the collapse of the USSR, in the mid-1990s - both in connection with acute crisis phenomena in the energy sector of the Russian economy, which threatened serious violations of normal, sustainable energy (electricity, heat, fuel); and in connection with the general increase of attention to different aspects of national security in the conditions of breaking-up of a new social and economic system. By this time, the Federal Law "On Security" (from 05.03.92) was adopted, in which security (actually - national security) is defined as "the state of protection of the vital interests of the individual, society and the state from internal and external threats". In the Institute of Energy Systems of L.A. Melentiev of the Siberian Branch of the Russian Academy of Sciences (ISEM) proposed the definition and interpretation of the concept of "energy security" in the following formulation [2-4]: ES - the state of protection of citizens, society, the state and the economy as a material basis for their existence and progressive development, from threats of shortage in providing their energy needs with economically accessible energy resources of acceptable quality, threats of disruption the continuity of energy supply. The state of security is thus understood as a condition that, under normal conditions, provides for fully justified energy demand, and in extreme conditions, guaranteed assurance of the minimum required volume of needs. Thus, ES is lack of deficit (physical accessibility), continuity (reliability) and economic accessibility of quality energy supply. ES is characterized by three main factors:

- by the ability of the fuel and energy complex (FEC) to provide a sufficient supply of economically available quality energy resources;
- by the ability of the economy (as a system of consumers of energy resources) to rationally (carefully) spend them and, accordingly, to limit their demand, realizing an economically feasible potential for energy saving;
- by a sufficiently high level of sustainability of energy systems, the fuel and energy sector in general, the economy as a whole to disturbing impacts of potential threats to ES.

2 Scheme of energy security research

The main content of research in the field of ES and justification of measures to ensure is to predict the conditions of operation and development of energy and fuel and energy complex in general, taking into account the realizing of ES threats; in assessing the state under these conditions and identifying bottlenecks in fuel and energy supply systems consumers, in the choice of possible directions and specific measures to reduce the negative impact of threats in these systems.

* Corresponding author: nata@isem.irk.ru

This requires appropriate methodological, model and software development. There are many foreign developments in the field of energy modeling.

The main representatives of imitation models of foreign researchers in the field of energy are MARKAL (MARKet ALlocation) [5, 6], EFOM (Energy Flow Optimization Model), MESSAGE (Model for Energy Supply Strategy Alternatives and their General Environmental Impacts).

MARKAL is a numerical model used to carry out economic analysis of different energy related systems at the country level to represent its evolution over a period of usually of 40–50 years. The word MARKAL was generated by concatenating two words (MARKet and ALlocation). Various parameters such as energy costs, plant costs, plant performances, building performance and so on, can be input and the software will choose an optimal technology mix to meet that demand at minimum cost. It is available from the International Energy Agency. TIMES is an evolution of MARKAL and both energy models share many similarities.

TIMES (The Integrated MARKAL-EFOM System) is a bottom-up optimization model generator that allows accurate energy and environmental policy analysis at the local, national or multi-regional level. This tools combines a technical engineering approach and an economic approach to represent technologies, fuels, emissions and their effect on all economic sectors.

The MESSAGE program [7, 8] was developed by the International Institute for Applied Systems Analysis (IIASA) for planning and forecasting the development of energy systems. In fact, this is a modeling system for medium- and long-term planning of energy systems development, energy policy analysis and development scenarios development. This is a flexible tool that allows you to compare alternative energy technologies and build the most appropriate scenario for the development of the energy system. The criterion of expediency is the minimum of the given system costs. To find the optimal solution, the linear programming method is used. Optimization is carried out on the whole considered prediction interval.

MESSAGE allows you to compare alternative scenarios for the development of the energy system in terms of impact on the environment.

The information-model complex NEMS (National Energy Modeling System) is developed by the US Department of Energy.

NEMS [9] - a system that models the US economy with the allocation of energy to its individual sector. This complex builds forecasts for the production, import, processing, consumption and prices of fuel and energy resources, taking into account macroeconomic and financial indicators, the world market of fuel and energy resources, the choice of certain technologies, the quantitative and qualitative characteristics of the chosen technology, etc. The complex is used to determine the strategy of energy development, the impact of energy on the economy, the environment of the country.

Energy security problems in Russia have a number of features that require the development of original methods, models and software.

To study the energy security problems of Russia, a two-level technology is proposed that integrates the stages of qualitative analysis (using semantic modeling tools) and quantitative analysis (using linear economic-mathematical models and traditional software systems).

The general scheme of the tasks to be solved and the modeling methods used in assessing the effect of ES threats on the state of energy security are shown in Fig.1.

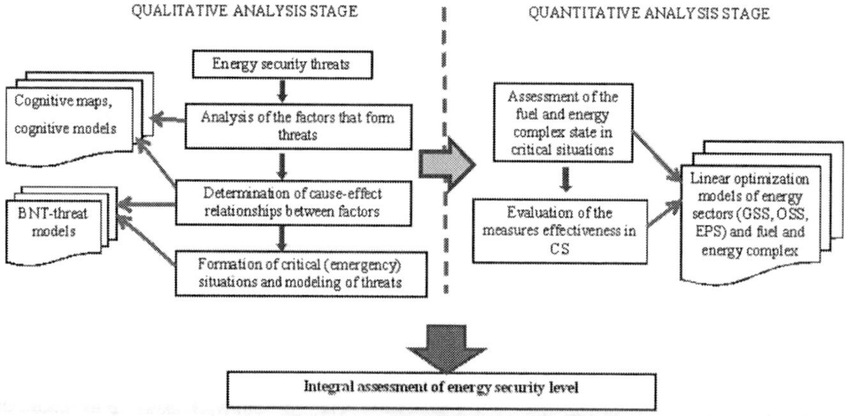

Fig.1. General scheme of studies of ES problems.

Based on the methodological approaches created in the ISEM of the SB RAS, a scheme has been developed for obtaining an integrated assessment of the level of energy security in the region, based on an indicative analysis of the region's energy security and model studies (Fig. 2) [10].

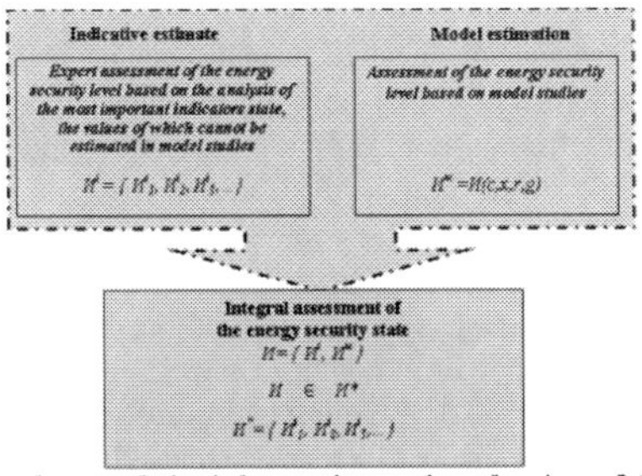

Fig. 2. Scheme of obtaining an integral evaluation of the regional level

To date the most developed methodically the stage of quantitative analysis, which solves the problems of researching the functioning and development of energy, taking into account the requirements of the ES. The basis for research is the technical and economic characteristics of energy facilities and reporting data on the state of energy systems, the results of research on the development of the fuel and energy sector, that justifying the choice of long-term strategy and the formation of energy policy. Based on the socio-economic program for the development of the country's economy, which determines the demand for fuel and energy resources, an analysis and assessment of energy consumption levels taking into account energy conservation is carried out.

Within the qualitative analysis phase, based on the above characteristics and the analysis of the ES threats, the calculation conditions for the computational experiment are formed, which is carried out at the stage of quantitative analysis.

A two-level system of models was proposed for carrying out experimental studies of energy security problems [4].

To harmonize the models of different levels of the hierarchy in the studies under consideration, a methodical approach is applied that allows solving the corresponding problems in the studies of the fuel and energy complex using a multilevel hierarchy of optimization studies. The interrelation of models and the main solved tasks are shown in Fig. 3.

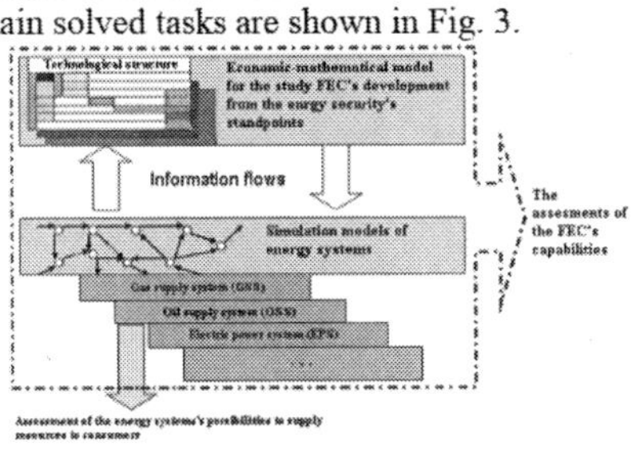

Fig. 3. Interrelation of models of two-level technology and the main tasks to be solved

In this approach, upper-level models are constructed by aggregating low-level models in accordance with conditionally optimal solutions previously obtained for them. The lower-level models are simulation models of energy systems for analyzing the options for their development, assessing the state and identifying "bottlenecks" when operating under different conditions. Aggregate solutions found using top-level models are passed to lower-level models and used by them as boundaries within which a detailed solution is determined.

Using a two-level system of models and methods of combinatorial modeling, studies were conducted on the selection of possible trajectories of the development of the fuel and energy complex with subsequent analysis of these trajectories from the point of ES [11].

3 Energy security threats and their modeling

When analyzing the current state of the country's energy sector, the main strategic threats to ES were identified at the present stage, that is, threats that can form today and in the near future a sustained long-term deficit in the country's energy balance:
- insufficient level of investments in the fuel and energy sector;
- energy wastefulness of the economy;
- iinsufficient increments of explored hydrocarbon reserves;
- the dominant role of natural gas in the fuel and energy balance of European regions of Russia;
- reduced opportunities to increase gas production due to the economic risk of developing the gas resources of the Yamal and the shelf of the northern seas;
- high deterioration and low rates of equipment renewal in the fuel and energy sector.

A special place among ES threats is the threat of reducing opportunities to increase gas production, as today the gas in the fuel balance of the country occupies 70-72%, and in some (European) regions of the country reaches 90% and higher.

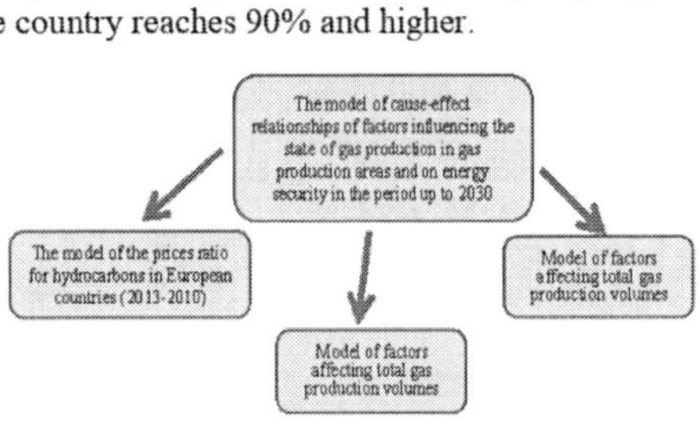

Fig. 4. Models of worsening / improvement for the situation for the strategic threat of ES "Decrease in opportunities to increase gas production"

Using the methods of cognitive modeling for this threat, a system of models of deterioration / improvement of the situation is constructed (Fig. 4), allowing to visualize and link all the factors that form a threat at various time stages [12].

DOI: 10.1051/e3sconf/20172501007

To analyze energy security threats and their consequences, it is suggested to create special models using an approach that allows, based on the use of Bayesian networks of trust, to describe the cause-effect relationship between the factors that form threats and lead to the emergencies. This approach makes it possible to take into account the probability of occurrence of the factors underlying the threat being analyzed, or the probability that they will assume a certain state.

The models constructed with the help of Bayesian networks allow taking into account a lot of factors and dependencies between them. A methodology for modeling energy security threats has been developed. With its use, models have been developed [13, 14] to assess the consequences of such threats to ES, such as:

- low rates of renewal of power generating equipment;
- lack of investment (for example, the oil industry);
- cool weather with a corresponding increase in the consumption of boiler-furnace fuel and heat and power generation

As an illustrative example, the results of the simulation of the threat "Low rates of renewal of power generating equipment" are shown (Fig. 5).

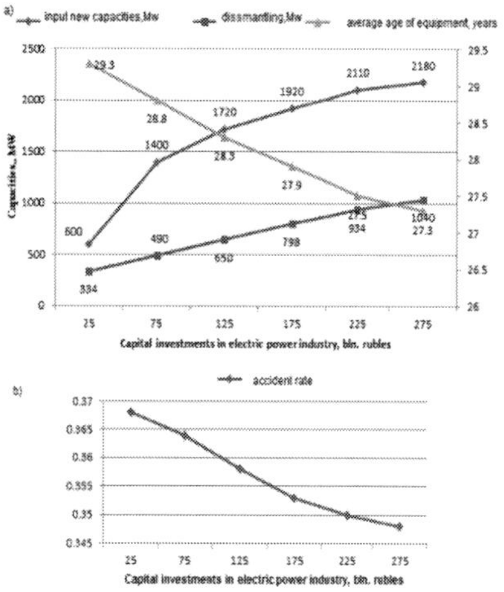

Fig.5. a) volumes of dismantling of old and commissioning of new equipment (MW) and mean age of equipment (years); b) changes in the accident rate in response to increase the volume of capital investments

The increase in investment volumes leads to increased levels of dismantling and equipment commissioning, while upgrading at a faster pace (compared to dismantling) can lead to a slight increase in accident rate due to the introduction of completely new equipment, but it reduces the accident rate sufficiently, compared with initial values.

The performed experimental calculations showed the possibility and expediency of using Bayesian networks of confidence for modeling the process of development of energy security threats with the possibility of assessing the likelihood of realizing the threats themselves (ie, the occurrence of emergencies), the likelihood of the consequences of their implementation and the possibility of assessing the effect of the corresponding compensation and liquidation measures.

4 Dynamics and transformation of energy security threats

It is also important to account the possible dynamics and transformation of threats to energy security at different time intervals. The conducted studies show possible changes both in the composition of threats and in the importance (priority) of individual threats at different time stages. Based on the real state of affairs with the provision of Russia's EBC in the period from 2011 to 2016, and the expected state of the same condition for 2020, the analyzed period is divided into two intervals: from 2021 to 2025 and from 2026 to 2030. Table 1 shows the change in the composition of threats in the future up to 2030 for different time intervals, as well as their possible influence on the energy sector and the economy of the country.

Table 1. Assessment of the direction of transformation of strategic threats to Russia's energy security for the period from 2021 to 2030.

N / N	The essence of the strategic threat of Russia's ES	The direction of the transformation of the threat by time intervals	
		2021-2025	2026-2030
1	Lack of investments in energy sectors	Significant increase in threat	Threatening the threat
2	Low rates of renewal of OPF in the energy sector	Some increase in the threat	Some weakening of the threat
3	Energy-saving economy of Russia	Some weakening of the threat	Weakening the threat
4	Domination of gas in the FER balances of the European regions of the country and the Urals	Some weakening of the threat	Some weakening of the threat
5	Delay with the development of new areas of gas production	Strengthening the threat	Enhancing the threat
6	Reduction of gas production capacity in the country	Increased threat	Increased threat
7	Reduction of oil production capacity in the country	-	Significant increase in the threat
8	A large share of electricity and heat produced at large thermal power plants and large boiler houses	-	Increased threat

The carried out researches on transformation of threats have allowed to determine the separate indicators of energy development. So, in the implementation of strategic threats to the ES without carrying out cardinal measures to overcome them by

DOI: 10.1051/e3sconf/20172501007

2025-2030. The country may face significant problems for its economic development. The analysis of prospective opportunities of the energy industries production in comparison with domestic needs for fuel and energy resources in this period shows that the tension will increase to meet external obligations for the export of oil and gas. Unless drastic measures are taken, the volume of oil production in Russia by the year 2030 will be 450 million tons (in 2016, the volume of production was 547 million tons), and gas - at 620 billion cubic meters (in 2016 - 653 billion cubic meters). It is especially alarming that the opportunities for exporting Russian gas (without taking into account the conjuncture of foreign markets and provided that the country's domestic requirements for gas are fully provided) are drastically reduced and, without taking drastic measures, by 2030, can be only 80 bcm versus 180 bcm in 2016.

5 Conclusion

To date, the ISEM SB RAS has carried out a complex of studies to develop an appropriate scientific and methodological base for studying the problem of energy security. In particular:

- methodical bases of an estimation of a level of ES of the country and its regions, tested at an estimation of level of ES of some regions (subjects of the Russian Federation and federal districts) are developed;
- the role of the gas industry (as dominant in the balance of primary fuel and energy resources) in providing the country's ES is assessed and requirements for the future development of this sector are determined from these positions;
- a system of decision support for the provision of ES, relevant information technology and software are developed;
- the process of conducting model studies of the interconnected work of the fuel and energy complex and the energy systems in the context of the implementation of strategic threats to the ES was prepared and implemented;
- methodical provisions were developed for the formation of directions for adjusting options for the development of the fuel and energy complex of Russia from the ES's point of view.

In addition, with the participation of the Institute, the draft of the Doctrine of Energy Security of Russia (protocol approved by the President of Russia on November 29, 2012, No. 3167) and the Methodology for Monitoring Energy Security in Russia and Regions were developed [15-17].

This work was carried out with the partial support of the Program of the Presidium of the Russian Academy of Sciences I.33P.

References

1. *The Routledge Handbook of energy security / ed. by B.K. Sovacool.* (London and New-York, Taylor and Francis Group, 2011, 436 p).
2. V.V. Bushuev, N.I. Voropay and others, *Energy security of Russia* (SO RAN Publ., Novosibirsk, 1998)
3 *Energy security. Terms and definitions* (Moscow: Energia, 2005)
4. N.I. Pyatkova, V.I. Rabchuk, S.M. Senderov and others, *System research in power engineering: Retrospective of scientific directions of SEI – ISEM. Research in the field of energy security of the country and regions (*SO RAN Publ., Novosibirsk, 2010)
5. A. J. Seebregts, G.A. Goldstein, K. Smekens, Energy/Environmental Modeling with the MARKAL Family of Models 2001 Adobe Acrobat Reader. URL: ftp://ftp.ecn.nl/pub/www/library/report/2001/rx01039.pdf
6. MARKAL 2011. URL: http://www.iea-etsap.org/web/Markal.asp
7. MESSAGE. URL:http://www.iiasa.ac.at/web/home/research/researchPrograms/Energy/MESSAGE.en.html
8. E.V. Fedorova *Nuclear energy* **4**. (2004).
9. *The National Energy Modeling System: An review 2009* http://www.eia.doe.gov/oiaf/aeo/overview/index.html
10. N.I. Pyatkova, S.M. Senderov, E.V. Pyatkova *Izvestia RAN. Energetika [Proc. of the Russ. Acad. of Sciences. Pow. Engin.]*, **2**, 7. (2014).
11. S.M. Senderov, V.I. Rabchuk, N.I. Pyatkova *Izvestia RAN. Energetika [Proc. of the Russ. Acad. of Sciences. Pow. Engin.]*, **5**, 7. (2009).
12. N.I Pyatkova, L.V. Massel, A.G. Massel, *Izvestia RAN. Energetika [Proc. of the Russ. Acad. of Sciences. Pow. Engin.]*, **4**, 8. (2016).
13. E.V. Pyatkova, L.V. Massel, *Methodological issues of reliability research for large energy systems,* **61**, 8. (2011).
14. E.V. Pyatkova *Methodological issues of research of large energy systems,* **63**, 12. (2013).
15. V.V. Bushuev, Voropai N.I., Senderov S.M., *Economika regiona*, **2**, 11 (2012).
16. S.M. Senderov, *Methodological recommendations for assessing the state of energy security of the Russian Federation at the federal level* (Irkutsk: ISEM SB RAS, 2013).
17. S.M. Senderov, *Methodology for monitoring the state of Russia's energy security at the regional level* (Irkutsk: ISEM SB RAS, 2014).

 DOI: 10.1051/e3sconf/20172501008

Analysis of the reserve generating capacity in unified power system of Russia in terms of adequacy assessment

Dmitry Krupenev[1*], *Ludmila* Lebedeva[1], *Gennady* Kovalev[1], *Nikolay* Belyaev[2], *Andrey* Egorov[2], *Roman* Gromov[3]

[1]Melentiev Energy Systems Institute, 130 Lermontov str., Irkutsk, Russia
[2]Ministry of Energy of Russian Federation, Moscow, Russia
[3]Association "NP Market Council", Moscow, Russia

Abstract. The paper is devoted to a topical issue of determining the required total capacity reserve for a given level of adequacy in the operation of the Interconnected Power System of Russia. This paper places an emphasis on a close interdependence between adequacy and total capacity reserve. The higher the adequacy, the larger the reserve should be. We propose that the adequacy indices, including the probability of failure-free operation, be assessed and calculated based on the requirement to adopt one or another adequacy standard. Moreover, the calculated reserve should be considered complete and necessary, but not normative, to maintain the given level of adequacy.

1 Introduction

For modern conditions of functioning and development of the Unified Power System (UPS) of Russia, the problem of reserving the generating capacity has become particularly acute and this was facilitated by the following main reasons:
- the current situation in the electric power industry, when, due to the instability in the Russian economy, large surpluses of established generating capacities in a significant part of the out-of-date resources have emerged in the UPS, and therefore are of low reliability, but are on the balance sheet and require maintenance costs;
- despite the presence of excess capacity in the whole of the UPS, reliability in some regional energy nodes is characterized as insufficient;
- flaw introduction of modern equipment diagnostics for repair organization according to the technical condition in power companies;
- the discrepancy between the adopted model of the capacity and electricity market to the existing realities of the functioning of the UPS of Russia, including the lack of incentive mechanisms for replacing obsolete capacity, etc.

Taking into account the above, the question of what level of reservation of generating capacity and its location in the system are most effective is multifaceted. The answer to this question depends on factors such as the technical characteristics of the generating equipment used, the capacity of power transmission networks and electric power, the requirements of consumers to ensure the required level of reliability of their electricity supply, etc. The question of sufficient reserves of electric power systems (EPS) is essential for the economy of each region and the country as a whole, since the development of industries, especially energy-intensive industries, largely depends on the reliability of their electricity supply.

The overestimated level of reliability of EPS leads to additional costs that are shifted to electricity consumers, and inadequate reliability leads to damage for both consumers and energy companies. Therefore, the solution of the problem of optimal reservation of EPS should be based on a sound scientific approach, using specialized software products for analysing the reliability of EPS.

From 2010 to 2016, the utilization factor of installed capacity in the UPS of Russia declined from 53 to 47 percent [1]. Excess capacity in power systems reaches up to 30–40 percent, from the maximum load. The so-called forced generation in recent years has led to a twofold increase in costs, which have to be compensated by the end user. Only in 2015, they were transferred an additional 19 billion rubles.

In this paper, using the example of the United Power System (UPS) of Siberia, the levels of capacity redundancy in it have been analysed from the standpoint of standard of adequacy standards.

2 EPS adequacy assessment

Let's consider the basic moments of the procedure for estimating adequacy of EPS. For EPS, the most acceptable method for estimating the adequacy is a method based on statistical modeling (the Monte Carlo method) [2, 3]. AS a rule, the adequacy estimation is carried out for a one-year period, while every hour of the EPS operation is modeled. In this case, only the basic structure of EPS is considered: generation capacities and

* Corresponding author: krupenev@isem.irk.ru

the backbone network, as the basis for the formation of adequacy (system). To assess the power supply systems of specific consumers, the reliability of the local distribution network must be taken into account.

Let N – a given number of a rally calculated states of EPS. Each state of the EPS is characterized by a set of random states: the working generating capacity, the load value $\bar{y}_i^k$ in the i-th node, the capacity of the transmission lines between the nodes i and j, $i=\overline{1,n}$, $j=\overline{1,n}$, $i \neq j$, $k=\overline{1,N}$. The values of these random variables are played out using the Monte Carlo method. To assess the capacity deficit of the generated EPS states, the following problem is solved.

Task variables: x_i – the used generation power in node i, y_i – the load covered at node i, – the capacity flow from node i to node j, $i=\overline{1,n}$, $j=\overline{1,n}$.

When assessing the capacity deficit of the k-th state of EPS, $k=\overline{1,N}$, it is required to find:

$$\sum_{i=1}^{n} y_i \to max \tag{1}$$

taking into account balance limits

$$x_i - y_i + \sum_{j=1}^{n}(1-a_{ji}z_{ji})z_{ji} - \sum_{j=1}^{n} z_{ij} = 0, \quad i=\overline{1,n},\ i \neq j, \tag{2}$$

and linear restrictions-inequalities on variables.

$$y_i \le \bar{y}_i^k,\ i=\overline{1,n}, \tag{3}$$

$$x_i \le \bar{x}_i^k,\ i=\overline{1,n}, \tag{4}$$

$$z_{ij} \le \bar{z}_{ij}^k,\ i=\overline{1,n},\ j=\overline{1,n},\ i \neq j, \tag{5}$$

$$y_i \ge 0,\ x_i \ge 0,\ z_{ij} \ge 0,\ i=\overline{1,n},\ j=\overline{1,n},\ i \neq j, \tag{6}$$

where a_{ij} is a given positive coefficients of specific capacity losses when it is transferred from node i to node j, $i \neq j$.

After determining the deficits in EPS, during the accounting period such basic indexes of adequacy as:
- probability of failure-free (no-deficit) operation of EPS;
- the mathematical expectation (m.e.) of undersupply of electricity to consumers;
- the ratio of electricity supply to consumers;
- the values of the estimated reserves of various types for the nodes and the system as a whole;
- energy-reliability characteristics of ties;
- dual (objectively determined) estimates of the deficiency of the main resources (capacity of generators by nodes and transfer capabilities of ties).

3 The practice of EPS adequacy indexes rating

To date, there are no approved standards for adequacy indexes for the UPS of Russia. There are recommended values of the probability of failure-free (no-deficit) work of consumers, prescribed in the "Energy Strategy of Russia 2030". In accordance with this document, UPS of Russia at the level of 2030 should ensure the probability of failure-free operation of electricity consumers at the level of 0.9997. Moreover, there are two intermediate stages – 2015 and 2020, for which the probability of failure-free operation should be at the level of 0.999 and 0.9991, respectively. At the same time, it is not specified how to interpret this standard for the UPS as a whole or for the calculated nodes of the UPS.

The standard indicator of adequacy at the level of the main structure of the UPS of Russia (the calculated energy nodes of the UPS and interconnections) is well known – the probability of a no-deficit work, equal to 0.996. This indicator is justified in [4] for the conditions of functioning and development of the energy system of the USSR.

In the practice of controlling the power systems of a number of foreign countries, to ensure their reliability and the required level of reserves of generating capacity, a probabilistic approach is applied when assessing the adequacy and the standards of probability reliability indexes are used [5–10]. Here are some examples:
– in France, the criterion of adequacy is the values of the reliability index LOLH (loss of load hours) – the expected number of days in a year during which there is a capacity deficit. It is believed that EPS works reliably, if LOLH does not exceed 3 hours per year;
– in the Netherlands, the standards of value LOLH is 4 hours per year;
– in the Scandinavian countries, the reliability criterion of the capacity system is LOLP (loss of load probability) – the probability of a capacity deficit in a small time interval, usually during the maximum load period, which is 0.001 % per year;
– LOLE (Loss of Load Expectation) = 0.1 day / year is adopted in the USA.

4 Analysis of UPS Siberia adequacy at the level of 2015

Investigation of the level of reservation of generating capacity was carried out at the UPS of Siberia.

The study consisted of the following stages:

1. Estimation of adequacy of the UPS of Siberia.

2. Based on the probability of failure-free operation at the nodes of the UPS of Siberia and the accident rate of the generating units, an alternate shutdown of generating units with the highest accident rate at nodes with the highest probability of failure-free operation was carried out.

3. Improving the probability of failure-free operation at the nodes of the UPS of Siberia to levels of 0.9997; 0.999; 0.996.

DOI: 10.1051/e3sconf/20172501008

The calculations were carried out using the 12-node scheme of the UPS of Siberia, shown in fig. 1.

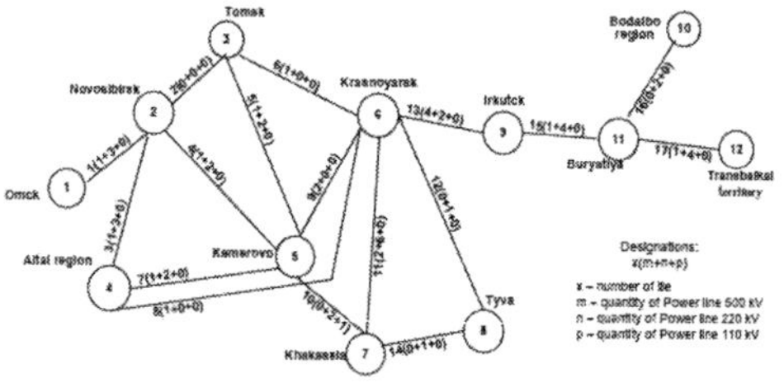

Fig. 1. Calculated schema UPS Siberia

The presentation of the UPS of Siberia on energy units was made on the basis of regional UPS. The external ties of the UPS of Siberia are accepted open, and all the initial data for the calculation were adopted at the level of 2015.

In table 1 shows a brief description of the nodes of the UPS of Siberia [11–13].

Table 1. Brief description of the nodes.

№	Node name	Absolute maximum of load, MW	Available capacity, MW	Total own reserve	
				MW	%
1	Omsk	1782	1479	-303	-17
2	Novosibirsk	2690	2730	40	1,49
3	Tomsk	1302	918	-384	-29,49
4	Altai region	1884	1444	-440	-23,35
5	Kemerovo	4535	5028	493	10,87
6	Krasnoyarsk	6235	12006	5771	92,56
7	Khakassia	2155	5430	3275	151,97
8	Tyva	152	40	-112	-74,01
9	Irkutsk	7570	12550	4980	65,79
10	Bodaibo	90	20	-70	-77,78
11	Buryatia	945	898	-47	-4,97
12	Transbaikal territory	1260	1156	-104	-8,25
Siberia UPS		30298,6[a]	43699	13400	44,23

[a] According to the UPS of Siberia, the combined maximum of the load is indicated (the sum of annual maxima for the nodes is 30,600 MW).

As can be seen from table 1, the total own reserve in the UPS of Siberia is at the level of 44 %, which exceeds the recommended standard (12 %) [14] by 32 %. The data on the transmission capacities of the UPS of Siberia are given in table

Table 2. The values of the transmission capacities of the UPS of Siberia.

Tie number	Nodes adjacent to tie (node number and name)	Transfer capabilities of tie (MW)
1	Omsk – Novosibirsk	1305
2	Novosibirsk – Altai region	1440
3	Novosibirsk – Kemerovo	950
4	Tomsk – Kemerovo	1170
5	Tomsk – Krasnoyarsk	780
6	Altai region – Kemerovo	950
7	Altai region – Krasnoyarsk	850
8	Kemerovo – Krasnoyarsk	1560
9	Kemerovo – Khakassia	1650
10	Krasnoyarsk – Khakassia	3400
11	Krasnoyarsk – Tyva	135
12	Krasnoyarsk – Irkutsk	3630
13	Khakassia – Tyva	135
14	Irkutsk – Buryatia	885
15	Bodaibo – Buryatia	66
16	Buryatia – Transbaikal territory	410

When assessing the reliability, each tie is represented by a set of power lines entering into it with characteristic accidents and capacity loss factors. The transmission capacities of individual transmission lines was taken from [15] with allowance for given permissible limitations on throughput capacities in sections.

In table 3 shows the results of calculations of the adequacy indexes.

Table 3. Dynamics of the change in the probability of failure-free operation and the mathematical expectation of an undersupply of electricity at the nodes of the UPS of Siberia.

Node name	Variants							
	initial		$\mathcal{P}$=0,9997		$\mathcal{P}$=0,999		$\mathcal{P}$=0,996	
	$\mathcal{P}$	E_{und}, MW·h	$\mathcal{P}$	E_{und}, MW·h	$\mathcal{P}$	E_{und}, MW·h	$\mathcal{P}$	E_{und}, MW·h
Omsk	0,999977	16,8	0,999748	336,1	0,999097	1271,9	0,996104	5993,2
Novosibirsk	0,999999	0	0,999931	11,3	0,999682	66,5	0,998381	329,9
Tomsk	0,999999	0	0,999926	20,8	0,999634	171,6	0,998197	804,7
Altai region	0,999999	0	0,999927	22,4	0,999652	109,7	0,998156	682,1
Kemerovo	0,999999	0	0,999933	9,3	0,999665	78,1	0,998233	460,3
Krasnoyarsk	0,999999	0	0,999933	20,7	0,999637	117,1	0,998200	634,2
Khakassia	0,999999	0	0,999946	5,3	0,999791	92,1	0,999002	391,5
Tyva	0,999980	11,5	0,999893	42,3	0,999584	134,6	0,998123	551,1
Irkutsk	0,999999	0	0,999936	9,4	0,999647	134,5	0,997985	2106,4
Bodaibo	0,8	7495,7	0,8	7670,3	0,8	8071,1	0,8	9776,0
Buryatia	0,999989	1,2	0,999736	175,8	0,999074	675,6	0,996158	3045,0
Transbaikal territory	0,991265	5607	0,990816	6222,7	0,989769	7796,8	0,985308	13719

As can be seen from table 3, the probability of failure-free (no-deficiency) operation at the nodes of the UPS of Siberia in the initial variant is at a high level, practically at all nodes, except for the Transbaikal territory and the Bodaibo power node. Reliability of power supply in problem nodes can be increased either by strengthening ties with other nodes or by building generating capacities directly at these nodes. After applying the above-described approach to lowering the generating capacity for different standard values of the probability of failure-free operation, it was possible to significantly reduce the generation capacity reserves and finally it was found that in the UPS of Siberia, it is sufficient to maintain a reserve capacity R = 4293 MW for reliable operation (ensuring the probability of a no-deficit operation at 0.996) , having carried out for 9107 MW connection of additional load power in general for UPS of Siberia.

In table 4 shows the values of power reserves in the nodes of the UPS of Siberia and for the system as a whole, which were obtained at each step.

Table 4. The values of power reserves of UPS Siberia for different levels of adequacy.

Node name	Variant							
	initial $\mathcal{P}$		$\mathcal{P}$=0,9997		$\mathcal{P}$=0,999		$\mathcal{P}$=0,996	
	MW	%[b]	MW	%	MW	%	MW	%
Omsk	-303	-17	-303	-17	-303	-17	-303	-17
Novosibirsk	40	1,49	40	1,49	40	1,49	40	1,49
Tomsk	-384	-29,49	-384	-29,49	-384	-29,49	-384	-29,49
Altai region	-440	-23,35	-440	-23,35	-440	-23,35	-440	-23,35
Kemerovo	493	10,87	493	10,87	493	10,87	493	10,87
Krasnoyarsk	5771	92,56	1223	19,62	788	12,64	788	12,64
Khakassia	3275	151,97	2870	133,18	2870	133,18	2870	133,18
Tyva	-112	-74,01	-112	-74,01	-112	-74,01	-112	-74,01
Irkutsk	4980	65,79	1894	25,02	1908	25,2	1261	16,66
Bodaibo	-70	-77,78	-70	-77,78	-70	-77,78	-70	-77,78
Buryatia	-47	-4,97	-47	-4,97	-47	-4,97	-47	-4,97
Transbaikal territory	-104	-8,25	-104	-8,25	-104	-8,25	-104	-8,25
UPS Siberia [c]	13400	44,23	5361	17,7	4940	16,31	**4293**	14,17

[b] % of maximum load

[c] The values of capacity reserves in the system as a whole are related to the combined maximum of the load (30298.6 MW), and the nodes to their own.

Fig. 2 graphically shows the dependence of the required reserve to ensure the required level of reliability of the EPS under study. It follows from the figure that the sharp increase in the required reserve begins with the probability of a no-deficit work of 0.9997.

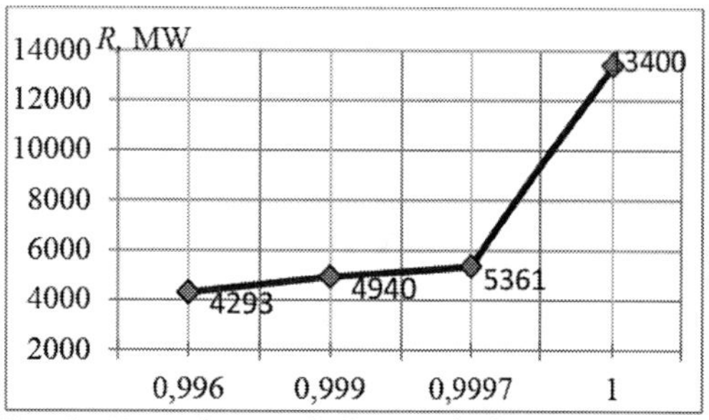

Fig. 2. Dependence of the required reserve on the reliability index.

5 Problems in the collection and processing of initial information

To assess the adequacy of EPS, a vast amount of information characterizing the EPS under investigation is needed. The completeness and reliability of the initial information depends on the reliability of the received indexes of the adequacy of EPS. Below is an analysis of the basic principles of the collection of initial data and problems that arose when collecting information on the UPS of Siberia for this work.

A. The design scheme of EPS (equivalent power nodes and ties between them). The design scheme should be designed in such a way that each power node represented in it is concentrated, i.e. within the node there should be no restrictions on capacity flows in any likely regimes.

In this study, the breakdown of the calculation scheme was carried out according to the areas of responsibility of the regional dispatch offices. The only exception is the Tomsk and Tyva power systems, as well as the Bodaibo power node.

B. Characteristics of generating equipment in the calculation units of EPS. For the assessment of adequacy EPS it is necessary to have the following information: the composition of generating units in EPS nodes, the unit available capacity, the standard for scheduled (major, mid-life and current) repairs, the failure rate of each unit.

During the research, the following problems were identified: information on the total available capacity of the node in different sources is different. There are no data on the accident rate of generating units. Thus, in the reporting data the indicator "Total time spent in repair (planned, unplanned) for the reporting year, hours" is given. From this indicator it is difficult to find out what time the unit was in routine repair, therefore it is necessary to oblige generating companies to correctly separate planned and unplanned repairs in the reporting data. In this study, the issue of the failure rate of generating units was solved as follows:

1. From the indicator "Total time spent in repair (scheduled, unscheduled) for the reporting year, hours" for 4 years from 2012 to 2015, the mathematical expectation (m.e.) of this indicator for each unit was found.

2. Using the data of SB 34.04.181-2003 according to the standards of scheduled repairs of generating units, the average duration of idle time of each unit in scheduled repairs per year was found.

3. The difference between m.e. "The total time spent in repair (scheduled, unscheduled) for the reporting year, hours" and the average duration of idle time of each unit in scheduled repairs per year is taken as the total time of the emergency idle of generating units for the year.

When processing the data on the accident rate of the generating equipment in the UPS of Siberia, the units were identified which, in emergency repairs (idle times), are more than 2500 hours per year, which indicates either a critical condition of these units or inaccurate data provided by generating companies.

C. Information on loading graphs in the nodes of EPS. To conduct the study, it is necessary to have the characteristic daily and annual graphs of the monthly maximum loads in each of the nodes, the root-mean-square deviations of the loads from the predicted graphs. When compiling a design scheme for EPS based on clustering, the boundaries of some nodes will not correspond to the boundaries in which regional power systems are located, therefore additional information on load schedules of these nodes is needed to estimate the adequacy for such schemes.

6 Conclusion

In the current conditions of the functioning of the UPS of Russia, there are problems that affect, namely, reduce the efficiency of its operation. One of the problems is unreasonable excess of generating capacity, which lead to economically inefficient operation of the UPS, including the increase in the cost of electricity for consumers and other negative consequences. To determine the values of the excess generating capacity, we propose an approach based on the use of ambiguous estimates obtained after solving the problem of minimizing the power deficit at the nodes for given standard values of the adequacy indexes of EPS.

Within the framework of the conducted studies, the level of redundancy of generating capacity in the UPS of Siberia was analysed. For the analysis of the dynamics of changes in the reservation level, several standard values of the probability of failure-free (no-deficit) operation of the UPS were accepted: 0.9997; 0.999; 0.996; As a result, it was found that the growth of the necessary reserve of generating capacity is moderate to 0.9996–0.9997, after which the growth rate of its growth is significantly increased.

It should also pay special attention to the fact that the accuracy of calculations is largely determined by the quality of the source information, the completeness and reliability of which at the current stage remain problematic. It takes a lot of work to create at the federal level an infrastructure for the collection and processing of primary information on the parameters of the operation of EPS and equipment. This will also help to increase the validity of decisions taken to ensure adequacy.

DOI: 10.1051/e3sconf/20172501008

References

1. E. Voskanyan, Newspaper "Ene. & Ind. of Rus.", No. 08, **292** (2016)

2. D.S. Krupenev, S.M. Perzhabinsky, Contr. of lar. sys., issue 54, Pp. 166-178. (2015)

3. G.F. Kovalev, L.M. Lebedeva *Reliability of electric power systems,* Nauka, **224** (2015)

4. I. M. Wolkenau, A. N. Zeiliger, L. D. Khabachev, *Economics of the formation of electric power systems,* Energia, **319** (1981)

5. Yu. N. Kucherov, Yu. G. Fedorov, *Electro. Electr. eng., elec. pow/ ind., electrotech. ind.* **№ 6**, p. 2–11. (2010)

6. *The concept of determining the reserves of installed (available) capacity in the UES of Russia, necessary to meet the public needs for reliable and sustainable supply of electrical energy* (CRA Project No. D16525) / (Charles river Associates, John Hancock Tower 200, Clarendon Street, I-33, Boston, Massachusetts, 0211, USA,) and Melentiev Energy Systems Institute of Siberian Branch of the Russian Academy of Sciences (ESI SB RAS), Lermontov Street, 130, 664033, Irkutsk, Russia. (2011)

7. K. Uhlen D. Cirio, L. Haarla and oth., *IEA ENARD CIGRE 43-th Session, pub.* C4-305 (2010)

8. *Grid master plan*, Nordel, Access at: http://www.pfbach.dk/firma_pfb/historien/data_files/nordic_grid_master_plan_2008.pdf, (2008).

9. *Generation adequacy report on the electricity supply demand balance in France, RTE.* Access at: http://www.rte-france.com/en/article/forecast-assessment-electricity-supply-demand-balance, (2009)

10. *Reliability Standarts for the Bulk Electric Systems of North America. NERC,* **1010** http://www.nerc.com/files/Reliability_Standards_Complete_Set.pdf, (2008)

11. *Scheme and program for the development of the Interconnected Power System of Russia for 2016–2022 /* Order No. 147. of the Ministry of Energy of Russia from March 1, (2016)

12. *Report on the functioning of the Interconnected Power System of Russia in 2015 /* System Operator of the Interconnected Power System, **40** (2015)

13. *Scheme and program for the development of the electric power industry of the Irkutsk region for 2017–2021. /* Approved by the decree of the Governor of the Irkutsk region No. 257-ug. of October 17, (2016)

14. *Methodical recommendations on designing the development of power systems /* CO 153-34.20.118-2003, Approved by Order of the Ministry of Energy of Russia of 30.06.03 № 281e., (2003)

15. *Handbook on the design of electrical networks /* Ed. D. L. Faibisovich. – 4-th ed., alteration and additional. – Moscow: ENAS, **376**, 2012

E3S Web of Conferences **25**, 01009 (2017)
RSES 2017 DOI: 10.1051/e3sconf/20172501009

Effect of duration of the pause single-phase auto-reclosing on electro-power transmission capacitance

Tatyana Krasil'nikova[1], and *Saiyod* Jononaev[1*]

[1]Novosibirsk State Technical University, 630073, Novosibirsk, Russian Federation

Abstract. This paper discusses the problem associated with accidents in the aerial line (AL) ultra-high voltage (UHV) due to its big length. In lines with a voltage of 500-1150 kV the overwhelming proportion of trips (98%) is caused by single-phase short circuit (SPSC). A substantial portion (70%) single-phase short circuits is erratic arc accidents which can be successfully eliminated in a high-speed auto-reclosing (HSAR) or single-phase auto-reclosing (SPAR). Success single-phase auto-reclosing (SPAR) at liquidation by single-phase short circuit (SPSC), on the one hand, is determined by the characteristics of the secondary arc current, and on the other hand the effectiveness of ways to reduce secondary arc current and recovery voltage development. The minimum dead time, at a HSAR it is usually taken as 0.5 s., at single-phase auto-reclosing (SPAR) it depends on the current value of the arc support is in the range of 0.5-3.0 s. The article shows high efficiency of use single-phase auto reclosing (SPAR) at liquidation SPSC in a single-chain AL voltage of 500 kV, the dependence of the bandwidth of transmission in maintaining the dynamic stability from the length of the pause SPAR.

1 Introduction

The prevailing type of accidents in the aerial line (AL) of ultrahigh voltage (UHV) are single-phase short circuits (SPSC), the proportion of which exceeds 98% [1-4]. At the same time, a significant part of the SPSC represents unstable arc faults, which can be successfully eliminated in the cycle of automatic high-speed reclosing (AHSR) or single-phase automatic reclosing (SPAR).

As a design scheme we can use the classic scheme, representing the station, working through the line on the tires of infinite capacity (Fig. 1).

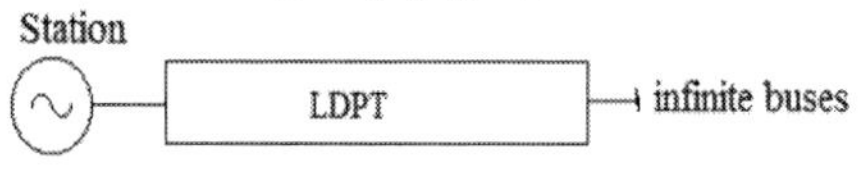

Fig. 1. The calculation scheme

If the minimum no-current pause at AHSR is usually taken of the order of 0.5 sec, in the case SPAR it depends on the current value of the arc support is in the range of 0.5-3.0 sec.

It is of interest to analyse the nature of dynamic transitions when using AHSR and SPAR of different duration and to assess the level of bandwidth long distance power transmission (LDPT) UHV by the condition of dynamic stability. Features of the dynamic transition at the estimated failure can be traced with the help of angular characteristics of a normal mode, mode, SHSC, AR (AHSR or SPAR) and post-accident [5].

$$P_V(\delta) = \mathrm{Re}\left(\overline{Y}_{11V}E_V^2 - \overline{Y}_{12V}E_VUe^{j\cdot\delta}\right) \quad (1)$$

where $\overline{Y}_{11V}, \overline{Y}_{12V}$ – is conjugate values of self and mutual conductances of the circuits for the v - th mode; E_V – electromotive intensity of the station in the v - th mode; U – voltage on the infinite bus.

In the analysis of dynamic stability of SHSC is modeled in the scheme of direct sequence shunt composed of a resistance of zero and negative sequence relative to the injury site, which usually races was a matter of the beginning of the line. It is useful to note that the ratio of the maxima of the angular characteristics in the SHSC in normal mode weakly depend on the circuit parameters, namely the length of the line and the resistance of the starting system (Fig. 2). Therefore, when lengths of over 200 km can be taken

$$P_{SHSC}(\delta) = K_{SHSC} \cdot P_{норм}(\delta) \quad (2)$$

where K_{SHSC} – is a coefficient varying within narrow limits, the value of which with reasonable accuracy can adopted $K_{SHSC} = 0.64$.

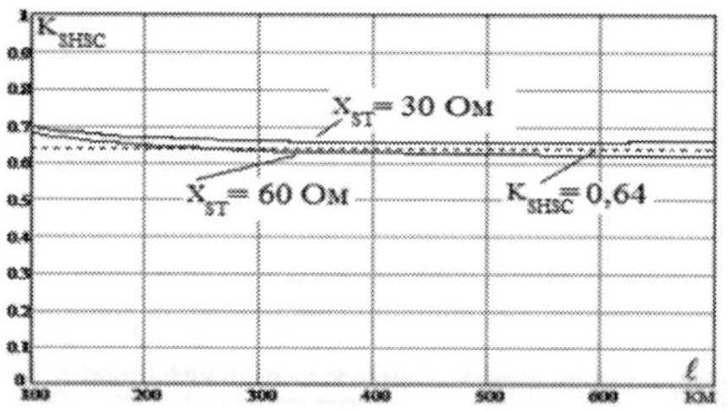

Fig. 2. The ratio of the maxima of the angular characteristics at SHSC in the normal mode depending on circuit parameters

* Corresponding author: jononaev.87@mail.ru

 DOI: 10.1051/e3sconf/20172501009

As in the case of the SHSC, the ratio of the maxima of the angular characteristics in the pause SPAR in the normal regime depends weakly on the parameters of the circuit (Fig. 3).

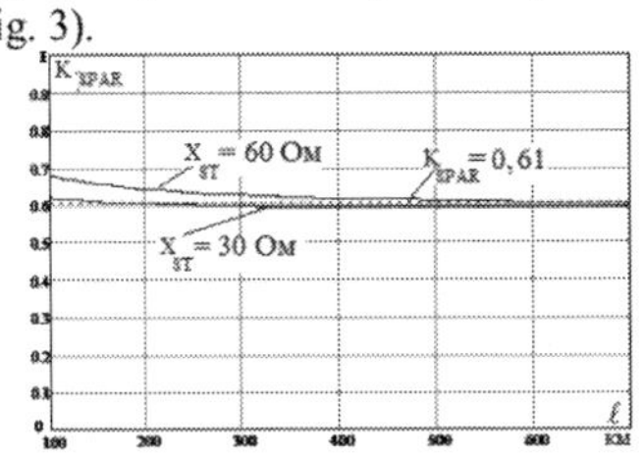

Fig. 3. The ratio of the maxima of the angular characteristics in SPAR in the normal mode depending on circuit parameters

Therefore, as in the case of the SHSC, it is possible to take

$$P_{SPAR}(\delta) = K_{SPAR} \cdot P_{норм}(\delta) \quad (3)$$

where K_{SPAR} – is a coefficient varying within a very narrow range, the value of which may be adopted $K_{SPAR} = 0.61$.

Angular characteristics allow it to perform the dynamic nature of the transition, using the method of squares. To determine the angle change of the generators of the power plant in time requires the solution of the equations of motion [6]

$$\frac{d^2\delta}{dt^2} = \frac{18000}{T_j \cdot S_{nc}}\left[P_T(t) - P_{epg}(\delta, t)\right] \quad (4)$$

where T_j – a constant of inertia of the station; S_{nc} – is the nominal capacity of the plant; $P_{TP}(t)$ – shaft power turbines, which generally varies in the dynamic process of transition, for example, by the discharge pulse of the turbines or the power runs out; $P_{epg}(\delta, t)$ – electromagnetic shaft power of the generators.

Compare the efficiency AHSR and SPAR with the elimination of SHSC in single-circuit overhead line 500 kV. Analysis will hold in relation to the scheme in which when transmitting 1000 MW the requirements for static stability (the safety factor is 0.25).

2 The use of AHSR

As analysis shows, in the case of transfer of such power dynamic stability is violated, i.e. the limiting factor for the capacity of the power transmission (PT) is the dynamics. Thus, by the condition of the dynamics, a reduction in the maximum transmitted power in the normal mode is required, i.e. throughput capacity of the PT, so that when the accident is eliminated, dynamic stability is maintained. The solution of the equation of motion of the generator (4) with the duration of the SHSC 0.1 sec and the 0.5-second AHSR pause showed that the dynamic stability is retained only when power transmission is no more than 420 MW (Fig. 4). The deviation of the angle to the end of the AHSR pause when the transition to the angular characteristic of the post-emergency mode occurs is $\delta_{AHSR} = 98°$.

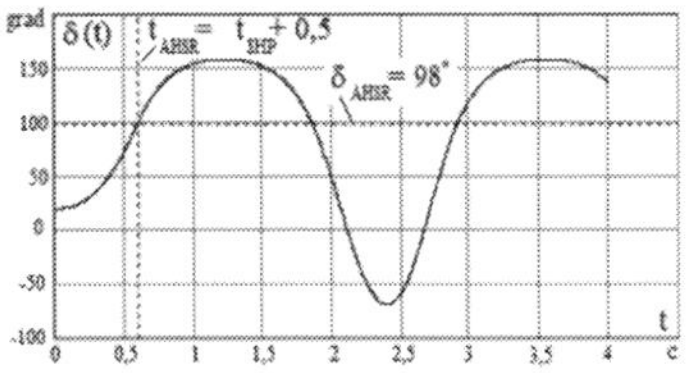

Fig. 4. The angle change in the process of dynamic transition

The condition of preservation of dynamic stability is illustrated using angular characteristics in Fig. 5, which shows the equality of the sites of acceleration and braking.

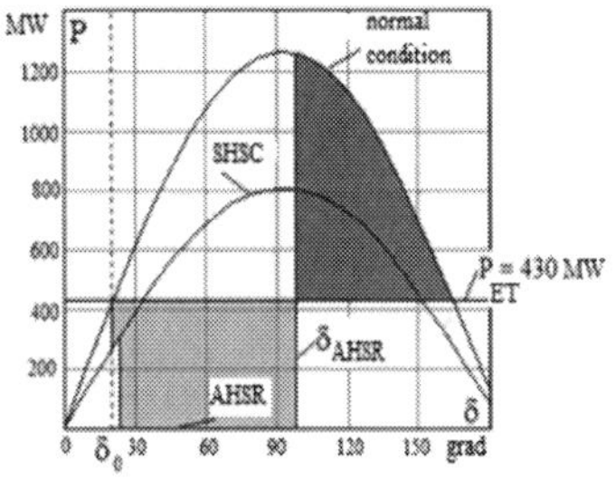

Fig. 5. Angular characteristics of acceleration and deceleration when using AHSR

3 The use of SPAR

The solution of equation (4) with the duration of the SHSC 0.1 sec and the pause of the SPAR 0.5, 1.0 and 2.0 sec is illustrated in Fig. 6.

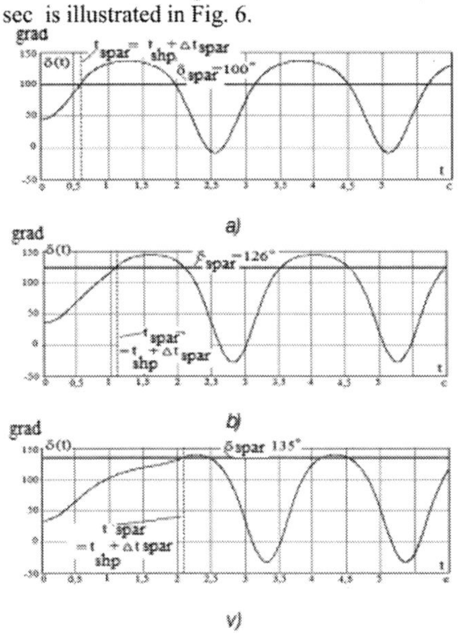

Fig. 6. The angle change in the process of dynamic transition a – $\Delta t_{SPAR} = 0.5$ sec ; b – $\Delta t_{SPAR} = 1.0$ sec ; v – $\Delta t_{SPAR} = 2.0$ sec

Table 1. The characteristic parameters of the dynamic transition at SPAR

Pause duration SPAR, sec	transmission capacity, MW	The deviation of the angle to the end of the pause SPAR, hail
0.5	863	100
1.0	724	126
2.0	679	135

The conditions of preservation of dynamic stability under various pauses SPAR are illustrated using angular characteristics in Fig. 7, which shows the equality of the respective sites of acceleration and braking.

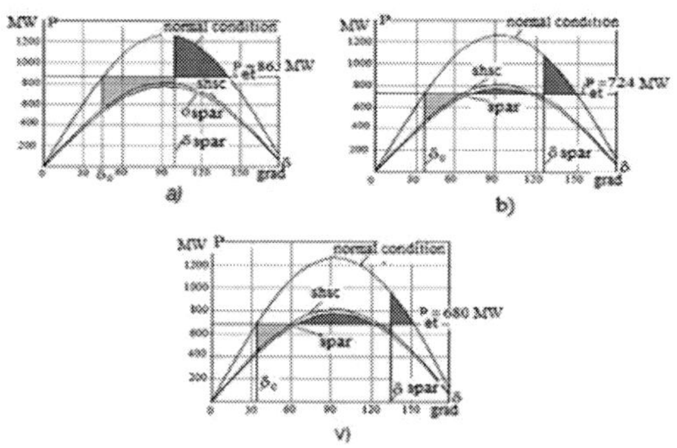

Fig. 7. Angle characteristics and areas of acceleration and braking when using the SPAR: a – $\Delta t_{SPAR} = 0.5$ sec ; b – $\Delta t_{SPAR} = 1.0$ sec ; v – $\Delta t_{SPAR} = 2.0$ sec

Using the angular characteristics in Fig. 8, we find the transmission capacity of the power transmission, in which the duration of the SPAR pause by the condition of the dynamics can be arbitrarily large.

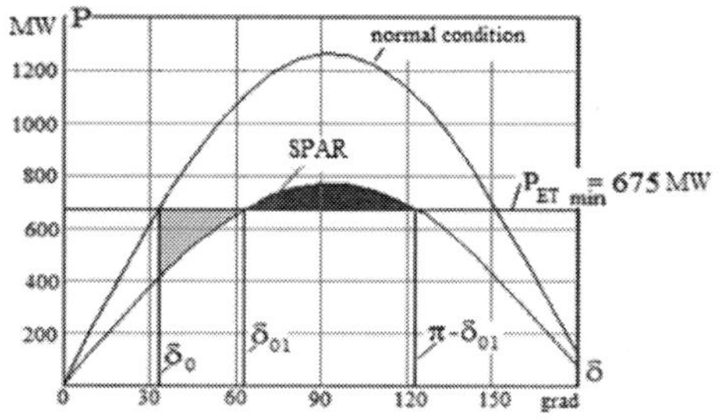

Fig. 8. The definition of transmission capacitance with unlimited pause SPAR

This capacity is found from the condition of equality of the acceleration and deceleration areas in the SPAR mode (the angular characteristic for SHSC differs little from the angular characteristic for the SPAR and therefore they can be assumed to be coincident)

$$\int_{\delta_o}^{\pi-\delta_{01}} \left(P_{ET\,min} - P_{SPAR\,max} \cdot \sin\delta\right) d\delta = 0 . \quad (5)$$

After the transformations, we obtain the following equation for determining the minimum throughput that can be achieved with an unlimited pause of the SPAR

$$P_{et\,min}\left(\pi - arc\sin\frac{P_{et\,min}}{P_{n\,max}} - arc\sin\frac{P_{ET\,min}}{P_{SPAR\,max}}\right) - \frac{P_{SPAR\,max}}{P_{n\,max}}\sqrt{P^2_{н\,max} - P^2_{et\,min}} - \sqrt{P^2_{SPAR\,max} - P^2_{et\,min}} = 0 \quad (6)$$

The solution of this equation with the received data is given by $P_{ET\,min} = 675\,MW$.

Now, the dependence of transmission capacity on the condition of maintaining dynamic stability on the duration of the SPAR pause can be constructed (Fig. 9). In Fig. 9 also shows the size of the transmission capacity of the transmission by the condition of the dynamics when using the AHSR with a pause duration of 0.5 sec.

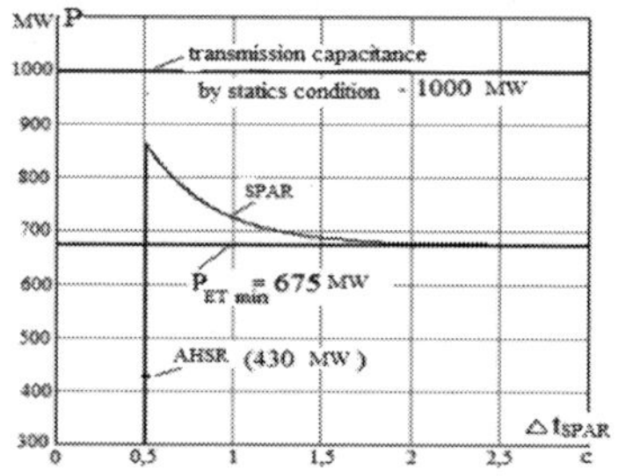

Fig. 9. The dependence of the transmission capacity of the transmission in condition for the dynamic stability from the length of the pause SPAR

On the basis of the above algorithm, we analyze the conditions for the extinction of the secondary arc during the pause of the SPAR in the projected AL of the Kyrgyz Republic-Republic of Tajikistan with a voltage of 500 kV and a length of 477 km. Lines of this length, as a rule, are equipped with three reactors, which provides a degree of compensation of the charging power of 100% (Figure 10).

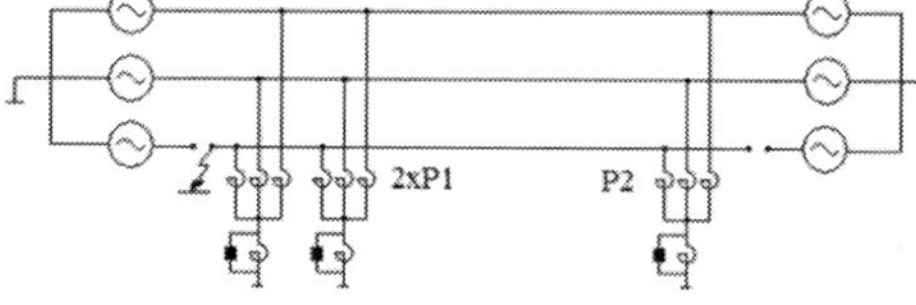

Fig. 10. Scheme of long power line with different number of reactors at the ends of the line

We will focus on the necessary conditions to ensure a successful SPAR. The duration of successful SPAR is projected in the range of 0.5-1.0 seconds, if the secondary arc current (SAC) will satisfy the condition

$$I_{\partial} \leq 15-46A_{\,ampli}$$

The second circumstance to consider is the creation of acceptable conditions of work overshoot suppression (OS) pause SPAR. Recovering voltage (RV) to pause SPAR in the place of installation of the arrester should be limited to $U_{RV} \leq U_{\text{cons.bea}}$, where $U_{\text{cons.bea}} = 230\kappa B$ for overhead lines with a voltage of 500 kV.

In Fig. 11 shows the distribution of RV in pause SPAR for different phases of the resistance of the neutral reactor, and Fig. 12 the dependence of RV at the end of the line from the angular shift δ when SPAR in phases a, b, c in the real transposition. In the case of actual transposition of RV depends on which phase is SPAR.

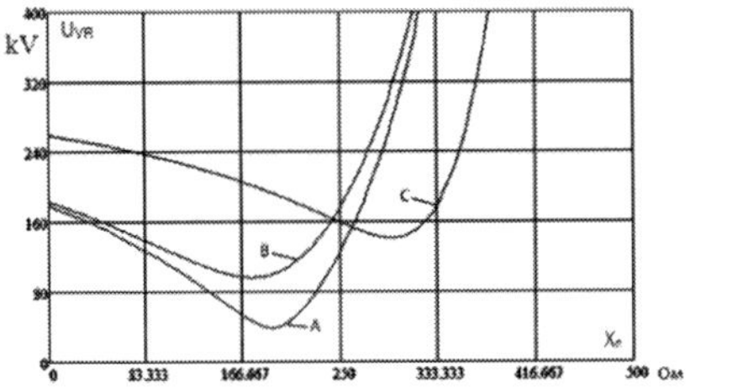

Fig. 11. The distribution of RV in the phases of AC in pause SPAR in real transposition

As we can see from figure 11, the circuit is resonant, and the voltage level at the critical phase exceeds the permissible increasing the voltage on the OS located at the ends of the line.

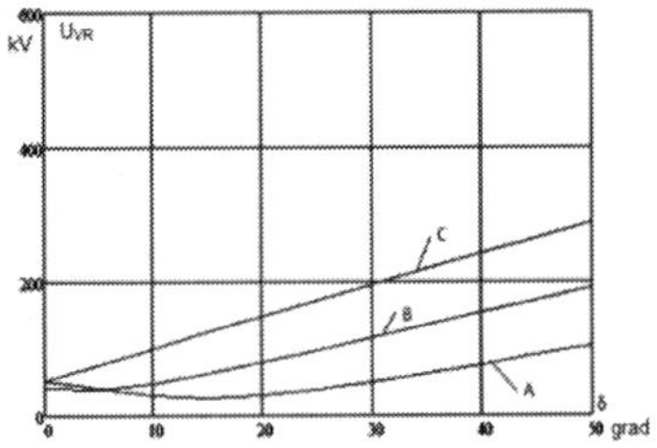

Fig. 12. The dependence of RV on phases at the end of the line from the angular shift between the voltage in a pause SPAR in real transposition

A possible solution to the problem of successful SPAR is to install in series with the shunt reactor (SHR) at the end of the line in the critical phase from the neutral conclusions of the capacitor bank (CB), shunted in normal operation the switch (Fig. 13).

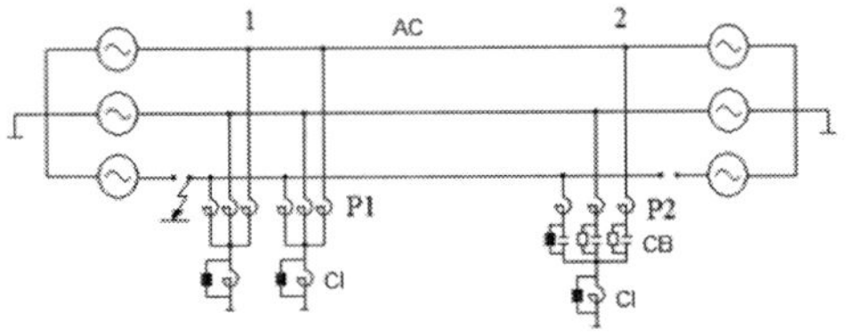

Fig. 13. Device to improve the efficiency SPAR CB in all phases

The calculation shows that the maximum SAC is estimated by the value $I_{arc} = 35A_{ampli}$, and the corresponding pause duration SPAR - 0.8 sec.

4 Conclusion

The using SPAR compared AHSR allows to increase the throughput of long-range transmission condition of dynamic stability is 1.5 – 2.0 times. The greatest effect is the use of SPAR if the pause duration SPAR is in the range 0.5-1.0, which requires limiting the secondary current to arc (30-45) A (peak value). If you pause more (1.5 to 2.0) sec with its duration has no significant effect on the bandwidth condition of the dynamics, and in this case, the pause duration is determined only by the condition of quenching secondary arc current (SAC).

When calculating RV in the end of the line with a short-circuit on phase "a" resonance is not observed, therefore no need to connect the capacitor banks to this phase. When a short-circuit on phases "b" and "c" require the installation of capacitor banks.

References

1. A.S. Zelichenko, B.I. *Smirnov Proektirovanie mekhanicheskoi chasti vozdushnykh linii sverkhvysokogo napriazheniia [The planning of mechanical part of overhead lines of extra high voltage]*. (Moscow, Energoizdat Publ., 1981. 336 p).

2. I.G. Barg, V.I. Edel'man *Vozdushnye linii elektroperedachi: Voprosy ekspluatatsii i nadezhnosti [The overhead transmission lines: Issues of operation and reliability]*. (Moscow, Energoatomizdat Publ., 1985. 248 p).

3. N.N. Beliakov, V.S. Rashkes, M.L. Levinshtein, M.I. Khoroshev *Perspektivy primeneniia OAPV v elektroperedache 1150 kV [The prospects of application of the Single-Phase Auto-Reclosing in power transmission 1150 kV]. Elektroperedachi 1150 kV [1150 kV transmission]*. (Moscow, Energoatomizdat Publ., 1992, bk. 1, pp. 129-158).

4. T.G. Krasil'nikova, S.G. Dzhononaev Sbornik nauchnykh trudov Novosibirskogo gosudarstvennogo tekhnicheskogo universiteta, **no. 2 (84)**, pp. 116-130. (2016).

5. Iu.P. Ryzhov, ed. Rezhimy *dal'nikh elektroperedach v primerakh [The modes of long-distance transmission in the examples]*. (Moscow, MPEI publ., 1994. 88 p).

6. V.A. *Venikov Perekhodnye elektromekhanicheskie protsessy v elektricheskikh sistemakh [The transient Electromechanical processes in electrical systems]*. (Moscow, Vysshaya shkola Publ., 1985. 536 p).

Efficiency of using cables on 6-110 kV with sticked polyethylene isolation

Michail Korotkevich[1*], *Sergey* Podgaisky[2], *Andrey* Golomuzdov[2]

[1] Belarusian National Technical University, Minsk, Republic of Belarus
[2] JSC "Energokomplekt" Industrial Association, Vitebsk, Republic of Belarus

Abstract. The expediency of using more expensive 6-10 kV cables with XLPE insulation was proved by us on the basis of the method of multi-purpose optimization, which took into account both the resulted costs for the construction and operation of cable lines, and the reliability of its operation, as well as the convenience of installation. Without compensation of capacitive currents, permissible difference in the height of the cable laying, the effect of cable lines on the environment.

1 Introduction

The expediency of using more expensive cables with XLPE insulation in comparison with cables with paper-and-oil insulation will be considered for cable lines made with cables:

- three-core cross-sectional area of conductors 35-240 mm^2 with a voltage of 6-35 kV;
- single-core cross-sectional area of conductors 35-240 mm^2 (cables with XLPE insulation) and three-core with the same cross-sectional area of wires (cables with paper-and-oil insulation) with a voltage of 6-35 kV;
- single-core cross-sectional area of strands 150-630 mm^2 (cables with XLPE insulation) and single-core cross-sectional conductors (oil-filled cables with paper-oil insulation) with a voltage of 110 kV.

To assess the indicated expediency of using power cables with XLPE insulation in relation to cables with impregnated paper insulation, we will use both the reduced cost method and the multi-purpose optimization method, which allows taking into account not only the cost but also the quality indicators.

2 The resulted annual costs for the construction and operation of cable lines

Consist of the costs for the construction of the cable line $К_{к.л}$ and the cost of its operation И, i.e.

$$З = Е_к К_{к.л} + И \qquad (1)$$

where $Е_к$ - coefficient of efficiency of capital investments (with refinancing rate of the National Bank of the Republic of Belarus equal to 14%, value $Е_к = 0{,}14$).

Capital costs for the cable line construction $К_{к.л}$ are comprised of the cost of purchasing the cable $К_к$ and the cost of its installation $К_л$. In the total cost of building cable lines, the cost of cables is on average 48% for 6-10 kV lines; 57.5% for 35 kV lines [1], 23-52%, or an average of 37,5% for lines with a voltage of 20 kV [2] and 33% for lines with a voltage of 110 kV [1], i.e. $К_к/К_{к.л}$= 0.48; 0.375; 0.575; 0,33 - respectively for cable lines with voltage 6-10, 20, 35 and 110 kV.

The cost of laying three-core cables with viscous or artificial insulation can be considered equal, and equal to 52; 62.5; 42.5; 67% of the total cost of the cable line, respectively, for lines with a voltage of 6-10; 20; 35 and 110 kV. Here the cost of laying was estimated as

$$К_п = (1 - \frac{К_к}{К_{к.л}}) К_{к.л} \qquad (2)$$

The ratio of the cost of building a cable line with cables with XLPE insulation $К^п_{к.л}$ to the cost of building a cable line with cables with paper-oil insulation $К_{к.л}$ can be represented as

$$\frac{К^п_{к.л}}{К_{к.л}} + 1 + \frac{\alpha_2 - 1}{\alpha_1} \qquad (3)$$

where the coefficient taking into account the number of times the cost of building a cable line exceeds the cost of the cable itself; the value of $\alpha 1$ = 2.08; 2.67; 1.74; 3,3 - respectively, for lines with a voltage of 6-10; 20; 35; 110 kV; a factor that takes into account the increase in the cost of a cable with XLPE insulation compared to the cost of a cable with combined insulation.

It is established from the expression (3) that the increase in the cost of a cable with XLPE insulation in comparison with the cost of a cable with paper-oil insulation in 1,2, ..., 2,0 times (i.e., $\alpha 2$ = 1,2-2 , 0) will lead to an increase in the cost of building a line with XLPE insulated cable in 1,1, ..., 1,48; 1.07, ..., 1.37; 1.11, ..., 1.57; 1.06, ..., 1.3 times, respectively, for lines with a voltage of 6-10; 20; 35; 110 kV.

When comparing the capital costs for the construction of a cable line with single-core cables with XLPE insulation and with three-core cables with paper-and-oil

* Corresponding author: elsyst@tut.by

insulation, one must bear in mind both an increase in the cost of rolling three cables and a reduction in the cost of laying cables with XLPE insulation at the expense of :

- a smaller mass of drums with a cable, exported to the track;
- the need to install a smaller number of couplings due to large cable lengths;
- smaller diameter of each of the single-core cables compared to a three-core paper-insulated cable;
- a smaller permissible bending radius in comparison with three-core cables with an aluminum shell;
- there is no need to warm up the cable when installing in a negative temperature environment.

When assessing the cost of cables with XLPE insulation, it is necessary to take into account the fact of increasing the cable capacity due to the increase (up to 90 C) of the permissible heating temperature of the cores [3, 4].

The increase in the long-term permissible current corresponds, as it were, to an increase in δF of the required cross-sectional area of the cable cores, i.e.

$$\delta F = \frac{1}{j_н}(I_д^п - I_д) \quad (4)$$

where $j_н$ - is the normalized current density; for cables with paper insulation with aluminum or copper conductors; take 1.4 or 2,5 A/mm^2 (with the number of hours of use of the maximum load of 3000-5000 h [5]); $I_д^п, I_д$ - long-lasting current for cables with polyethylene and paper insulation.

The values of the proportion of the increase in the cross-sectional area of cables with XLPE insulation as compared to cables with traditional insulation are:

- when comparing three-core cables with a voltage: 10 kV (0,07-0,21) (copper cores); (0,11-0,28) (aluminum core); (the lower values for the section are 240 mm^2, the upper ones are 70 mm^2); 20 kV (0,17-0,27) (copper cores); (0,25-0,45) (aluminum core); (the lower values for the section are 185 mm^2, the upper ones are 70 mm^2); 35 kV (0,20-0,24) (copper conductors); (0,28-0,36) (aluminum core); (the lower values for the section are 150 mm^2, the upper ones are 120 mm^2);
- when comparing three-core cables with paper insulation and single-core cables with XLPE insulation: 10 kV (0,13-0,34) (copper conductors); (0,09-0,45) (aluminum core); 20 kV (0,24-0,42) (copper cores); (0,34-0,61) (aluminum core); 35 kV (0,27-0,33) (copper conductors); (0,39-0,46) (aluminum core); here the upper and lower values refer to the same sections as above;
- when comparing single-core cables at a voltage of 110 kV (0,12-0,32) (copper cores); 0,19 - aluminum conductors; (lower values for section 630 mm^2, upper 150 mm^2).

The cost per unit length of cable with polyethylene insulation $K_к^п$

$$K_к^п = K_{из}^п + F(1-\sigma)\gamma c \cdot 10^{-6} \quad (5)$$

where $K_{из}^п = G_{из}c_{из}$ - the cost of polyethylene insulation; $G_{из}$ is the mass of insulation, kg; $c_{из}$ - the cost of one kilogram of insulation; is approximately equal to 209 s.m.u/kg; c - the cost per unit of mass of the core (equal to 1,94 and 5,37 s.m.u./kg respectively for aluminum and copper, respectively); γ - specific volumetric weight or density of the vein material (for copper is equal to 8890 kg/m^3, aluminum - 2703 kg/m^3); F - cross-sectional area of the cable core, mm$^{2;}$ σ coefficient less than 1, taking into account the increase in the capacity of cables with polyethylene insulation by increasing the permissible temperature of heating the veins; (the limits of the change in the cross-sectional area of the wires are given above).

The cost of thecopper core material exceeds the cost of polyethylene insulation in (1,8-19,7) times, the cost of the aluminum core material - in (1,1-1,9) times. The cost of an aluminum core material with a cross section of 150-630 mm^2 for 110 kV cables is approximately equal to the cost of a cross-linked polyethylene insulation. With a cross-sectional area of aluminum veins up to 70 mm^2, the material cost of veins is less than the cost of insulation. The ratio of the cost of material and fabrication of cores to the cost of material and the manufacture of cable insulation can lie within (0,1-1.0).

Through cable lines with cables with XLPE insulation, compared to paper-and-oil-insulated cable lines, operating costs are reduced by unnecessary of:

- mapping of soil corrosion zones;
- coating with anticorrosion varnishes and paints of non-zinced steel armor cables laid in cable structures;
- periodic replacement of vertical sections of cable lines due to dehumidification;
- monitoring the pressure and sampling of oil from oil-filled cable channels and couplings to determine the characteristics of the oil and the presence of undissolved air in it; monthly inspections of the feeding points (for oil-filled lines with a voltage of 110 kV);
- periodic (on average, once every 1,8 years) isolation tests with an increased rectified voltage (here it is recommended to conduct a measurement of the dielectric permittivity of insulation against the applied voltage and frequency every 5 years);
- replacement of end couplings after 20 years of operation.

As a result, the cost of maintaining cable lines with XLPE insulation can be reduced compared to the associated costs of servicing paper-oil-insulated cable lines to:

9% for cable lines with a voltage of 35 kV;

17% for cable lines with a voltage of 110 kV.

Nevertheless, the costs for the construction and operation of XLPE insulated cable lines exceed the same costs for cable lines with paper-oil insulation in (1,03-1,43) times.

3 Integrated assessment of the feasibility of using cables with XLPE insulation

A comprehensive assessment of the feasibility of using cables with XLPE insulation is given using the multi-target optimization method.

It should be noted that single-core oil-filled cables with paper insulation for 110 kV are currently not manufactured by industry and their presence in this report is of only methodological importance. In terms of the method of multi-purpose optimization, the problem is formulated as follows: it is necessary to evaluate the feasibility of using cables with XLPE insulation for a voltage of 6-110 kV in comparison with cables with paper-oil insulation when achieving the following objectives:

- the minimum of the resulted costs for the construction and operation of cable lines (target № 1);
- maximum reliability of work (target № 2);
- maximum ease of installation (goal № 3);
- the maximum length of the line without charging power compensation (target № 4);
- the maximum permissible difference in the height of the cable laying (target № 5);
- minimal impact on the environment (target № 6).

The expediency of using more expensive cables with XLPE insulation is ensured when the maximum optimization criterion E → max, [6] is reached, i.e.

$$E = \sum_{i=1}^{n} v_i e_i \to \max \tag{6}$$

where v_i - is an estimate of the importance of the i-th goal, i = 1, ..., n; e_i - is the relative effectiveness of the i-th goal.

A point estimate of the values of vi was obtained on the basis of the ranking of these goals by experts-specialists in the design, construction and operation of networks, the total number of which was 32.

The results of determining the importance of goals: v_1 = 0,174; v_2 = 0,281; v_3 = 0,185; v_4 = 0,1377; v_5 = 0,1155; v_6 = 0,1066.

To assess the trouble-free operation of cable power lines, the failure flow parameter λ and the mean time between failures (inversely proportional to λ) are used. Experience in the operation of cable lines with cables with XLPE insulation showed that the failure flow parameter for such lines is an order of magnitude smaller, and the operating time for failure is an order of magnitude greater than for cable lines with paper-oil insulation.

Convenient installation of cable lines depends on the following factors, typical for cables with traditional insulation and insulation from cross-linked polyethylene: the mass of drums with cable, which must be taken out to the track for the subsequent installation of one kilometer line; outer diameter of the cable; permissible bending radius, allowable negative temperature of outside air, at which the cable is laid without its preheating.

We established [7] that the mass of drums with a three-core cable with aluminum conductors, lead sheath and paper-oil insulation impregnated with a non-sticking compound for a voltage of 10 kV is greater than the weight of the drums with a three-core cable for the same voltage with XLPE insulation of approximately 1, 3 times.

The weight of drums with a three-core cable with paper and oil insulation with aluminum cores and aluminum shells for voltage 6 and 10 kV is less, and with a lead sheath greater than the mass of drums with a three-core cable with aluminum conductors and XLPE insulation approximately 1,12 times.

The weight of drums with a single-core cable with aluminum conductors with XLPE insulation is less than the weight of a three-core cable with paper insulation and with the same core material approximately in 1,09-1,5 times.

The weight of drums with single-core copper oil-filled 110 kV cable is 1,36 times higher than with copper single-core cable with XLPE insulation.

The diameter of three-core cables with a voltage of 10 kV and paper insulation is less than the diameter of the same cables with XLPE insulation of approximately 1,23 (ААБ and АПвП cables) and 1,16 (cables of the ЦАСБ and АПвП type) times or an average of 1 , 2 times.

The diameter of single-core oil-filled cable with paper insulation for 110 kV is more in 1,1-1,15 times the diameter of the cable at the same nominal voltage with XLPE insulation.

The diameter of three-core cables with a voltage of 10 kV and paper insulation is greater than the diameter of a single-core cable with XLPE insulation in 1,75 (AAB and APVP cables) and 1.83 (cables of the CASB and APVP type) times or an average of 1,8 times.

Three-core cables with polyethylene insulation have a smaller permissible bending radius in comparison with three-core cables with paper insulation in:

1,5 times - for cables with a lead sheath;

2,5 times - for cables with aluminum sheath.

Single-core cables with polyethylene insulation have a smaller (1,7 times) permissible bending radius in comparison with three-core cables with paper insulation and aluminum shells.

The possible temperature range at which the cable is laid without its preheating:

- cables with paper-oil insulation - 41°C (from 0 to + 40°C);
- cables with polyethylene insulation - 61°C (from minus 15 or 20°C to + 40°C, including 0°C). Therefore, the maximum temperature range at which the cable is laid without its preheating, for cables with polyethylene insulation is 1,4 times greater than for cables with paper-oil insulation.

Estimation of the level of convenience of installation will be given on the basis of the method of multi-purpose optimization with the same importance for all purposes (vi = 0,25), when it is necessary to ensure:

- minimum weight of drums with cable;
- the minimum outer diameter of the cable;
- the minimum radius of bending;
- the maximum permissible temperature range when laying the cable without preheating it.

Relative values of targets for cables with paper-oil insulation are taken as one; for cables with XLPE insulation were determined relative to the ones taken for the unit.

The relative effectiveness of the ei goals was calculated using the formulas:

- for minimized targets

$$e_i = \frac{\min x_i}{x_i}, \tag{7}$$

- for maximized goals

 DOI: 10.1051/e3sconf/20172501010

$$e_i = \frac{x_i}{\max x_i}, \quad (8)$$

where x_i is the current value of goal indicators; $\min x_i$, $\max x_i$ - minimum and maximum values of target indicators. With the same importance of all goals, the value of the optimization criterion E (formula (6)) is calculated as

$$E = \frac{1}{4}\sum_{i=1}^{4} e_i \quad (9)$$

For three-core cables with XLPE insulation (E = 0,95), the installation of the cable is large by 1,2 times (compared to the installation of a cable type ААБ and ЦАСБ) convenience.

Installation of a single-core cable with XLPE insulation for 10 kV voltage is 1,4 and 1.6 times more convenient, respectively, compared to the installation of three-core cables of the ААБ and ЦАСБ type having paper-oil insulation.

Installation of a line of single-core cables with XLPE insulation for 110 kV voltage is 1,4 times more convenient than installing a single-core oil-filled cable with paper insulation for the same voltage.

It is known [8, 9] that the value of the charging current is lower for cables with polyethylene insulation (because of the lower value of the working capacity compared to cables with paper-oil insulation), the value of the dielectric power losses taken into account for 110 kV cables is almost an order of magnitude lower kV.

Compensation of capacitive currents of a single-phase earth fault in electrical networks with a voltage of 6; 10; 20; 35 kV should be provided when the value of these currents exceeds 30; 20; 15; 10 A respectively.

The limiting length of cable lines with cables with polyethylene insulation is 1,5 times the permissible length of lines with cables with paper-oil insulation by the condition of operation without compensation for capacitive earth-fault currents and is equal to the ratio of the relative dielectric permeability of paper-oil insulation equal to 3,5 , to the relative dielectric permeability of cross-linked polyethylene equal to 2,3.

Cables with XLPE insulation allow an unlimited difference in height of the gasket, while the allowable difference in cable heights with paper-and-oil insulation is 15 m.

The unlimited difference in height of the cable laying with polyethylene insulation in comparison with cables with paper-oil insulation can be numerically described as follows

$$\delta h = A\, \mathbf{lg}\frac{h^{п}}{h} \quad (10)$$

where δh - is an estimate of the increase in cable routing height with cross-linked polyethylene insulation; $h^{п}$, h - is the permissible difference in cable routing heights with XLPE insulation (for the conditions of the Republic of Belarus more than 1000-1500 m is not required) and the permissible difference in cable heights with paper-and-oil insulation; (equal to 15 m).

Given the extreme values of $h^{п}$ and h, their ratio is 100 and the value of δh = 2A. For A = 1, δh = 2.0.

The operation of cable lines is accompanied by the influence of the environment on them and the influence of cable lines on the environment (land, water, air).

Nature and external factors limit the conditions for laying cables:

- with XLPE insulation and polyvinyl chloride outer sheath - the value of soil moisture;
- with paper impregnated insulation - the values of corrosive (acidic, alkaline) soil activity, vibration.

In general, it can be said that the environment equally affects the cable lines, regardless of the insulation material of the cables and this effect is reflected in the reliability of cable lines.

Consider the influence of underground cable lines on the environment.

On the lines of cable lines, agricultural work is allowed, but the construction of houses and planting of large trees is excluded.

The electrical and magnetic fields around the cable line even at 500 kV are insignificant. However, when the cable is operating, its surface is heated to a temperature of 50-55°C or 70-75°C (cables with paper-oil and polyethylene insulation, respectively), which leads to soil drying and has negative effects on living organisms living there. the surface of cables with XLPE insulation is heated 1,4 times more than the surface of paper-insulated cables.

The peculiarity of using a cable with XLPE insulation, compared to cables with paper-oil insulation, is that they do not have a liquid dielectric that can flow into the soil when performing repair work on 6-35 kV lines or leak out of pressure tanks , pipelines, end, connecting and locking clutches on lines with oil-filled cables with a voltage of 110 kV [10].

The absence of oil leakage from the cross-linked polyethylene insulation of the cable in terms of environmental impact will be estimated by the coefficient 2 (as well as by formula (10)), i.e. Damage to the environment will be 2 times less than for cables with traditional insulation.

When operating cable lines with cables that have a toxic lead sheath, additional costs are required for the rehabilitation of workers in contact with lead. Impact of lead on the environment refers to the high (third) level.

Cables with XLPE insulation can have a weakly toxic PVC jacket, the decomposition products of which cause irritation of the upper respiratory tract of the person and mucous membranes of the eye and relate to the average (second) level of environmental impact.

The foregoing allows us to establish the correlation of the factors of the effect of cable lines with paper and artificial insulation on the environment, i.e.

$$\frac{И^{п}_{окр}}{И_{окр}} = \frac{И^{п}_1}{И_1}\frac{И^{п}_2}{И_2}\frac{И^{п}_3}{И_3} = +1{,}4\frac{1}{2}\cdot\frac{2}{3} = 0{,}47 \quad (11)$$

where $И^{п}_{окр}$, $И_{окр}$ - factors affecting the environment of cable lines with polyethylene and paper-oil insulation; $И^{п}_1$, $И_1$ - annual expenses for the reproduction of natural resources associated with heating the soil with cables with polyethylene and paper-oil insulation; $И^{п}_2$ and $И_2$ - annual expenses on reproduction of the natural resources connected with leakage of oil oil in soil; $И_з^п$, $И_з$ - annual costs associated with the work of personnel with toxic (lead and polyvinylchloride) materials. In the absence of polyvinyl chloride shell $И_{окр}^п = 0{,}35 И_{окр}$.

It follows from expression (11) that the effect of cables with polyethylene insulation on the environment will be less than that of cables with paper-oil insulation more than 2 times.

For a comprehensive assessment of the efficiency of using 6-110 kV cables with XLPE insulation, we use the method of multi-purpose optimization and the above formulation of the problem.

The value of targets in relative units for three-core and single-core cables with paper-oil insulation will be taken as one. The cost of a cable with XLPE insulation is more than twice the cost of a cable with paper-and-oil insulation.

The values of the relative effectiveness of goals are determined by the formulas (7), (8), and the optimization criterion - according to the formula (6) with the values of the importance of goals.

The values of the optimization criterion E for cables with paper insulation and XLPE insulation are shown in Table 1.

Table 1 - Values of the optimization criterion

Name of cables	Optimization criteria for cables		
	ААБ	ЦАСБ	АПвП
1. Three-core cables for 10 kV	0,558	0,615	0,959
2. Three-core cables with paper-oil insulation and single-core cables with XLPE insulation for 10 kV	0,536	0,579	0,959
3. Single-core oil-filled cables with paper-oil insulation type МНС and single-core cables with XLPE insulation type ПвП for 10 kV	0,536 (МНС)	-	0,967 (ПвП)

As can be seen from the data in Table 1, the value of the optimization criterion characteristic of cables with XLPE insulation exceeds the corresponding value of the optimization criterion for cables with paper-and-oil insulation in 1,6-1,8 times, which means the unconditional efficiency of using cables with insulation from cross-linked polyethylene in comparison with cables with paper-oil insulation.

4 Conclusion

The expediency of using more expensive 6 ... 110 kV cables with XLPE insulation was proved by us on the basis of the method of multi-purpose optimization, which took into account both the resulted costs for the construction and operation of cable lines, and the reliability of its operation, as well as the convenience of installation. Without compensation of capacitive currents, permissible difference in the height of the cable laying, the effect of cable lines on the environment.

References

1. V. A. Kozlov *The Power Supply of Cities*. (Leningrad, Energiya Publ. 280. 1977).

2. A. V. Maiorov, A. V. Shuntov. *Methodical Problems of Research of Reliability of Large Energy Systems. Issue 66: Topical Problems of Reliability of Energy Systems*. Minsk, BNTU, p.125–132 (2015)

3. *Rules for Eleecrical Units Installation*. (Moscow, Energoatomizdat Publ. 646 p. 1987).

4. *The Cables and Wires. The Product Catalog*. (Vitebsk, "Energocomplekt" Production Association. 58 p. 2015).

5. *Handbook on the design of electric power systems*. (Moscow: Energoatomizdat, 1985. - 352 p).

6. Yu. B. Guk *Reliability Analysis of Electrical Installations*. (Leningrad, Energoatomizdat Publ., Leningrad Branch. 224 p. 1988).

7. M. A. Korotkevich *Operation of Electric Networks*. (Minsk, Vysheishaya Shkola Publ. 350 p. 2014).

8. *A guide to high voltage electrical installations*. (Moscow: Energoatomizdat, 1989. - 768 p).

9. B. Widy *High voltage cable lines* (Moscow: Energoatomizdat, 1983. - 232 p).

10. E.T. Larina *Power cables and cable lines* (Moscow: Energoatomizdat, 1984. - 368 p).

On low-resistance neutral earthing mode in 20 kV overhead and cable networks

Andrey Mayorov[1], *Kirill* Osintsev[2,*], *and Andrey* Shuntov[2]

[1]Public company «United Energy Company», 115035 Raushskaya Naberezhnaya, 8, Russian Federation
[2]National Research University «Moscow Power Engineering Institute», 111250 Krasnokazarmennaya 14, Russian Federation

Abstract. The problems of justification and selection of the required single-phase short-circuit current values in cable and overhead networks of 20 kV with low-resistance neutral earthing are considered. It is shown that the desired values of the short-circuit current can be determined on the basis of harmonization of conflicting influencing factors: reliability of the relay protection and automation devices and required resistances of the earthing devices of electrical installations, including personnel safety. In this case, the main influencing factor is the electrical network structure and parameters

1 Introduction

The advantages of low-resistance neutral earthing mode in medium-voltage networks are well-known. They are conditioning for almost complete elimination of high frequency arc overvoltage and transition of single-phase short circuits to interphase (multiplace) short circuits (SC), avoidance of the personnel injury at the single-phase ground fault (SFGF), selective operation of relay protection and automation devices (RPA) at the SFGF and some other advantages.

2 SFGF current selection

The cable 20 kV electrical network with low-resistance neutral earthing has massively been introduced in Moscow since the beginning of the 2000th. At the first stages of decision-making Russian specialists took into account the experience of the west-european countries (primarily – France), where 20 kV networks were widespread from the second half of the last century [1].

By copying the 20 kV network construction scheme of Paris, at all main substations (MS) in the 20 kV neutrals of 110–220/20 kV supply transformers there were installed resistors with the resistance of Rr=12 Ohms. At the same time the SFGF current (in fact – the current running through the resistor) $I_r \approx U_{nom}/(1.73 \cdot R_r) = 20/(1.73 \cdot 12) = 0.96$ kA, where U_{nom} – nominal network voltage. The reason for choice of the current at the level of 1000 A was earlier mentioned in [2]: "Historically the value of 1000 A was adopted due to low sensitivity of the previous protection systems against ground faults...with the minimum possible setting of 0.5A (on the secondary side) which switched on phase current transformers of 1000/5 A... Respectively with the setting of 0.5 A on the secondary side the minimum possible setting on the primary side was 100 A. According to the long familiar foreign rule of relay protection for reliable operation of protection the SFGF current shall exceed the setting by 10 times. Therefore the value of the resistor current shall be nothing less than 1000 A".

At the same time nobody paid attention to foreign 20 kV network structure [1, 3]. 20/0.4 kV transformer substations (TS) commute with the 20 kV main line basically on branch lines or by load interrupter switches. The switches are installed only on connections of the 20/0.4 kV transformers. With the architecture of that kind all the electrical network is deemed to be the distribution one.

Ever since the times of the USSR, for decades the electricity supply systems of the cities in our country have been formed differently, according to the so-called two-tiered architecture. The first tier is network feeders, i.e. cable lines (CL) from the MS to distribution centers (DC), and the second – distribution networks, i.e. CL from DC to TS. From them, at 0.4 kV voltage, end consumers are powered (Fig. 1).

* Corresponding author: kirmosh@mail.ru

 DOI: 10.1051/e3sconf/20172501011

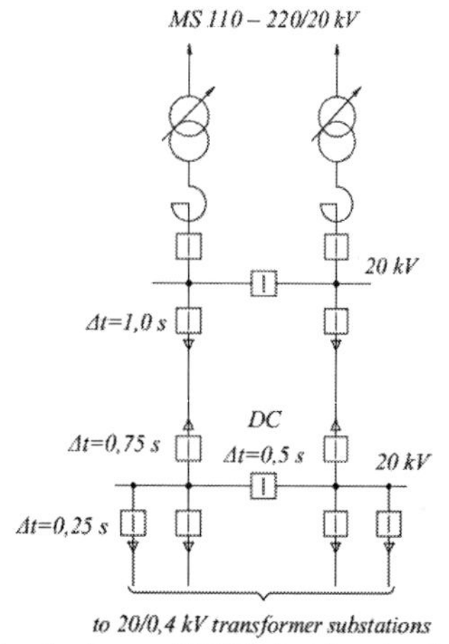

Fig. 1. Two-tiered scheme of 20 kV electric network.

With the two-tiered architecture, there appear three additional intervals of the RPA devices selectivity: on the main switch and bus section breaker of the DC, as well as on the switches that go to the TS lines; total $\Delta t=3\cdot0.25=0.75$ s. As a result, abroad for current protections, departing from the MS lines it is sufficient to have time delay of 0.3–0.4 s, and for the two-tiered architecture – not less than 1.0 s. The latter extremely stiffens nominal conditions for feasibility and selection of conductors, machineries and earthing devices of electrical installations.

In Russia, there is no regulatory framework for development of the 20 kV electric network. According to [4], operation of networks up to 35 kV could provide for different neutral earthing mode, including through a resistor. At the same time, the requirements for earthing devices of electrical installations above 1 kV, which is one of the basic criteria of the electrical safety, are standardized only for networks with effectively earthed and isolated neutral. Therefore, when providing electrical safety in a 20 kV network with low-resistance neutral earthing, there remains to take into account the standards for contact voltage and step potential [5, 6].

Thus, according to the well-known permissible fault voltage curve (on contact) $U_f(t)$ for the time of fault (trip) t given in [6], it is not difficult to estimate the required resistance of the earthing device $Re(t) = U_f(t)/I_r$ depending on the SFGF current $I_{sfgf} = I_r$, created by the resistor (Table 1).

Table 1. Values of the permissible resistances of the earthing devices.

SFGF trip interval, sec	up to 0,1	0,2	0,5	0,7	1,0	1,0–5,0
Fault voltage, V	500	400	200	130	100	70
Permissible resistance of the earthing device, Ohm						
I_r = 1000 A	0,49	0,39	0,20	0,13	0,10	0,07
I_r = 800 A	0,6	0,49	0,24	0,16	0,12	0,09
I_r = 600 A	0,80	0,63	0,32	0,21	0,16	0,11
I_r = 400 A	1,12	0,89	0,45	0,29	0,22	0,16

Note: The resistances of the earthing devices were evaluated under the condition of imposing capacitive current with the value of 200 A on the resistor current.

From Table 1 it follows that with the adopted on all MS of $I_r\approx1000$ A, two-tiered net-work architecture and the SFGF trip time delay of 0.75 s (see above), the resistance of the earthing devices for the 20 kV DC in the cable networks should be slightly more than 0.1 Ohm, which is unattainable. In default of regulatory requirements for earthing devices of 20 kV electrical installations in the country, engineering companies took the easy route by adopting as the target value the minimum possible from [4] – 0.5 Ohm for electrical installations with effectively earthed neutral. However, in the urban setting it is extremely difficult to achieve even such resistances for widely used compact DC and TS, provided that high priced special depth electrodes are used.

The requirements for the earthing devices resistance are mitigated (see Table 1) as the SFGF current decreases. The latter is limited to reliable operation of RPA devices, namely the minimum permissible sensitivity coefficient Cs, which for cable networks is adopted as [4] $C_s > 1.25$, and for overhead – $C_s > 1.5$.

Overcurrent SFGF protection in cable networks of the above-mentioned region is made by undirected residual current protection (RCP). The method of choosing the parameters of their operation is known (see, for example, [7, 8]) and therefore is not described in detail.

In the switchgear cubicles there are installed single-phase current transformers. Based on them, zero sequence current filters (ZSCF) are configured. In this case, the operative current I_{op} of the RPA devices is graded from the unbalance current of the I_{ub} of the current transformers at the short-circuit (short-circuit currents, as a rule, are limited to 12 kA). As seen from Table 2 it turns to be impossible to provide the required sensitivity of the RCP here.

Table 2. Calculated values of the RPA sensitivity coefficients.

SFGF protection	Operative current alternative	Sensitivity coefficient $C_s=I_{sfgf}/I_{op}\geq 1.25$	Current/ resistor resistance, A/Ohm
Undirected residual current protection with ZSCF	$I_{ub}=C_{unif}\cdot\varepsilon\cdot I_{sc}$; $I_{op}=C_r\cdot I_{ub}$	I_{sc}=12 kA C_s=0.29–0.73	1000/12
As above, but with time grading from short circuits between phases	$I_{ub}=C_{unif}\cdot\varepsilon\cdot I_{load}$; $I_{op}=C_r\cdot I_{ub}$	I_r=400–1000 A C_s=6.7–16.7	90/130
Undirected residual current protection with CBCT	$I_{op}=C_r\cdot C_{ck}\cdot I_c$	I_c=68–85 A C_s=1.25–1.50	230/50

Legend: C_{unif} – coefficient of current transformers uniformity; ε – error of the current transformer winding; C_r – coefficient of reliability; C_{ck} – coefficient of capacitive kick; I_{sc} – short-circuit current.

In order to provide the required sensitivity for protection against the SFGF with the current I_r=400–1000 A, there could be introduced an additional time delay, exceeding that for protection from short circuits between phases by a selective interval (0.2–0.3 s), or blocking of the protection against the SFGF at the start-up of the overcurrent protection (OP). In this case, the RCP is graded from the load currents I_{load}. Here, at I_r=400–1000 A and theoretically possible I_{load} = 1000 A, the calculated sensitivity coefficient of the protections shall be (see Table 2) C_s= 6.7–16.7, which significantly exceeds the required value. Therefore, in order to ensure the normalized Cs≥1.25 for cable networks, it is sufficient to adopt the minimum current of the resistor I_{rmin}=90 A (130 Ohm). With that current, the sensitivity of the RCP shall be guaranteed oversized. However, introduction of an additional time delay is not advisable due to toughening of the requirements for the earthing devices resistance.

To ensure sensitivity of undirected RCP, responsive to the fundamental current harmonic $3I_0$ with an isolated cable core balance current transformer (CBCT), as the condition for choosing the current of the resistor there is adopted the offset from the peak capacitive current I_c of the connection. For the cable with a core section of 500–630 mm^2, with the capacitance C = 0.42–0.46 mfd/km and the maximum possible CL length of 15–17 km in the megapolis conditions, I_c=68-85A. Whence the required value of I_{rmin}=230A. It is provided by the resistor resistance of only 50 Ohm (see Table 2). Thus, for the real power system diagram of 20 kV, the value of the resistor current could be four times smaller than that accepted at the moment (about 1 kA). The latter is fundamentally important from the standpoints of acceptable resistance of earthing devices and ensuring the personnel safety.

Thus, when selecting the SFGF current in the networks with low-resistance neutral earthing, it is necessary to harmonize the conflicting influencing factors: reliability of RPA devices operation, required resistances of earthing devices of electrical installations, including the personnel safety. In this case, the main influencing factor is the structure and parameters of the electrical network, namely, its configuration, the schemes of distribution centers of the electrical installations, equipment parameters. Unlike the 20 kV cable networks (there have already been built more than 1000 km only in Moscow), the decisions on construction of the overhead networks of the voltage class and the neutral mode under consideration are only being made. In view of the above, it is critical to avoid previously made mistakes.

When choosing the SFGF current in the overhead networks, it is necessary to take into account their specific preferred configuration (Fig. 2) [9]. It is represented by a well-known loop circuit with connection from two geographically separated MS of 110/20 kV, sectionalized by reclosers – automatic distribution stations of the overhead line (OL). Transformer substations 20/0.4 are connected to the main line on branches with installation of a disconnecting device or recloser (depending on the branch length). Transformers 20/0.4 are protected by 20 kV fuses. In the normal operation mode the circuit layout is opened at one of the reclosers using automatic load transfer (ALT). The number of branches to the TS between switching devices of the main line is shown in Fig. 1 conventionally (usually there are 5–7 of them). Mainly there are installed single-transformer package substations (PTS) of the column (nominal transformer capacity S_{nom}=16–100 kV·A), mast (S_{nom}=160–250 kV·A) and kiosk (S_{nom}=400–1000 kV·A) types.

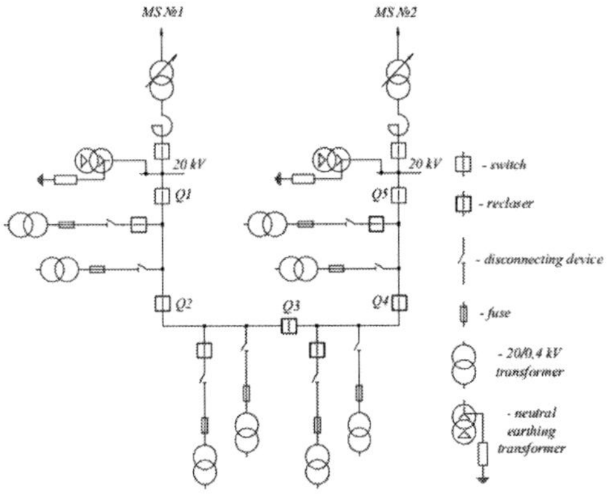

Fig. 2. Typical configuration of the overhead network.

The scheme shown in Fig. 2, in real conditions may even be more cumbersome and have a greater number of distribution stations. The minimum selective time interval of the RPA terminals of modern reclosers, guaranteed by the manufacturer, is only 0.1 s. The latter is less than that of the RPA devices installed in the switchgear cubicle (SC) of the MS and DC (0.2–0.3 s). However, even in the simplest scheme in Fig. 1 in the repair mode, let us assume, when the Q5 switch is turned off (in which case Q1 to Q4 shall be switched on), the time delay of the protection from the SFGF in the head sections will exceed 0.5 s, inclusive of the fuses blowing time (of the fuse links). Herewith, the cost of the earthing device may exceed (and noticeably) the expenses for the PTS itself, which is unreasonable. Therefore, interval grading of the RPA devices in the overhead networks is not applicable.

The operation of undirected residual current protection shall be based on the minimum possible time Δt of unselective circuit breakers operation at the SFGF and then their sequential single acting autoreclosing (AR) with acceleration of protection, starting with the MS circuit breaker. The protection acceleration time t_a in the overhead networks is usually taken to be not less than 0.1 s. The minimum possible time for the protection grading t_{gr} of the MS and the "downstream" recloser is 0.2 s, i.e. the selective time interval, guaranteed by the manufacturer for the RPA devices (see above). Therefore $\Delta t = t_a + t_{gr} = 0{,}1+0{,}2=0{,}3$ s.

For protection against the SFGF in the overhead networks, due to obvious constructional features, it is necessary to focus on the use of zero sequence current filters (rather than zero-sequence current transformers, as in cable networks). The operating current is assumed to be the highest, based on the three conditions of grading from:

• The unbalance current;
• The own capacitative current of the connection;
• The time-current characteristics of the fuse links melting.

The first two conditions of grading in the overhead networks, as a rule, are not determinative. When grading from the unbalance current, premised on the three-phase fault current $I^{(3)}$ at the entrance of the protected zone, usually $K_s<1.5$. Therefore, in order to increase the sensitivity of the RPA devices, it is necessary to provide for blocking of the protection against the SFGF when starting the overcurrent protection (OP). In this case, operating current of the OP instead of $I^{(3)}$ is taken into account. As for the grading from the capacitative currents, the latter in the overhead networks, all other conditions being equal, in comparison with the cable networks are about the next lower order and are not a weighty influencing factor from the positions under consideration.

The third condition (see above) is the determining factor when choosing the actuation data of the RPA devices. The time Δt = 0.3 s shall be sufficient for blowing in the first place the fuse of the protected connection if the latter is damaged. Therefore, the tripping currents of the RPA devices should be adjusted from the time-current characteristics of the fuse links melting in the following manner [8]: $I_{tr} = (1+\varepsilon)\cdot I_f(t)$, where ε=0,15 is the coefficient, making allowance for errors of the RPA terminals and the current transformers (for reclosers 0.05 is sufficient); $I_f(t)$, – the melting current of the fuse-link of the fuses, depending on the time, taking into account the standard 20% variation of their time-current characteristics.

When selecting the resistance of the earthing devices of the PTS 10(6) kV, i.e. in the networks with the isolated neutral, more stringent requirements to their values are specified not by the side of 10(6) kV of the electrical installation (where $R_e \leq 250/I_{SFGF}$, and not more than 10 Ohm [4]), but the side of 0.4 kV. For it R_e=4 Ohms. The implementation of such earthing device is not burdensome, both financially and materially. It would be advisable that these values shall be preserved for the 20 kV electric network, for example, not 4, but at least 2 Ohms. As it was already noted earlier, R_e=0.5 Ohm is not acceptable, since the costs for it may exceed those for one particular PTS, i.e. the electrical installation for the wide use, which should be as economical as possible. For example, at R_e=4 Ohms, in the conditions of the Moscow region, it is sufficient to install six electrodes on the area of 5×10 m, and for R_e=2 Ohms – 15 on the area of 10×20 m. At the same time, the estimated cost of the earthing device according to the current prices shall be about 20 and 60 thousand rubles respectively.

Table 3 represents the characteristics of interrelations among the basic influencing factors (such as the transformer power, the operative current of the RPA devices, the resistance of the earthing device from the electrical safety position, the sensitivity coefficient of the RPA devices), at the SFGF current change from 100 to 400 A and its unselective circuit breaker operation at Δt=0.3 s.

In the calculations of Table 3, there were taken into account the actual parameters of the 10 kV network, which was supposed to be "switched" to the voltage of 20 kV. The length of the main line between the MS was 26 km with the total length with the branches of about 60 km. The SFGF current of 100–400 A is the current at fault close to the buses of the MS, the minimum SFGF current – the current at the end of the protected zone in the network maintenance diagram, i.e. at the temporary power supply from the single MS. The possible 40% overload of the oil transformers was taken into account. The shadows in Table 2 indicate the zones in which the required sensitivity coefficient of the RPA devices is provided. For the record, the RPA operative current at the unbalance current grading (with the allowance for blocking of the protection from the SFGF during the start-up of the OP) is 31 A, and in case of grading from its own capacitative current, it is only 21 A.

Table 3. Alternatives of the earthing device resistance at the simultaneous unselective operation of the RPA devices within 0.3 s

Transformer capacity, kV·A	Operative current of the RPA, A	SFGF current, A			
		100	200	300	400
		Minimum SFGF current, A			
		90	163	222	275
		Earthing device resistance, Ohm			
		4,25	2,13	1,42	1,06
		Sensitivity coefficient of RPA, rel. unit			
Up to 75	37	2,4	4,4	5,9	7,4
160	84	1,0	1,9	2,7	3,3
250	101	0,9	1,6	2,2	2,7
400	145	0,6	1,1	1,5	1,9
630	248	0,4	0,7	0,9	1,1
1000	386	0,2	0,4	0,6	0,7

3 Conclusion

Thus, in the opinion of the authors, for the time being, the SFGF current at the level of 200 A is the most compromise for domestic 20 kV overhead electric networks with low-resistance neutral earthing (the resistor resistance being 60 Ohm) under the following restrictions: the resistance of the earthing devices is not more than 2 Ohm, and the power of the step-down transformers – not more than 250 kV·A. With the capacity of above 250 kV·A, it is necessary to provide for unselective operation of RPA devices with regard to grading from the time-current characteristics of the fuses. Any SFGF shall be eliminated by unselective circuit breaker operation within 0.3 s, followed by the circuit restoration in the series of the sequential single-shot AR with acceleration of protections starting from the power center switch. The identified limitations of the transformers power may have a positive effect in terms of creation incentives for disaggregating of 20/0.4 kV transformer substations. This makes it possible to significantly simplify and reduce the cost of the 0.4 kV networks, which is an alternative to the tendency of the recent years – the introduction of a new intermediate voltage level equal to 0.95 kV.

Touching upon the 20 kV cable networks with the neutral earthing mode under consideration, it has to be recognized that only by means of the circuit designs and selection of the required minimum SFGF current, it is possible to achieve acceptable resistances of the earthing devices and the required electrical safety.

References

1. R. Pelisier, *Energy systems* (Higher School, Moscow, 1982)

2. A. Nazarychev, S. Titenkov, A. Pugachev, EEPIR **3**, 14 (2016)

3. V. Kozlov, *Municipal Power Supply* (Energoatomizdat, Leningrad, 1988)

4. *Electrical Installations Code* (KNORUS, Moscow, 2010)

5. GOST 12.1.038-82, *Occupational safety standards system. Electric safety. Maximum permissible levels of pick-up voltages and currents* (IPC Publishing House of Standards, Moscow, 2001)

6. GOST R 50571-4-44-2011 (IEC 60364-4-44: 2007), *Low-voltage electrical installatios. Part 4-44. Safety Requirements. Protection against voltage and electromagnetic disturbances* (Standartinform, Moscow, 2012)

7. V. Shuin, A. Gusenkov, *Protection against ground faults in 6-10 kV electrical networks* (RTC "Energoprogress", "Energetik", Moscow, 2001)

8. M. Shabad, *Calculations of relay protection and automation in distribution networks* (PPEIAT, St. Petersburg, 2003)

9. A. Vologin, EEPIR **4**, 76 (2016)

About loading of autotransformers in main electrical grids and conclusions for designing

Nikita Vasilenko[1], *Vladislav* Plotnikov[1], and *Andrey* Shuntov[1]*

[1]National research university "MPEI",Institute of Electrical Power Engineering, 14 Krasnokazarmennaya St, Moscow, Russia

Abstract: The justification issues and the capacity selection of autotransformers in the main electrical grids, which require the accounting of the operation regimes of power units for the future of 10 or more years are considered. Appropriate methodological provisions for justification and choice in terms of generalized load schedules, load capacity of transformer equipment, as well as assessing the reliability of electrical installations in the design and development of the energy systems are proposed.

1 Introduction

Autotransformers rated capacity (S_{rated}) is selected with taking into account of kind of load characteristic, permissible normal cyclic loading and emergency loading. The first could have place systematically under uneven daily load-graph, the second – in emergency when maintenance of consumer power supply is needed regardless of autotransformers overloading.

Starting point of determination of load capability of autotransformer is its daily load-graph converted into equivalent two-stage load graph. Methods of conversion and definition of load capability are specified in Russian literature [1-2 etc] and international literature [3]; they are well known and are not considered in this paper. Accepted practice of justification and selection autotransformer capacity does not bring methodical problems in the design of power-supply system of particular consumers. There is understanding of normal load graph shape for them.

However, it is alternately for electrical installations in main electrical grids. Particularly for large system substations with autotransformers 220 – 750 kV related to Unified National Power Grid. Selection of their capacity is performed on 10 and more year's long term in electrical energy system development scheme and main electrical grids development scheme, i. e. under conditions of material uncertainties. In this case, substation load graphs are defined not by particular consumers, but interconnection operational states. On practice project engineer operate not with load graphs, but with predicted values of main electrical grids power flows in the period of winter and summer maximal load.

Moreover, scope of regulation [1] does not apply to lion's share of autotransformers installed in main electrical grids – installations 220 kV and above with capacity from 125 to 800 MVA with OF cooling system. Therefore, project engineers do not have methodical base for making decision in the problem domain.

Finally, consideration of virtually any possible rated condition in electrical energy system becomes a regular practice.

2 Loading of autotransformers

Modern software applications that calculate steady-state regimes allow calculating and ranging fast thousands of electrical energy system conditions.
Exhaustive search of any possible regimes in electrical energy system not only with n-1 principle (failure of one network element), but also with failure of two concurrent network elements (n-2 principle) becomes compulsory in designing substations. Thus, for substations – failure of any overhead power transmission line (OHL) 110 kV and above in electrical energy system under planned outage of AT in considering electrical installation. A number of regimes can come to hundreds. Regime with AT loading higher than in regime with N-1 principle always can be found. It is harder regime, but half-proven for capacity selection.

As a result of all listed factors of main electrical grids a low load of expensive transformer installations reveals itself. By authors calculations in the real network map 110 kV and above in the period of maximum load of 2015 year in n-1 regime about 90% of AT are not loaded higher than their rate capacity. In addition, their average loading in considering regime is equal about 60% S_{rated} and it is independent of year of commissioning of equipment.

As mentioned before, AT loading in main electrical grids is determined by interconnection operational states: balance power flows, flows connected with realization of interconnection effect and system reliability, high-water period passage conditions, etc.

Values of characteristic equivalent two-stage load graphs of substations 220-500 kV from different regions of the country are given in the Table 1. There were processed real daily check measurements of check measurement days in 2015 year for more than 500 substations according with their voltage type. Defined:

– f_{max} factor, calculated as the winter peak load ($S_{max(w)}$) divided by the summer peak load ($S_{max(s)}$);

* Corresponding author: sh280859@gmail.com

– load factor f_{load} (the average load divided by the peak load);

– load curve irregularity factor f_{irreg} (the minimum load divided by the peak load).

The defined average value of f_{max} = 1,2. However, the summer peak of about 30% substations is higher than the winter peak, i.e. f_{max} < 1,0. These substations exist in all regions. It is harder design condition, because rated AT capacity conform to ambient temperature of 20°C, i.e. permissible loading is decreasing while growth of temperature and conversely.

Average country values are f_{load}=0,825 and f_{irreg}=0,627. Therefore, average pre-load K_1 of equivalent two-stage load graph is in comparably narrow range 0,627<K_1<0,825 (arithmetical averages were taken nextly).

Table 1. Characteristics of Equivalent two-stage load graph

Prevalent type of load	f_{irreg}, pu.	f_{load}, pu.	K_1, pu.	K_2, pu.	h, h	T_{max}, h
The first group – industrial (regions of Siberia)	0,920	0,834	0,877	1,0	8,4	7547
The second group – zone of large industrial centers and cities (regions of Ural)	0,844	0,659	0,751	1,0	9,0	6719
The third group – mixed (Central regions)	0,820	0,611	0,716	1,0	8,8	6413
The fourth group – mixed with developed agricultural industry (South regions)	0,762	0,497	0,630	1,0	8,6	6128
Country average	0,825	0,627	0,726	1,0	8,7	6685

Characteristics of daily load graph structured in the Table 1 by four formal groups from the perspective of type of substation load with subsequent calculation of average values. The first group – industrial, it is representative of oil production regions enhanced by high share of hydro power plants in the mix of generating plants in some cases. The second group – zone of large industrial centers and cities. The third group – mixed type of load with measurable percent of domestic consumer and service sector demand. And the fourth group – the same as the third group, but with developed agricultural industry.

Conversion of daily load graph into equivalent two-stage load graph was realized the following manner. For f_{load} and f_{irreg} from the Table 1 when K2=1 there was defined pre-load K1=(f_{load}+ f_{irreg})/2. We call duration of K1, as x. Therefore, duration of K2 is 24–x и 24firreg=K1x+K2(24–x). Thus for country average values (Table 1) we have: 24·0,825=0,726x+1,0(24–x). From which x≈15,3 h., and the duration of potential overload K2 h=8,7 h.

It will be seen from the Table 1 the equivalent load graphs could be differ in values of K1. They're varying in in the range of 0,63 to 0,88 with almost constant duration of overload h≈8–9 hours.

In the basis of Russian and international methods of definition of loading capacity of (auto)transformers there're thermal model of ageing of their insulation, top-oil and hot-spot temperature limitations and number of other boundary conditions [1, 2]. From the perspective of insulation ageing the known 6°C principle is used. However, permitted temperatures are different in [1, 2] (Table 2). The international standards are more sparing. Thus, maxi-mum permissible winding hot-spot temperature from [1] - 160°C, while in [2] – 140°C (Table 2).

AT loading examples are given in Figure 1: solid lines – normal cyclic loading, dashed lines - emergency loading. Curves in Figure 1 are given for equivalent ambient temperature of +20°C, curves in Figure 1,a are given for equivalent ambient temperature of –10°C and for +20°C in Figure 1,b. It is typical for Moscow region for winter and summer periods thoroughly. Calculations were settled by the procedure [2]. Normal cyclic loadings were calculated with daily loss of life is less than 1. Numbers with daggers in Figure 1 - duration (h) of overloading K2. The current limitation from Table 2, which is K2≤1,3, is disregarded in Figure 1.

Table 2. Maximum permissible characteristics of autotransformers

Characteristic	(auto)transformers capacity, MV·A		
	to 2,5	from 2,5 to 100	above 100
Normal cyclic load:			
Current, pu.	1,5/1,5/1,5	1,5/1,5/1,5	**/1,3/1,3
Winding hot-spot temperature, °C	140/140/120	140/140/120	**/120/120
Top-oil temperature, °C	95/105/105	95/105/105	**/105/105
Long-time emergency loading:			
Current, pu.	2,0/1,8/1,8	2,0/1,5/1,5	**/1,3/1,3
Winding hot-spot temperature, °C	160/150/140	160*(140*)/140/140	**/130/140
Top-oil temperature, °C	115/115/115	115/115/115	**/115/115

Note: the first number – from GOST 14209 – 85; the second – from IEC 354 – 91, the third – from IEC 60076 – 7:2005; * the first number for transformers up to 110 kV, the second – above 110 kV ** – not rated

From Figure 1 follows: AT loading for the real values K_1≈0,6–0,9 and h≈8–9 hours (Table 1) is almost constant. Thus, for equivalent temperature of –10°C the normal cyclic loading K_2≈1,2 and emergency loading – 1,5 (Figure 1, *a*). For normal cyclic loading it's explained by the limitation of defined (as above) daily loss of life while the winding hot-spot temperature is below permitted value – 120°C and for emergency loading – by the winding hot-spot temperature is below 140°C.

The regularity revealed in permitted loads of AT is essential in a context of uncertainty of initial data for making decisions on the period 10 years or above. From Figure 1 also follows that emergency loading is on 0,2–0,3 higher than normal cyclic loading with other factors being equal.

The above-accepted limitation for daily loss of life can be reduced. Usually there are two autotransformers installed on substations.

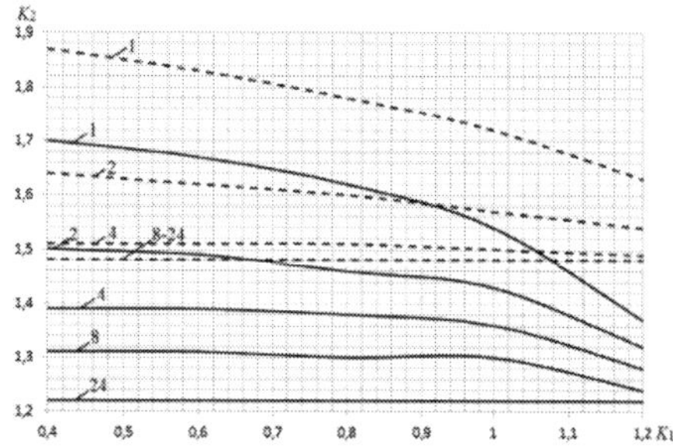

Fig. 1.a Loading of AT which rating over 100 MV·A with OF cooling system for –10°C.

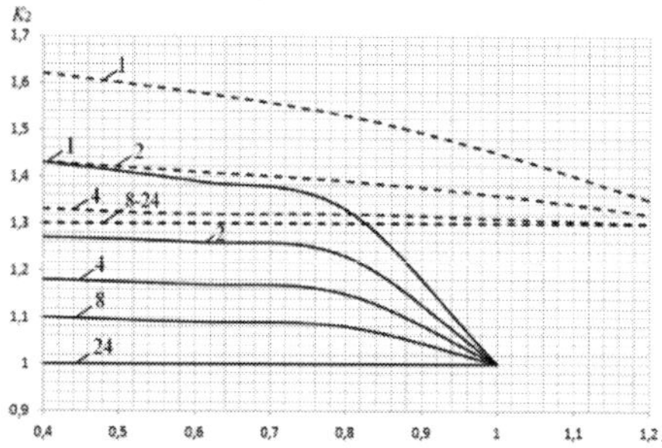

Fig. 1.b Loading of AT which rating over 100 MV·A with OF cooling system for +20°C.

Loading is divided proportional between them if possible in steady state regime. That's why most of the time they're underloaded and mid-annual duration of power system maintenance regime and postfault regime, as a rule, doesn't exceed 1/10 part of a year. In this case, normal cyclic loads can be limited not by daily loss of life, but the maximum permissible winding hot-spot temperature (120°C – Table 2).

For this controlling case permissible cyclic loading of AT which rating over 100 MV·A with OF cooling system is given in Table 3 for equivalent ambient temperature +20°C (numerator) and corresponding daily loss of life (denominator).

The comparison of data from Table 1 and Figure 1,b shows: with removing constraints from daily loss of life under other factors being equal the permissible loading K_2 is growing from 1,0 to 1,16. In this case daily loss of life for h=8 hours doesn't exceeded 3,98. Thus, in calendar day insulation is really «outmoded» on 3,98 days. For example, if annual average outage doesn't exceed, assume, a month then insulation will be outmoded on 30·3,98=120 days for this period.

In a steady state regime when two AT are in service and the loading of each is $(1{,}16/2)S_{\text{rated}}$, the daily loss of life is less than 0,01 (the winding hot-spot temperature is 57°C), so it can be neglected. This state is common to temperatures up to 80°C. Thus, determination of permissible cyclic loading based on winding hot-spot temperature is proven design condition for increasing of AT loading.

As it was mentioned above, n-1 and n-2 regimes are taken into account for justification and selection autotransformer capacity. The last also is harder designing condition. It is interesting to estimate its real severity.

There were calculated AT loadings $S_{n\text{-}1}$ and $S_{n\text{-}2}$ on 73 functioning two-transformer substations 220–500 kV in n–1 and n–2 in all regions of the country for peak loading scheme of 2015 year (Table 4). N-1 regimes were modeled with failure of one AT on a substation. In n-2 regimes all OHL 110 kV and above in a region were orderly and sequential failure in addition to n-1 regimes. For one particular substation, the number of design regimes was in the range of 100 to 400. Regimes were ranked depending on $S_{n\text{-}2}/S_{n\text{-}1}$ ratio (Table 4). From Table 4 follows:

– From 16073 design n–2 regimes about 30% (4919 regimes) lead to excess $S_{n\text{-}2}>S_{n\text{-}1}$;
– From chosen 4919 regimes 4718 design conditions (95,9%) lead to excess of autotransformer loading less than 0,1 in comparison with the same in n–1 regime and 4718+115=4833 (98,3%) – less than 0,2;
– The number of regimes, where $S_{n\text{-}2}/S_{n\text{-}1}>1{,}1$, decreases rapidly with increasing of power surge;
– Between 16073 design regimes there was found one extreme event with $S_{n\text{-}2}/S_{n\text{-}1}=1{,}81$.

Table 3. Cyclic loading and daily insulation loss of life for winding hot-spot temperature of 120°C

Overload duration, h	Permissible overload K_2 for K_1=0,4–1,2				
	0,4	0,6	0,8	1,0	1,2
1,0	1,43 / 0,60	1,39 / 0,64	1,33 / 0,77	1,25 / 1,89	–
2,0	1,27 / 0,61	1,26 / 0,67	1,23 / 0,78	1,20 / 1,90	–
4,0	1,19 / 1,05	1,18 / 1,06	1,18 / 1,35	1,17 / 2,47	–
8,0	1,16 / 2,44	1,16 / 2,57	1,16 / 2,85	1,16 / 3,98	–
24,0	1,16 / 11,10	1,16 / 11,20	1,16 / 11,46	1,16 / 11,77	–

As was shown before, emergency loading is on 0,2–0,3 higher than normal cyclic loading with other factors being equal. Therefore, if AT capacity is selected according to n-1 regime, then while verify it in n-2 regime we can exclude 98,3% design conditions (Table 4) from consideration, $S_{n-2}/S_{n-1}<1,2$. Next, we will do reliability evaluation of n-1 regimes for the real example.

In detailed analysis of all variety regimes, (Table 5) authors could not found even one for reliability measurement. As a fact capacities of AT in main electrical grids in the most cases are exceed power flows through AT not only in n-1 regime, but also in n-2 regime. In other words, rated capacities of given installations due to various reasons turn out to be such uprated, that it's impossible to estimate real limitations for doing reliability evaluation. For example, let's take into consideration the substation № 6 from Table 5 with almost twice power surge in n-2 regime (S_{n-2}/S_{n-1}=1,81). It has two installed groups of single-phase autotransformers 500/220 kV with capacity of 3·167 MV·A each group. In a steady state regime the peak loading of AT is $0,16S_{rated}$, in the n-1 regime is $0,22S_{rated}$, and in the n-2 regime is $0,39S_{rated}$. That is why regime reliability factor is analyzed from a perspective of hypothetical reasonability of decreasing of installed AT loading.

In this context the substation № 5, which is located at the interface of Ural and West-Siberian regions, is demonstrative example.

It has two installed AT 220/110 kV with S_{rated}=250 MV·A. Real characteristics of equivalent two-stage load graph of the substation - K_1=0,726, h=9,6 h; $S_{max(w)}/S_{max(s)}$=1,32; T_{max}=6522 h. Maximal power flow in n-2 regime in the period of winter peak loading is 200 MV·A (max($S_{n-2}/S_{ном}$)=0,8) and in n-1 regime is 145 MV·A (S_{n-2}/S_{n-1}=1,38). It is caused by emergency failure of OHL 500 kV with length l=113 km. The second estimated event – failure of another OHL 500 kV with length l=60 km (S_{n-2}/S_{n-1}=1,32). The third – failure of OHL 110 kV with length l=101 km (S_{n-2}/S_{n-1}=1,21). Then a logical question appears: could we use autotransformers with rated capacity of 125 MV·A instead of 250 MV·A on the substation?

AT loading in n-1 regime will be 145/125=1,16. Even for +20°C such cyclic loading is permitted (Table 3). We can consider that emergency loading in n-2 regime could be 0,2 higher than permitted cyclic loading; total K_2=1,16+0,2=1,36. Estimated event is harder with taking into account of current limitation 125·1,3≈163 MV·A according with [2]. Therefore, the loading of substation in n-2 regime should be limited to 200–163=37 MV·A (ΔP≈37 MW). Limitations for the second and third estimated events are 28 and 12 MW. Other n-2 regimes from ranges of 0-10 and 10-20% from Table 4 are neglected, because they do not lead to exceeding of permissible emergency loading of AT.

Table 4. AT loading on substations in n-2 regime

№	Amount of regimes, ea	Among them regimes, ea, leading to $S_{n-2}/S_{n-1}>1$ within the ranges, %										max (S_{n-2}/S_{rated}) pu
		0–10	10–20	20–30	30–40	40–50	50–60	60–70	70–80	80–90	0–90	
						500 kV						
1	113	50	0	0	0	0	0	0	0	0	50	0,79
2	140	58	1	0	0	1	0	0	0	0	60	0,70
3	352	108	4	0	0	0	0	0	0	0	112	0,79
4	176	81	2	3	1	0	0	0	0	0	87	0,67
5	352	119	5	2	1	1	0	0	0	0	128	0,35
6	352	23	2	0	0	1	0	0	0	1	27	0,39
7	390	304	4	2	0	0	0	0	0	0	310	0,63
8	390	138	0	0	1	2	0	0	0	0	141	1,11
…	…	…	…	…	…	…	…	…	…	…	…	
27	254	127	1	0	0	0	0	0	0	0	128	1,39
1–27	6733	2467	56	15	5	5	1	0	0	1	2550	
						220 kV						
28	132	32	0	0	0	0	0	0	0	0	32	0,56
29	113	48	2	4	5	0	1	0	0	0	60	0,70
30	140	54	4	1	0	0	1	0	0	0	60	0,31
31	254	115	1	0	0	0	0	0	0	0	116	1,12
32	113	30	2	0	0	1	1	1	0	0	35	0,60
33	390	46	0	0	0	0	0	0	0	0	46	0,86
34	230	152	2	6	0	0	0	0	0	0	160	0,50
35	230	73	2	1	2	0	0	0	0	0	78	0,80
…	…	…	…	…	…	…	…	…	…	…	…	…
73	230	40	0	0	1	0	1	0	0	0	42	0,62
28–73	9340	2251	59	35	14	3	6	1	0	0	2369	
						220–500 kV						
1–73	16073	4718	115	50	19	8	7	1	0	1	4919	

Table 5. For n–2 regimes reliability estimation

Index	Estimated event		
	l=113 km, ΔP=37 MW	l=60 km, ΔP=28 MW	l=101 km, ΔP=12 MW
ω_Σ, 1/year	$4{,}1 \cdot 10^{-3}$	$2{,}3 \cdot 10^{-3}$	$9{,}3 \cdot 10^{-3}$
m_t, hours	9,8	9,9	7,2
$1/\omega_\Sigma$, year	244	442	108
f_{of}, pu	$4{,}6 \cdot 10^{-6}$	$2{,}6 \cdot 10^{-6}$	$7{,}7 \cdot 10^{-6}$
ΔW, MW·h	2,2	0,9	1,2
Total ΔW=2,2+0,9+1,2≈4,0 MW·h			

Reliability characteristics of n-2 regimes for considering substation are given in Table 5. Agreed notations in Table 5: ω_Σ – resulting fault rate; m_t – mean time to scheme repair; f_{of} – scheme forced outage factor; ΔW – mathematical expectation of annual electricity sacrifice due to scheme unreliability. From Table 5 follows, that rate of occurrence of n–2 regimes is one time in 100 – 500 years.

Economic impact evaluation due to schemes unreliability has always caused great uncertainty because of known reasons. Setting aside them, please note that in recent years results of the most large-scale research in the area of values of specific damage by electricity sacrifice were published in [3, 4]. Thus, when selecting a scheme of main electrical grids aggregated data of specific damage is recommended at a level of 18–42 rub./(kW·h) at 2000 values. Therewith, weighted average value of specific damage in countries of Western Europe, the USA and Canada is 9,53 dol./(kW·h). For liquidation of power flows limitations in n-2 regimes it is possible to go towards a variant of increasing AT capacity from 125 to 200 MV·A.

Discounted costs estimation for considering variants is trivial. For maximum recommended value of specific damage 240 rub./(kW·h) the economic impact due to schemes unreliability for ΔW≈4,0 MW·h is up by more than decade of expenses, connected with increasing AT capacity. The same for taking the value of specific damage of countries of Western Europe, the USA and Canada.

Then a logical question appears: when n-2 regimes should be taken into account for technical and economic assessment of optimum level of probability of electrical installation? In analyzed substation scheme, it will be necessary, If power limitation is up by more than decade, i.e. not 37, but 370 MW, when S_{n-2}/S_{n-1}=(170+370)/145=3,7. Taking into account of factual data from Table 4 such a regime is unlikely possible.

3 Conclusion

Therefore, for selection of AT capacity in main electrical grids of electrical energy system one may recommend the following rules:

1. AT capacity is selected based on maximum power flow in n-1 regime with failure of one of AT on a substation.
2. There should not be power flows limitations in n-1 regime with taking into account a permissible cyclic loading of AT.
3. Cyclic loading is estimated based on non-exceedance of winding hot-spot temperature without AT insulation loss of life limitation.
4. Upon that, n-2 regimes are not considered, but they are considered for Emergency Control Automatics requirements definition.

References

1. *State Standard 14209 – 85. General-purpose oil-immersed power transformers. Permissible loads.* (Standartinform Publ., Moscow 2009)
2. B.N. Neklepaev *Electric Part of Power Stations and Substations* (Energoatomizdat 1986)
3. *IEC 60076 – 7 – 2005. Power transformers – Part 7: Loading guide for oil-immersed power transformers.* (IEC, 2005)
4. V.A. Nepomnyaschiy *Power market*, **9**, 36 – 47. (2014)

E3S Web of Conferences **25**, 01013 (2017) DOI: 10.1051/e3sconf/20172501013
RSES 2017

Development of express method for liquid dielectric quality control as a step to increase the reliability and effectiveness of oil-filled equipment exploitation

Svetlana Vysogorets[1*], *Alexandr* Nazarythev [1], *Alexey* Tadjibaev[1], *Ilia* Vysogorets [2]

[1]PEIPK, SPb, Russia
[2] NRU "HSE SPb", SPb, Russia

Abstract. Presented the results of researches, aimed to develop methods for express-analysis of transformer oil quality indicator "Acid Value". Developed an algorithm of preparation for and conducting measurement. Proposed implementation of mixed indicator. Outlined the valuation criteria for results of "AV" semiquantitative measurement via express-method.

1 Introduction

The aging equipment park requires medium and capital repairs, which are followed by regenerative work to restore the exploitational and resource characteristics of the liquid dielectric. The physico-chemical analysis of the oil is an extremely reliable tool that allows to influence the efficiency of these works: to control the technological cycle, to determine the point of the procedures completion. At the same time, carrying out control chemical measurements involves the need to transport samples to chemical laboratories with considerable time losses. The search for a minimum set of control parameters and the development of methods for their rapid analysis will improve the quality and efficiency of repair work in oil-filled equipment.

2 Determination of oil quality key parameter that allows to indirectly assess the change in the insulation resource

All the processes taking place in the insulation of oil-filled equipment affect, above all, the impregnating mineral oil and spread to solid cellulose materials (paper, cardboard). At the present time, there have been developed theoretical provisions on the fundamental aging processes of individual insulation components - oil and paper [1].

In [2], there is revealed a considerable influence of the oxidation processes occurring in the transformer oil and the aging products formed in it on the state of solid insulation, affecting, eventually, its electrical insulation characteristics and durability.

Considering the significant contribution of transformer oil in damage to the insulation of oil-filled equipment, the most important exploitational characteristics of a liquid dielectric are their chemical stability, i.e. the ability not to change their properties for a long time in the process of exploitation, as well as their chemical resistance - the ability for a long time not to destroy the solid insulation [3].

During the scientific analysis [2], it was established that these properties are in close connection with the processes of oil aging and the resulting oxidation products. Accordingly, timely and sufficient removal of aging products from the oil with subsequent restoration of its resource allows maintaining stable and reliable characteristics of both power transformers insulation and liquid dielectric.

The carried scientific experiment [2] statistically proved a significant relationship between the oil antioxidant stability (oil resource) and a number of quality indicators which are indicative also for aging products, among which are the indicators "Acid value" (further AV), "Water-soluble acids and alkalis" (further WAA), "Dielectric loss angle tangent" (hereinafter Tangent), "Antioxidant Additive Content" (hereinafter the Additive). In work [2] it is shown:

- The multicollinearity [4] of the indicators of AV and WAA, which means that there is a close relationship between them (high closeness);
- By the methods of correlation-regression analysis, it was revealed a significant direct relationship between AV and antioxidant stability (hereinafter Stability), a moderate direct relationship between Stability and Tangent and a weak inverse relationship between Stability and Additive.
- By the three non-parametric methods (trees of classifications, determinant analysis and logistic regression), the AV was found to be the main grouping factor of transformers in the indicator Stability.

* Corresponding author: s-151075@yandex.ru

 DOI: 10.1051/e3sconf/20172501013

Thus, AV is a key indicator of quality, allowing with a high degree of reliability to indirectly assess the change/deterioration of the oil resource. The obtained results of the studies [2] indicate that exactly the AV index is a reliable key parameter of the oil quality, which allows to empirically measure the efficiency of oil regeneration (restoration of the oil resource) and determine the work completion point with subsequent controlled regulatory measurements.

Due to the fact that a significant part of the oil regeneration work in the period of overhauls of power transformers is carried out at the place of their installation (at substations), there is an urgent need to develop a rapid method for estimating the AV, which will eventually result in:

- reducing the loss of time for the delivery of intermediate samples to the laboratory, and, accordingly, will affect the reduction in the duration of direct capital repairs;
- reducing the cost of capital repairs: reducing the idle time of crews before obtaining the results of control measurements, increasing the efficiency of regeneration machines by sorbent timely replacement based on the results of control measurements, reducing costs by eliminating the excessive carrying out of expensive laboratory measurements;
- improvement of the quality of directly regenerative work due to the provision of operational control without loss of organizational procedures, and accordingly to increase the period in-between overhauls of transformers, increase the reliability and durability of their isolation.

3 Development of a laboratory experiment on the search for the method of express analysis of transformer oil according to the AV index. Results of trial experiments

One of the recognized modern methods of measuring transformer oil AV is the method [5] providing for volume titration (titrimetric method) using an "Pigment Blue 61" indicator. In this case, unlike the more common method [6] of determination of AV using the indicator of nitrazine yellow, the method [5] does not contain boiling procedures.

Thus, method [5] of the author of this article is taken as a basis for the development of express analysis of the quality index of AV. The developed technique is an empirical (semi-quantitative) method for assessing the compliance of the AV of the analysed mineral transformer oil with a given initial value. To assess the results, a categorical assessment is used: "less than the set value" and / or "more than the set value". In the implementation of the rapid analysis of the AV, preliminary preparation of working solutions in laboratory conditions is provided, followed by carrying out test measurements at the facility (substation).

On the basis of mentioned [5] quantitative method for measuring AV, a list of additional solutions was determined, with the interaction of which with the defined volume of the transformer oil sample, according to the adopted rules, by the colour shifts the range of AV values characteristic for a given liquid dielectric is determined.

All experimental measurements are carried out on the basis of the chemical laboratory of PJSC "IDGC of the North-West" "Komienergo". During the installation of trial experiments, it was found that for a number of oils there is no more or less bright colour transition, which makes it difficult to determine the end point of titration: the category of evaluation.

4 Mixed indicator

Indicators are chemical compounds capable of changing the colour of a solution depending on environmental conditions without directly affecting the test solution and the direction of the reaction. Dissolved in water chemical compounds dissociate into positively charged ions - cations and negatively charged - anions. If the concentration of hydrogen and hydroxyl ions in the solution is the same, then such solutions are neutral and pH = 7. At a concentration of hydrogen ions corresponding to a pH from 0 to 7, the solution is acidic, if the concentration of hydroxyl ions is greater (pH from 7 to 14) alkaline. Qualitatively, the reaction of the solution can be determined with the help of special indicators (acid-based), changing their colour depending on the concentration of hydrogen ions. The advantage of acid-based indicators is cheapness, speed and demonstrativeness, however, due to subjectivity of colour determination and low accuracy, pH indicators are not always convenient. To improve the accuracy of pH measurement, the use of mixed indicators is recommended. For this purpose, two indicators with close intervals of the pH colour transition are selected, which have additional colours in this interval [7].

So, in order to provide a more noticeable transition of the color of the indicator during titrimetric analysis, a mixed indicator was chosen: "Pigment Blue 61" + curcumin. Justification of the choice of indicators:

1. "Pigment Blue 61" has a pH range of the color transition (in aqueous solutions): from pH 9.4 (blue) to pH 14.0 (red):

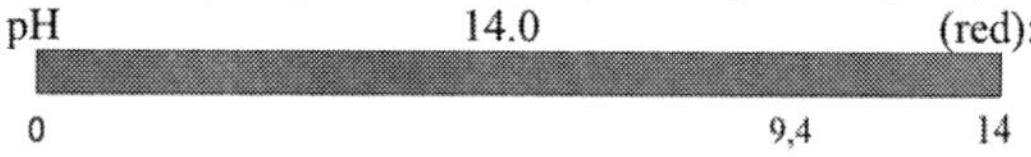

2. Curcumin has the first range of the pH colour transition (in aqueous solutions): from pH 7.4 (yellow) to pH 9.2 (brown-red); The second range of the pH colour transition (in aqueous solutions): from pH 10.2 (brown-red) to pH 11.8 (orange-yellow):

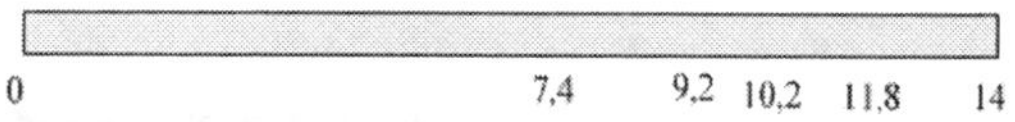

When using the above-mentioned pair of indicators at the same time, it is expected that the yellow colour of the curcumin indicator in an acidic region will not affect the

 DOI: 10.1051/e3sconf/20172501013

initial colour of the analysed mixture, given the original yellow colour of the oil. At the same time, in combination with the blue colour of the indicator of "Pigment Blue 61" in a mixture with oil, a strengthening of green shades is expected. When approaching the equivalence point, curcumin begins to give the red-brown colour of the mixture before the "Pigment Blue 61", and then strengthen the red colour it.

Thus, the use of an additional indicator made it possible to make the green tint of the colour more pronounced and hence the (more clearly observed) disappearance of the green tint is considered an achievement of the end point of titration.

5 Research results

The key objectives of the study are:

- Development of the methodology for express analysis of transformer oil AV: determination of conditions and algorithm for conducting measurements, a list of necessary reagents, the order of evaluation of measurement results;
- search for the dependence of the change in the colour component composition of the reaction mixture according to the RGB model from the measured value of the AV.

For the analysis of one working sample of transformer oil, a list of necessary solutions, an algorithm for preparing and conducting an express analysis of the AV quality index were developed (see Fig. 1).

Based on the results of the test measurements, it is envisaged to form a conclusion about the compliance of the AV to one of the set values. In the developed method of rapid analysis of the AV, the following interpretation of the results of the test analysis is provided without taking into account the numerical representation / change in the component composition of the colour of the solution during a chemical reaction:

- if the colour of the solution in the test tube with an aliquot of the sample retains the main colour / shade (red with tints) and significantly differs from the colour / shade of the test solution (green with tints), this means that the substances causing the oil AV are not sufficient to neutralize all the amount of KOH contained in Solution 2. Accordingly, it can be concluded that the AV of the oil sample does not exceed the preset AV value in Solution 2;
- if the colour of the solution in the test tube with an aliquot of the sample changed the main colour / shade became green and / or close to the colour / shade of the test tube with the reference solution, this means that the substances causing the oil AV in the mixture are contained in excess, all KOH is neutralized. Accordingly, it can be concluded that the AV of the oil sample exceeds the preset AV value in Solution 2.

The analysis of the colour component constitution of the solution in tubes is carried out according to the RGB model using the «ColorAnalyzer» software: the data is read in three channels - R (red), G (green) and B (blue), followed by its fixation. On the basis of the experimental data obtained, the dependence of the change in the component constitution of the solution colour during chemical reactions on the obtained test value of the AV is sought: the determination of the decision point.

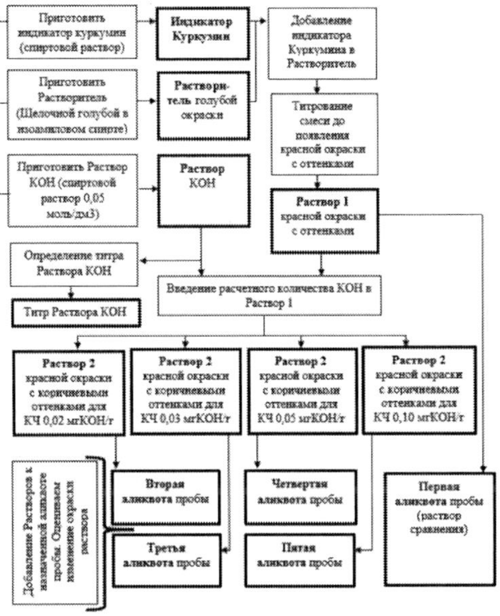

Fig. 1. Algorithm of preparation and conducting AV express-analysis

The evaluation of the reliability of the first developed rapid analysis method for AV is carried out by evaluating the ratio of the results of the analysis, obtained by the newly developed method to the results of the studied oil samples, the measurement of the AV in which was carried out using a control certified method [5].

According to the above algorithm, 40 samples of transformer oil with different values of AV were measured. Based on the results of the studies, it was found that in most of the analyzed samples, an acceptable visualization of the determination of the titration end point is found.

After carrying out in the laboratory conditions procedures of oil regeneration with the subsequent control measurement of the AV by the express-method, it also revealed the dynamics of the color change of analyzed ssolution, depending on AV change (decreasing of it). At the same time, it is continued to search for dependence of the change in the colour component composition of the reaction mixture according to the RGB model on the measured value of the AV.

6 Conclusions

1. The important properties of a liquid dielectric are chemical stability and chemical resistance, the restoration of which is mandatory during capital repairs of power transformers.
2. Substantiated the necessity of the development of express analysis of the AV oil quality parameter as a step

DOI: 10.1051/e3sconf/20172501013

allowing to increase the efficiency of capital repairs of power transformers.
3. The method of AV express analysis is developed, the algorithm of preparation and carrying out of measurements is presented. A mixed indicator is proposed. Developed a list of solutions necessary for preparation in laboratory conditions followed by application in "field" conditions. Proposed the criteria for estimating the results of a semi-quantitative measurement of the AV by the express method.

References

1. G.G. Lysakovskii, *Research of transformer oil aging in paper-oil insulation in close to exploitational circumstances* (*Liquid fuel and oil in energetics*: treatise collection. - M.: «Energoizdat», 1982)

2. S.P.Vysogorets, *Development of the new methods and quality valuation algorithm of 35-110kV power transformer exploitational oil. Dissertation for the degree of Candidate of Engineering Sciences* (SPb., 2012, 260 p).

3. *Enchiridion of repair and maintenance of power transformers*. (M.: IPKgossluzhby, 2008, 852 p).

4. D.E. Hank, *Business-forecasting* (translation from english – 7th edition. – M.: Publishing House «Williams», 2003, 665 p).

5. *MKHA 05-09 Measurement methods Determination of transformer oil acid number by titrimetric method.* - SPb.: Scientific and Production Association «Electrum» (2009 (2014 edition)) – 15 p.

6. *GOST 5985-79 (ST SEV 3963-83) Petroleum products. Method of acidity and acid number determination*. (M.: Izdatelstvo Standartov, 1994, 9 p).

7. *Reagents for technical analysis: B.R.* (M.: Metallugria, 1988, 384 p).

E3S Web of Conferences 25, 02001 (2017) DOI: 10.1051/e3sconf/20172502001
RSES 2017

New diagnostic methods aimed at increasing of reliability and safety of the main pipeline transport

Mikhail Sukharev[1,2], *Ksenia* Kosova[1,2,*]

[1]Gubkin Russian State University of Oil and Gas, 119991, Moscow, GSP-1, Leninsky prosp., 65, Russia
[2]LORES Engineering LLC, Moscow obl.,142700, Vidnoe, Zavodskaya ul., 2A, Russia.

Abstract. The paper considers the problems of detection of fluid leaks and calibration of measuring devices in main pipeline systems. On the example of the gas and oil products transport systems we propose new methods for solving these problems. The methods are based on the processing of measurements of standard measuring devices. The efficiency of the methods is verified by numerical calculations by processing real telemetry data.

1 Introduction

The pipeline transport systems for gas, oil and petroleum products play an extremely important role in the energy supply of the global community. For this reason, increasing the efficiency, safety and reliability of these systems are the most important tasks of their operation.

One way to obtain the desired results in these areas is monitoring the condition of pipeline systems (PS) using control and measurement instrumentation and telemetry systems. The growing level of IT penetration and computerization allows you to collect and process large amounts of information about fluid flow parameters (pressure, flow, temperature, etc.). At present, some information, received from standard measuring devices, is analyzed by specialized software complexes, which contributes to more justified adoption of operational decisions. However, the information potential of the PS is not limited to this.

Detailed data of fluid flow regimes allow us to develop existing and create new methods for increasing the reliability and safety of the main pipeline transport. Such methods are proposed in the present paper; they are based on processing the measurements of a standard measuring instrument and relate to the field of regime diagnostics. They are not expensive to implement and should be particularly attractive to operators.

Let us consider the two problems of increasing the reliability and safety of the operation of gas, oil and product pipelines. It is possible to advance in solving these problems involving new methods of regime diagnostics. The first of them is connected with the detection of leaks and illegal tie-in in the oil and product pipelines, the second - with the adjustment of a set of measuring instruments installed at the gas transmission system (GTS). Despite the fact that each of these tasks relates to a specific technology area, the approaches used to solve them can be developed and used also for other PS.

2 Detection of leaks and illegal pipelines taps

Rapid detection of leakages and illegal pipelines taps is one of the priority tasks of ensuring the safe operation of oil and product pipelines. Leakages [1, 2] and illegal taps [3] cause serious environmental problems and economic and reputation damage to operator companies.

The problem of the early detection of the leaks and their location in pipeline system has attracted the attention of researchers for more than a century. Beginning with the work of Zhukovsky N.E. (19th century), to identify leakages of oil and petroleum products a wide range of methods was developed. They are based on the registration and investigation of physical processes occurring during the formation of the leak.

Some methods, e.g., [4, 5] register front of the pressure wave that appears at the leak occurrence time. Others analyze the acoustic noise accompanying the leak [6, 7]. For identification of leakages [8, 9] also present balance methods: a leak signal is supplied when imbalance flows at the inlet and outlet of the pipeline occurs. In [10-12] for leaks detection the authors employ information about the mathematical models describing the flow regime of the fluid. In the paper [11] a leak detection procedure uses the stability criterion of Lyapunov. The approaches presented in [10, 12] are based on comparison operating parameters measurements with simulation results: the difference of the measured and calculated values indicates a leak. To take into account the influence of measurement errors of regime parameters on the results of leak detection methods the publications [9, 13] propose the methods of mathematical statistics. To improve effectiveness of the

* Corresponding author: kseniya_kosova@mail.ru

 DOI: 10.1051/e3sconf/20172502001

leak detection methods in [14] noise in the original data is removed by the wavelet analysis. In the standard [15] for leak detection wave acoustic emission is analyzed. Of course, these approaches do not represent the entire spectrum of the leaks identification methods. Additional information on this topic can be found e.g. in the Handbook [13] or paper [16]. [17] presents a review of pressure-based methods.

Modern information equipment of oil and product pipelines allows us to offer a new method for leaks and illegal taps detection that uses the apparatus of mathematical statistics and the theory of random processes (time series). The method described below belongs to the statistical class, but unlike the known methods [9, 13] does not require the recording of mass conservation and knowledge of the hydraulic model of fluid flow in the pipeline. The proposed method uses only regular pressure measurements in the nodes of the pipeline system and does not require knowledge of the flow rate along the pipeline. The developed procedure makes it possible to detect leaks in both stationary and non-stationary regimes of fluid flow.

2.1 Technological and mathematical formulation of the problem, method of solution

Depressurization of the pipeline is accompanied by a sharp drop in pressure. The wave propagates along the pipe. A leak can be detected by monitoring of pressure measured by standard instruments. Similar processes, however, can take place when stopping the pumping unit, opening/closing the line crane, etc. Therefore, the complexity of leak detection is due to the need to distinguish the response of flow parameters to the leakage occurrence and other technological causes.

To identify the leak occurrence time consider the sequence of pressure measurements $P(t)$ at the time points t, $(t = 1,2,...)$. The sequence $P(t)$ is a random process with discrete time – a time series*. In general, the process is non-stationary, which is due to technological reasons. At a random time T, a sudden pressure drop occurs due to leakage, the probabilistic characteristics of the random process $P(t)$ change. It is necessary to determine as soon as possible the time point when this change occurred.

In the statistics of random processes, the problem of detection whether a change in the probabilistic properties has taken place is called change-point problem, and the time T is the change-point [18]. To make a decision as quickly as possible whether the change-point has happened or not methods of sequential analysis have been developed [18, 19]. We have not a priori information on the distribution of the time T and the distribution of the observed sequence $P(t)$ after the change-point. Therefore, to identify the leakage in the pipeline, it is expedient to use nonparametric methods of sequential analysis [18, 19].

The idea behind these methods is that the sequence of observations $\xi(t)$ is used to calculate statistics $y(t) = y_t$. It defines the binary decision functions $d(y_t)$. If $d(y_t) = 1$, at the time t change has occurred, if $d(y_t) = 0$, then change was not.

Nonparametric change-detection methods assume that the observed sequence $\xi(t)$ is stationary. For technological reasons, in identifying leakages of oil and oil products the process $P(t)$ is non-stationary, therefore, before applying methods of sequential analysis it is necessary to transform the process $P(t)$ into a stationary form $\xi(t)$. To do this, receiving a new measurement $P(t+1)$ we consider the prehistory of the process $P(t)$, i.e. values at the time points $t-n,\ t-n+1,\ ...,\ t,\ (n \geq 0)$ when change was not detected. According to the observations $P(t-n),\ P(t-n+1),\ ...,\ P(t)$ we find a trend $f(t)$ [20]. Calculating $f(t+1)$ – the predicted value of the process $P(t)$ at the time $t+1$ we find the value $\xi(t+1) = P(t+1) - f(t+1)$. Big modulo value $\xi(t+1)$ indicates a sharp (abrupt) change in pressure – the appearance of change.

Thus, the proposed leak identification method sequentially checks incoming pressure measurements for the presence of change. If the incoming value $P(t+1)$ differs significantly from the forecast $f(t+1)$, then it is considered that there is a change, and t – the leak occurrence time. Whether or not the difference $P(t+1)$ from the predicted value $f(t+1)$ is great, the method determines using a non-parametric change-point method.

2.2 The results of calculations – the processing of actual measurements

We demonstrate the work of the proposed method using the example of a real product pipeline, at which pressure measurements $P(t)$ are carried out at a frequency of 0.5 s. In the calculations it was assumed that $n = 19$, as the nonparametric method of sequential analysis the method of cumulative sums (CUSUM) was used [19].

Statistics CUSUM has the form

$$y_t = \min\ y_{t-1} + \xi(t), 0\ ,\ y_0 \equiv 0,\ t = 1,2,..., \tag{1}$$

decision function –

$$d(y_t) = I(y_t < N), \tag{2}$$

where $N < 0$ – the critical value (parameter of the algorithm), I – indicator.

In relations (1), (2) it is assumed that the mathematical expectations of the sequence $\xi(t)$ up to the change point m and after the change point $h+m$ satisfy the inequalities

* It is possible to consider several time series - pressure measurements near place of leakages, but this is not done in this work.

DOI: 10.1051/e3sconf/20172502001

$$m > 0,\ h + m < 0\,. \qquad (3)$$

If conditions (3) are fulfilled before the change point $y_t = 0$, and after the change occurrence the values y_t decrease. The leak signal is supplied when y_t reaches a critical value N. The parameter N is selected by the method of computational experiment on the basis of processing of actual realizations of training time series. In the calculations it was selected $N = -0{,}01$, in order to satisfy inequalities (3), the value $b = 0{,}025$ that characterizes the maximum measurement error was added to the observed values $\xi(t)$.

The algorithm was tested on the following examples.

Example 1. Fig. 1 shows a graph of the pressure $P(t)$ measured by one of the sensors installed on the product pipeline. The period from the beginning of observation and until the time $t_1 = 8345 \cdot 0{,}5$ s is the response time. Then the leak occurs, which leads to a sharp drop in pressure.

As a result of the algorithm, the leak was detected at the time $t_2 = 8412 \cdot 0{,}5$ s, i. e. in 33.5 s after its occurrence. Part of the graph $P(t)$ (Figure 1) on a larger scale is shown in Fig. 2. There is also a graph of the function $f(t)$ (trend) obtained during the operation of the algorithm. Functions $f(t)$ and $P(t)$ are visually indistinguishable. Fig. 3 shows a stationary random process $\xi(t)$ at $t \geq 8000 \cdot 0{,}5$ s.

Example 2. Fig. 4 shows a graph of the wave process, the pressure $P(t)$ from the time $t = 58000 \cdot 0{,}5$ s. The actual leak occurrence time $t_1 = 58581 \cdot 0{,}5$ s in Fig. 4 is marked with a dotted line. The algorithm allowed identifying the leak occurrence time as $t_2 = 58582 \cdot 0{,}5$ s, that is, almost instantaneously, 0.5 s after its occurrence. In Fig. 4, the time points of leakage occurrence and its recognition are visually indistinguishable.

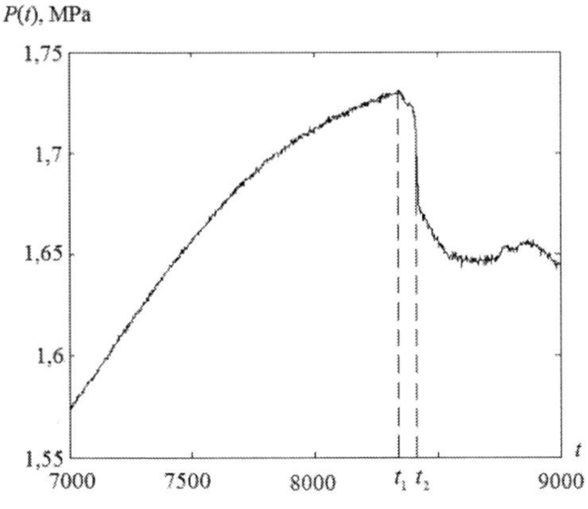

Fig. 1. Example 1. Pressure measurements $P(t)$.

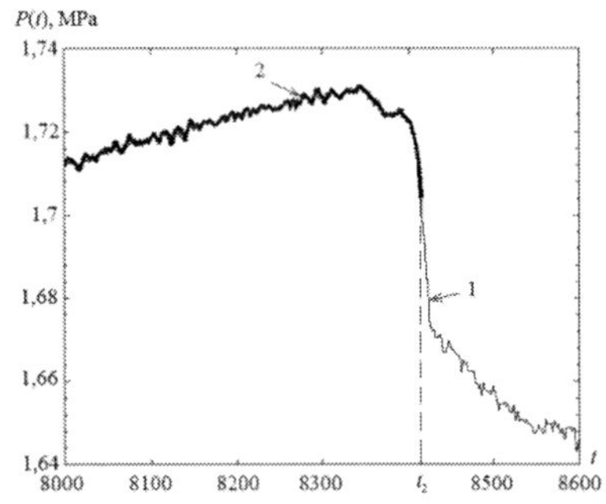

Fig. 2. Example 1. Graphs of functions. 1 – $P(t)$, 2 – $f(t)$.

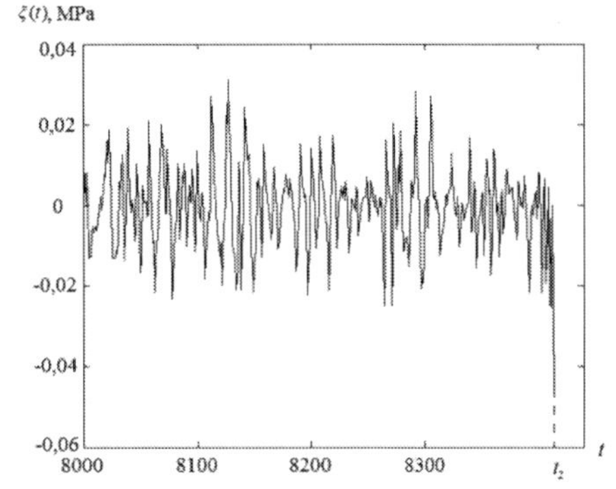

Fig. 3. Example 1. Graph of function $\xi(t)$.

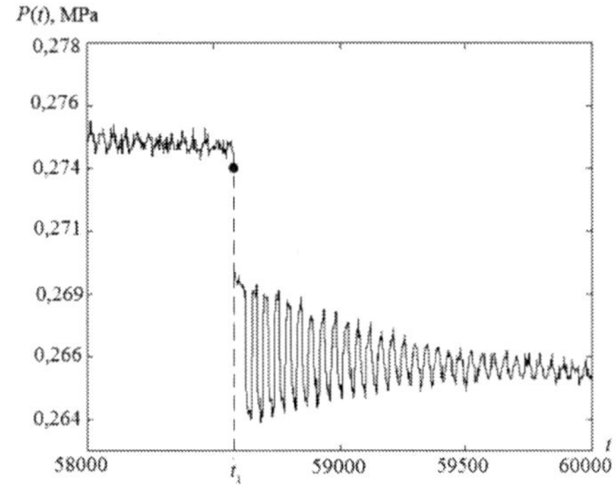

Fig. 4. Example 2. Pressure measurements $P(t)$.

The effectiveness of the method was also checked by examples without leaks.

Example 3. Fig. 5 shows one such case. The wave process that started at the time point $t_3 = 58700 \cdot 0{,}5$ s $= 489{,}2$ min is a consequence of the pumping unit shutdown (which took place before the time t_3). At the time $t_4 = 65020 \cdot 0{,}5$ s $= 541{,}8$ min the valve was opened. It would seem that there must be change-points in the same way as in examples 1 and 2.

However, the program does not record such time points. This means that there are not false alarms of the method when stopping the pumping unit and opening the valve.

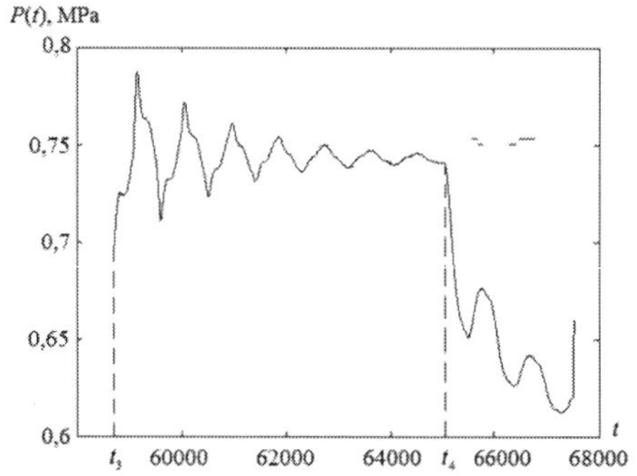

Fig. 5. Example 3. Pressure measurements $P(t)$.

Calculations show that processing of incoming measurements by the method can be performed in real time, a leak is quickly detected after its occurrence. The reactions of the method to leakage occurrence and to disturbances caused by other technological reasons: the valve opening and the shutdown of the pumping unit, differ from each other. Such a positive result is largely due to the fact that before the application of the CUSUM algorithm the trend is removed, that is, the model of the nonstationary random process is transformed into the model of stationary process. The successful choice of the parameter n – the length of the prehistory – allows us to predict adequately the further behavior of the process and prevent false alarms.

Of course, one cannot guarantee a priori that a method can always distinguish a leak from disturbances caused by other causes. However, we recall that the method uses standard measuring instruments and its implementation does not require additional equipment. This determines the expediency of further research and improvement of the proposed apparatus.

3 Verification of measuring instruments

Based on the improvement of information support, we offer one more way to improve the safety and reliability of the PS operation. The main source of information about fluid flow regimes is the instrumentation installed on the PS. Ensuring the quality of measurements is a necessary condition for the highly reliable and safe operation of equipment and PS as a whole. Errors in the initial data reduce the confidence level of calculations and can lead to the adoption of incorrect management decisions.

Increasing the quality of the information collected are periodic calibration and verification of measuring instruments. These procedures are an integral part of the maintenance of the PS. Their methodological basis has long been developed; the regulations for their conduct have been approved and have changed little since the adoption, over the course of decades. Qualitative changes in the field of PS computerization allow us to propose new methods of verification and calibration of devices that correspond to the current level of technology. One of these methods is described below. It is intended to correct the measures of the pressure sensors during the operation of the hydraulic system. The effect is achieved through the processing of interconnected measurements of several sensors. In the case of power systems, a similar approach has been developed, in particular, in the works [21-23].

Let us demonstrate the ideological side of the proposed methodology on the example of a gas transportation system (GTS) - a two-line technical corridor of main gas pipelines (Fig. 6). The outer diameters of the pipes are Ǿ = 1420 mm (line *I*) and Ǿ = 1220 mm (line *II*). The length of the route is 176.9 km. There are 9 measuring points on the object; the index i is used for their numbering $i = 1, \dots, 9$ (see Fig. 6). The measuring points are divided the object into 8 pipeline sections. Each section is assigned the number of the initial (in the gas flow direction) point $i = 1, \dots, 8$. The lines are in isolation regime, only the jumpers are open at the beginning and end of the corridor, all the jumpers along the route are closed.

Pressure, temperature and flow measurements were taken over the observation interval of 10 days. Manometers and thermometers are located at each jumper. The flow is measured at the end of the corridor, at the inlet of the compressor station (CS) 2. Fig. 7 and 8 show graphs of pressure measurements at $t \geq 5$ day in the line *I* (Fig. 7) and in the line *II* (Fig. 8). It was found that the temperature exerts a significantly less influence on the results of the calculation than the pressure, so the information on the temperature regime is omitted here.

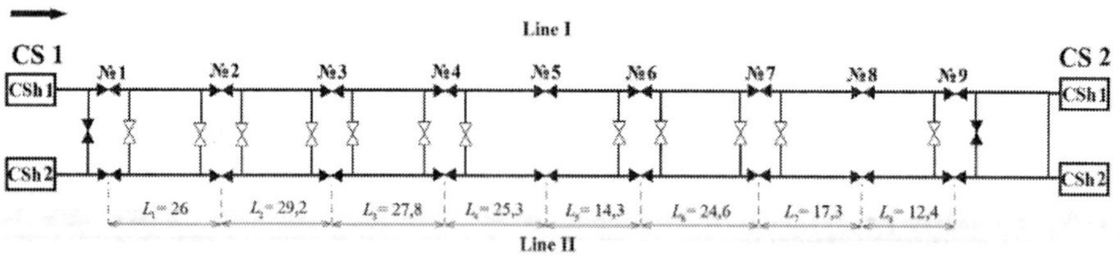

Fig. 6. Schematic diagram of the GTS. Section lengths, L_i, $(i = 1, \dots, 8)$ km. Unpainted jumpers are closed, painted jumpers are opened.

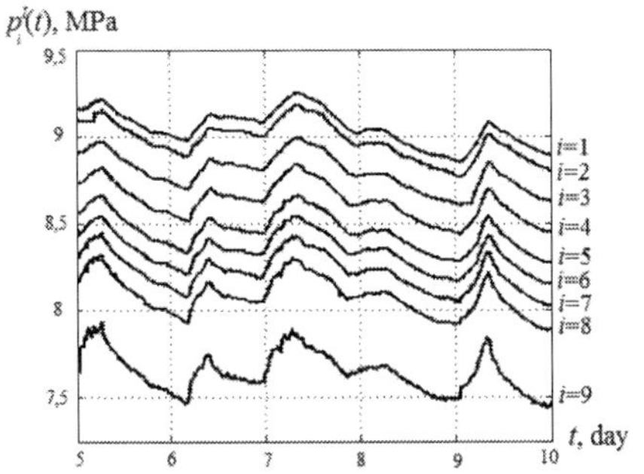

Fig. 7. Graphs of functions $p_i^I(t)$, $i = 1, \ldots, 9$ — pressure measurements in the line *I*.

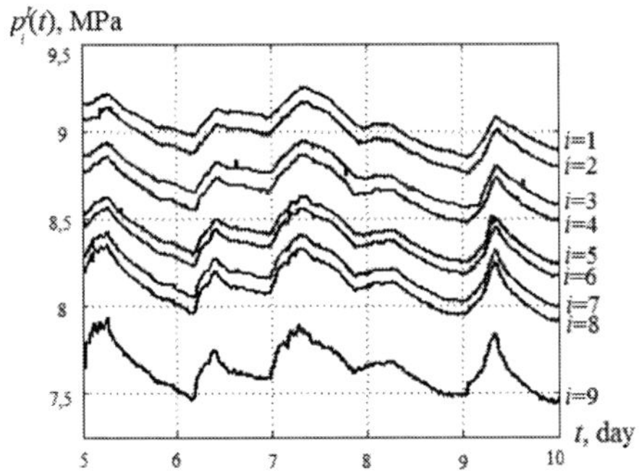

Fig. 8. Graphs of functions $p_i^{II}(t)$, $i = 1, \ldots, 9$ — pressure measurements in the line *II*.

3.1 Mathematical model

To describe the interdependence between the parameters of the gas flow, we will use the model of stationary quasi-isothermal flow. The considered corridor is horizontal, therefore the model looks like [24]

$$p_s^2 - p_f^2 = \Lambda D^{-5} L E^{-2} q^2, \quad (4)$$

where p_s, p_f – pressure values at the beginning and end of the pipeline; Λ – coefficient, depending on the characteristics of the pipe, the properties of the gas, the accepted units of measurement; D – internal diameter, L – length of the pipeline; E – coefficient of hydraulic efficiency (used to adapt the results of calculations to actual operating modes); q – commercial gas flow. The temperature is calculated in accordance with Shukhov's formula [24].

The formula (4) is applicable to each section of the corridor. For anyone $i = 1, \ldots, 8$, you can write

$$p_i^2 - p_{i+1}^2 = \Lambda_i D_i^{-5} L_i E_i^{-2} q_i^2. \quad (5)$$

Note that the values D_i and q_i depend on the line number. For all pipeline sections of the same line they are equal $q_1 = \ldots = q_8 = q$, $D_1 = \ldots = D_8 = D$. The differences of the coefficients Λ_i can be neglected. In addition, there is no reason to consider that the coefficients of hydraulic efficiency E_i are different. If this is so, then $E_1 = \ldots = E_8$ and the values $p_i^2 - p_{i+1}^2$ differ, mainly because of the lengths of the sections L_i.

3.2 Systematic errors in pressure measurements

Let us analyze the set of functions $p_i^I(t)$, $p_i^{II}(t)$, $i = 1, \ldots, 9$ (Fig. 7, 8). In accordance with the relation (5), the value $p_i^2 - p_{i+1}^2$ is proportional to the length of the section L_i, so, with growth L_i, the "distance" between the graphs of the functions $p_i(t)$ and $p_{i+1}(t)$ in Fig. 7, 8 should increase. It is even visually clear (Fig. 7, 8) that this condition is not valid. For example, the "distance" between functions $p_8(t)$ and $p_9(t)$ is more than other "distances" between adjacent graphs $p_i(t)$ and $p_{i+1}(t)$ in spite of the fact that the length of section 8 is minimal $L_8 = 12{,}4$ км. What is the reason for this discrepancy?

We did not find any physical grounds for explaining this effect. During the observation period, hydrate blockage, condensate or sludge deposits were not detected. The corridor was put into operation relatively recently, the flows are high, the gas in the pipeline is well-cleaned and drained. This means that there are no conditions for the appearance of local resistance. In the case of its appearance, the technical services would notice a deviation from the norm (partial failure) and take measures to eliminate it.

The only acceptable explanation for the observed effect is the presence of systematic instrument errors in the pressure measurements. Denote by Δp_i a systematic error and the measurement result by an asterisk from the top: $\overset{*}{p_i} = p_i + \Delta p_i$. Fig. 7, 8 show that the most convincing hypotheses about the existence of systematic errors are the following: for the line *I* $\Delta p_6^I \neq 0$ and $\Delta p_7^I \neq 0$, for the line *II* – $\Delta p_2^{II} \neq 0, \Delta p_3^{II} \neq 0$ $\Delta p_4^{II} \neq 0$, $\Delta p_5^{II} \neq 0$, $\Delta p_8^{II} \neq 0$. In addition, it is useful to evaluate systematic errors Δp_1 and Δp_9 (common for both lines).

We will search the required estimates using the parametric identification algorithm for gas supply systems [25, 26]. For this purpose, in the parameter identification model, in addition to systematic errors, we introduce the coefficients of hydraulic efficiency E_i^I, E_i^{II}, $i = 1, \ldots, 8$ (which can be considered equal). The identification procedure yields the desired estimates (Table 1).

DOI: 10.1051/e3sconf/20172502001

The results of calculations presented in the table 1 show that the absolute values of systematic errors are not small. Removing systematic errors from the measurements makes the plot of the pressure change more realistic (Fig. 9).

Thus, the proposed method makes it possible to carry out the procedure of verification and calibration of measuring devices during the operation of the GTS without removing the instruments and using any additional equipment. In calculating operating conditions systematic measurement errors can be properly taken into account reducing the likelihood of incorrect operational decisions. The application of the proposed method is not limited to the example considered. These ideas can be used not only in relation to the technical corridor of gas pipelines, but also to the GTS both tree-shaped and looped structure.

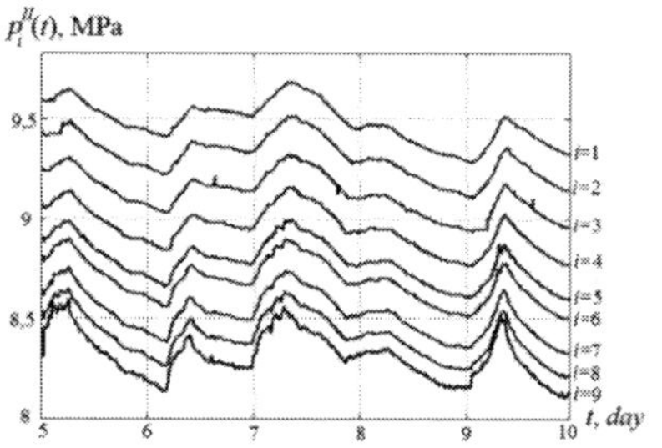

Fig. 9. Graphs of functions $p_i^{II}(t)$, $i = 1, \ldots, 9$ after elimination of systematic errors.

4 Conclusion

The paper demonstrates the additional possibilities of use of the regime parameters measurements for increasing the reliability and safety of the PS. To this end, we set and solved the following tasks: a) to detect the time of leakage and illegal tapping occurrence in the oil and product pipelines and b) to verify the measuring instruments installed at the GTS facilities. To solve them we propose new methods of regime diagnostics.

To detect the depressurization time of oil and product pipelines, a technique based on the use of nonparametric change-detection methods was developed. For verification of measuring instruments we presented a method based on algorithms of GTS parametric identification.

The processing of real telemetry data showed the effectiveness of the proposed approaches. Their implementation in production practice will require minimal costs of operators and will increase the safety and reliability of the PS operation.

Table 1. The estimations of the coefficients of hydraulic efficiency and systematic errors.

E_i^I, E_i^{II}	Δp_1, MPa	Δp_2^{II}, MPa	Δp_3^{II}, MPa	Δp_4^{II}, MPa	Δp_5^{II}, MPa	Δp_6^I, MPa	Δp_7^I, MPa	Δp_8^{II}, MPa	Δp_9, MPa
0,9929	–0,1032	–0,0171	–0,0439	0,0427	–0,0291	–0,0262	0,0283	0,0329	–0,3334

References

1. D. Sharkov. Russia wasting millions of tons of oil from leaking pipes. URL: http://www.newsweek.com /2015/04/17/russia-need-not-drill-arctic-if-it-repairs-oil-pipeline-leaks-321372.html
2. 'It's A Big One': Iowa Pipeline Leaks Nearly 140,000 Gallons Of Diesel. URL: http://www.npr.org/ sections/thetwo-way/2017/01/26/511636325/its-a-big-one-iowa-pipeline-leaks-nearly-140-000-gallons-of-diesel
3. Mexico's Pemex curbing pipeline thieves with unusable gasoline. URL: http://www.cbc.ca/news/world /mexico-s-pemex-curbing-pipeline-thieves-with-unusable-gasoline-1.2960890
4. P. Ostapkowicz. Maint. Reliab. **16**(1), 9–16 (2014).
5. S. Siebenaler, E. Tervo, P. Vinh, C. Lewis. IPC 2014. Paper no. IPC2014-33557.
6. J. E. Hough. Oil Gas J. **86**(47), 35-41 (1988).
7. B.-W. Yang, M. Recane. U.S. Patent no. 6668619 (2003).
8. H. H. Brons, H. Schaffhaussen. Pipe Line Ind. **5**, 50-53; **6**, 64-66 (1972).
9. J. Zhang. Pipeline Gas J. **228**(2), 42-45 (2001).
10. J. C. P. Liou, J. Tian. J. Energy Resour. Technol. **117**, 243-248 (1995).
11. K. E. Abhulimen, A. A. Susu. Chem. Eng. J. **97**, 47–67 (2004).
12. K. F. K. Oyedeko, H. A. Balogun. Int. J. Energ. Tech. Pol. **5** (1), 16–27 (2015).
13. M. Henrie, P. Carpenter, R. E. Nicholas. *Pipeline leak detection handbook* (Elsiever-Gulf Professional Publishing, United Kingdom, Oxford, 2016).
14. X. Lang; P. Li; Z. Hu; H. Ren; Y. Li. IEEE Access. **5**, 8659-8668 (2017).
15. ISO 18081: 2016. (2016).
16. T. R. Sheltami, A. Bala, E. M. Shakshuki. J. Ambient Intell. Humaniz. Comput. **7**(3), 347-356 (2016).
17. A. Abdulshaheed, F. Mustapha, A. Ghavamian. Renew. Sust. Energ. Rev. **69**, 902-911 (2017).
18. I. V. Nikiforov. *Sequential detection of changes in time series properties* (Nauka, Moscow, 1983) (in Russian).
19. B. E. Brodsky, B. S. Darkhovsky. *Nonparametric methods in change-point problems* (Kluwer Academic Publishers, The Netherlands, 1993).
20. M. G. Sukharev, *Forecasting methods*. (Russian State University of Oil and Gas, Moscow, 2009) (in Russian).
21. J. Zhu, A. Abur. IEEE Trans. Power Syst. **21**(2), 586-592 (2006).
22. L. Zhang, A. Abur. IEEE Trans. Power Syst. **28**(4), 3916-3923 (2013).
23. I. N. Kolosok, E. S. Korkina, L. A. Gurina. Methodological problems in reliability study of large energy systems. **66**, 231-237(2015).
24. M. G. Sukharev, R. V. Samoylov. *Analysis and control of stationary and non-stationary regimes of gas transport* (Gubkin Russian State University of Oil and Gas (National Research University), Moscow, 2016) (in Russian).
25. M. G. Sukharev, K. O. Kosova. Autom. Remote Control. **78**(5), 882-890 (2017).
26. M. G. Sukharev, K. O. Kosova. Methodological problems in reliability study of large energy systems. **67**, 110-119 (2016).

DOI: 10.1051/e3sconf/20172502002

Emergency and post-emergency control in the formation of micro-grids

Pavel Ilyushin[1*]

[1] Petersburg power engineering institute of professional development, Aviation str. 23, Saint Petersburg, Russia

Abstract. At the present time, there is a worldwide tendency to clustering the electrical power systems at LV and MV levels – creating the micro-grids. In this connection, numerous of technical issues appear in control approaches in general and in particular in emergency and post-emergency control. As micro-grids are meant to operate both in interconnected and islanded modes, the ability to steadily island and backward synchronize with an upper grid provides reduction of the negative impact of failure rate of electrical power system's elements, as well as power quality parameters influence toward operation regime of the loads. The consequence of previously described features is particular requirements for the following emergency automatics: overload-preventing automatics for overhead lines, cables and transformers; out-of-step automatics and under-frequency load shedding. Low inertia constants of some distributed generation units and low time constants of electromagnetic transients of generating units and ESS, interconnected with a grid by means of power-electronics, provide an ability to realize additional (fast) control signals in overload-preventing automatics, dispatching generating units and ESS with an aim to prevent considered elements tripping.

1 Introduction

Generation capacity decentralization forced by distinctive factors, including evolving of economically justified low-power energy generation and storage technologies (installed capacity less than 670 kW), added by electricity market liberalization, reflecting in price-responsive demand as a measure of consumer's involvement, end up with the formation of physical and virtual (consumption reduction) lump generation capacities in MV/LV distribution grids.

Being planned as passive one-port, distribution grids acquire steady-state regime control (steady-state stability) and transient control (transient stability) requirements, inherent for transmission networks, as cumulative install capacity of MV/LV generating units grows up.

Energy storage deployment is always coupled with automatic control system tying, which also can be used for implementation of demand response.

An aggregation of DG units, energy storage units and controllable loads, forming together so called DERs, by control integrity leads to the origin of controllable distribution grid's concept with the following objectives:

- Reduction of uncontrollable influence of commissioned and future DG objects towards both steady-state and transient regime parameters,
- Creation of the technological basis of non-discriminatory access for dispatched and non-dispatched «heterogeneous» DERs (comprising renewable energy sources),
- Creation of the technological basis for implementation of demand response,
- Creation of the technological basis for realization of ancillary services by DERs, which value in overall balance grows proportionally to the penetration level [1].

All abovementioned leads to a concept of micro-grid – an aggregation of electrically tied loads and DERs, having fixed electrical boundaries and operating as a single controllable element in both synchronized with an external grid and islanded modes.

In general, the appearing of this concept pursues economically effective local load supply. According to [2], micro-grid's electrical boundaries are the follows:

- A part of distribution grid limited by mains' circuit breaker (CB) (MV or LV),
- LV/MV feeder,
- Consumer's internal-supply electrical grid.

2 Emergency regimes in islanded micro-grids

Emergency regimes in islanded micro-grids are generally characterized by heavier burden from consumer's perspective than those under synchronized conditions, especially when accompanied by power deficit. Planning of emergency control encompassing electrical power supply normalization under micro-grid's islanded modes with part of generation cut off, load rise and etc. should be allocated as distinguished task due to the following:

* Corresponding author: ilyushin.pv@mail.ru

 DOI: 10.1051/e3sconf/20172502002

1. The field of those solutions is much wider in comparison with autonomous electric power supply systems. The likelihood of micro-grids, synchronized with the mains, to be islanded is inevitable, especially in case of equipment tripping in external grid under maintenance conditions (N-2). The problem generally worsens with deterioration of distribution grid's equipment, recruitment of skill-less manpower, related to emergency automatics (EA), and insufficient redundancy level.

2. Loss of a generator, a group of generators or an entire power plant cannot lead to significant active power deficit and another-reason supply interruption (voltage reduction and etc.) in an interconnected grid, whereas the same tripping events can cause full outage of reliable loads in islanded mode. Depending on micro-grid's islanding scenario, balance of generation and consumption can bias in diametrically opposite ways, such that it leads either to necessity of shedding a part of generating units or to 100% power deficit. Evidently, all the preceding should be considered while planning EA.

3. Technical features of modern generating units, such as gas turbines, reciprocating engines, wind energy conversion systems, photovoltaics, drastically differs from those of conventional steam turbines. Mainly, it relates to parameters, defining transient dynamics under external disturbances.

4. There is a tendency toward technical and economical characteristics enhancement of generating units, manufacturing abroad, which is on the strength of tripping times reduction and, thereafter, fastening of normal operating condition restoration due to applying of microprocessor-based protection.

Consequently, both islanding realization and ensuring continuous sustainable power supply in islanded mode should be evaluated while planning micro-grids, taking into account the aspects listed below:

- Frequency and active power control of generating units,
- Adjustments of generator voltage threshold U_{r0},
- Load transient-stability margins influence toward EA planning,
- Fault-voltage ride through characteristics of a micro-grid,
- Load shedding arrangement under unintentional islanding of a micro-grid,
- Out-of-step protection for generating units,
- Usage of energy storage devices.

3 Frequency and active power control

Commonly, generating units, operating in parallel with mains, acquire constant active power control, in autonomous grids – constant frequency control, whereas those tended to operate in both – combined control with switching between in the moment of islanding and synchronization respectively. The cornerstone here is an algorithm used as it can function incorrectly.

As a widespread practice, preselected CB's states (up to 5 depending on manufacturer) are used as flags for switching. Newer-the-less, in exploitation islanding can occur in neither of those CBs, non-causing algorithm to switch, resulting in frequency oscillations with possible violation of generating unit's frequency margins, ending up with this generating unit cut off.

Identical scenario can take place also when active power change is taken as the threshold. While generating units track fixed active power step, an islanding can be characterized by lower absolute value or an opposite sign.

The reasons of micro-grids to island comprise emergency equipment tripping, especially in maintenance schemes and due to EA trips, where capacity of remaining connected loads may vary drastically. If islanding happens with high load burden, load shedding takes place.

One of the load shedding algorithms applied on gas-turbine units is considered below. The latter adopts 50% or higher power loss during 200 ms as a threshold to reveal an islanding. It is assumed that generating unit has 80% active power in parallel mode. In calculations below 7 scenarios different in load value are depicted. Load varies from 5 to 100% of generating unit nominal power. Regime parameters, worth to be highlighted under question, are active power and frequency. The main calculation parameters:

- Active power control in synchronized regime: regulation actions keeping Pg = const,
- Kotating speed control: 0,5% droop with power increase and decrease time constants of 800 ms and 650 ms respectively,
- 60% of asynchronous rotating load.

Calculation results reveal small dependence on whether external fault is incepted in previous-to-islanding mode or not. Fig. 1 depicts frequency and active power curves for generating unit with constant active power for various load steps under fault and normal conditions respectively. It can be seen that differences take place only at initial stage of approximately two times fault duration.

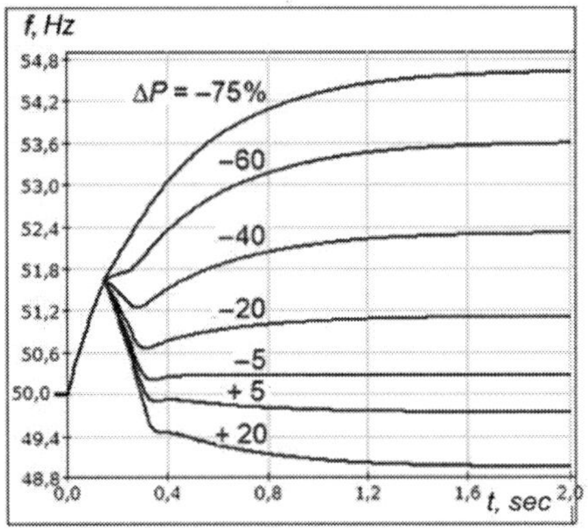

a)

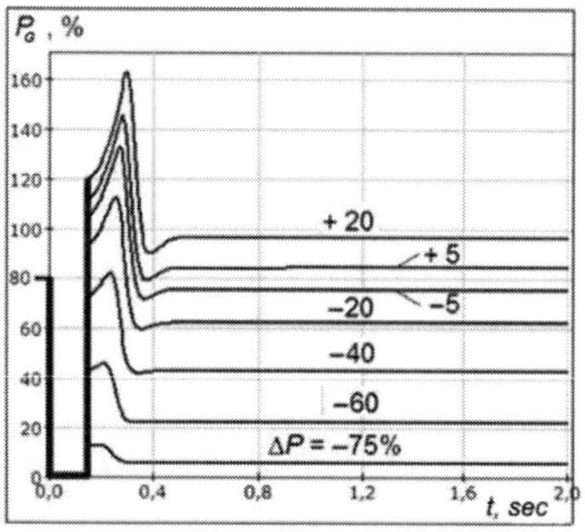

b)

Fig. 1. Frequency (a) and active power (b) under islanding conditions for no turbine fuel control and different load steps ΔP under three-phase short-circuit incepted in previous-to-islanding regime.

For load steps of 60% and 75% switching to rotating speed (frequency) control takes place, otherwise, generator continuously regulates its output active power. So, as load drops less than 50 % (ΔP of -40 to -5 % in Fig. 1) with respective frequency rise, the regulator approaches the preset power value, feeding more fuel to combustor, which provokes further frequency rise. Load surges (ΔP of +5, +20 % in Fig. 1) cause increment in frequency reduction.

Another principle, briefly mentioned above, consist in watching chosen CB's states. Whether a part of them are on, a regime is defined as parallel with mains, otherwise – islanded. This paradigm does not consider scenarios of grid splitting at different CBs, as shown in Fig. 2.

Based on preceding, the following findings are made:

1. The value of load steps cannot be used as a threshold, as it is strictly defined by previous-to-islanding regime: various in absolute value and sign load steps in various topological and parametrical situations.

2. Error penalization should be considered for combined regulators, as continuation to operate in power control mode in case of micro-grid islanding likely to lead to power outage, whereas, only inaccuracy of power dispatch occur in the opposite situation, when gas turbines operate with frequency droop in parallel regime.

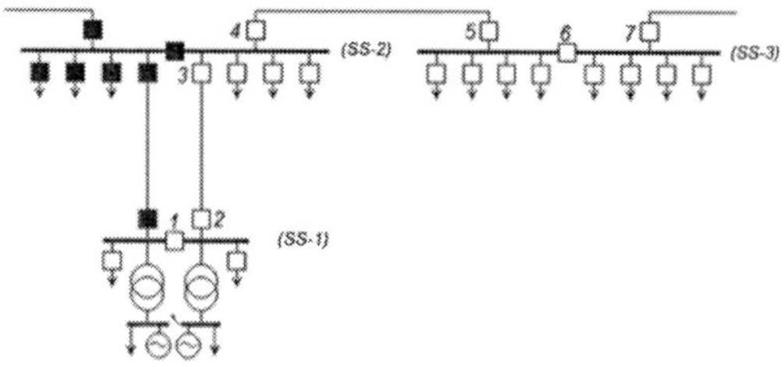

Fig. 2. For one busbar shutted down for maintenance tripping of CB 1-7 (3-7 are uncontrollable) will lead to mains loss.

A gas turbine with frequency droop of 4-5% without secondary control has power error of up to 5% nominal power for 0,1 Hz frequency oscillations.

As a solution for combined regulation issues, continuous frequency droop of 4-5% (Fig. 3) should be applied irrespectively. It can easily be done for low-power generating units by simply blocking the switching algorithm and setting the droop of 4-5%.

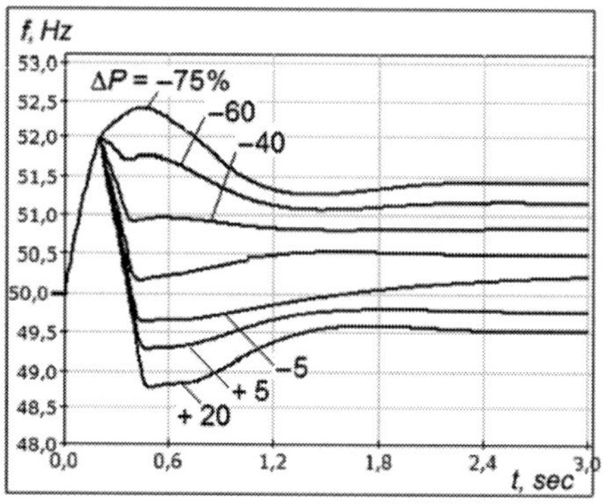

Fig. 3. All the above processes but with continuous speed (frequency) regulation.

4 Generator voltage threshold adjustments

For low generator voltage U_{G0} threshold values reactive power is chiefly drawn from an external grid. So, in case of islanding 10 kV busbar voltage drops (ΔU), biasing the voltage-regime point to up to voltage collapse. It alleviates active power deficit, where even frequency surge is possible. Fig. 4 depicts pre-fault generator Q value correlation with voltage ΔU and frequency at initial stage of islanded mode. Those parameters are expressed as generator voltage function in pre-fault regime with no under-frequency load shedding considered.

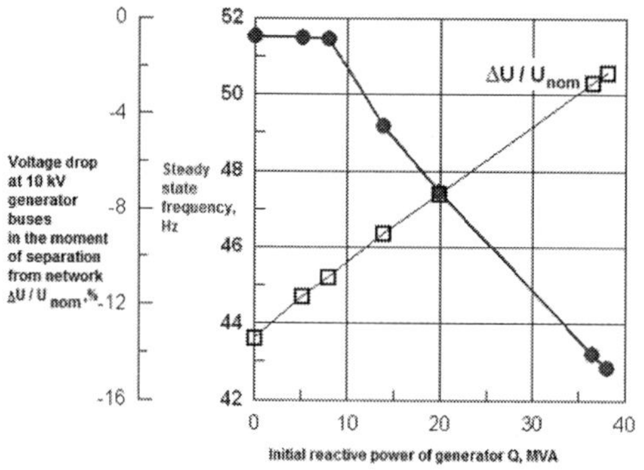

Fig. 4. The influence of generator voltage preset value toward islanding success.

If generator covers entire reactive load, the voltage drop is minimal (lower than 4% in absolute value); the loads are within their transient stability margins; the active power deficit is reflected in frequency reduction (Fig. 5a).

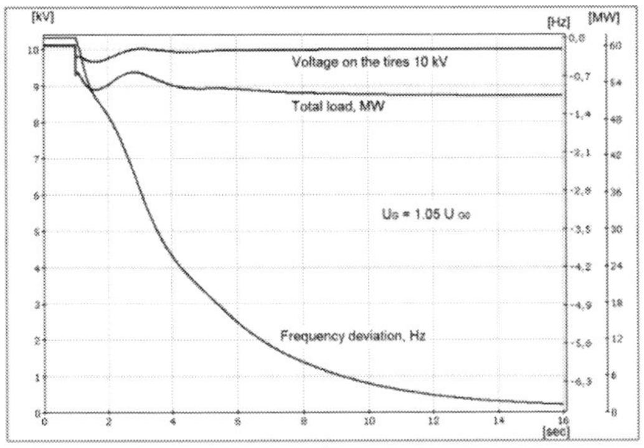

a)

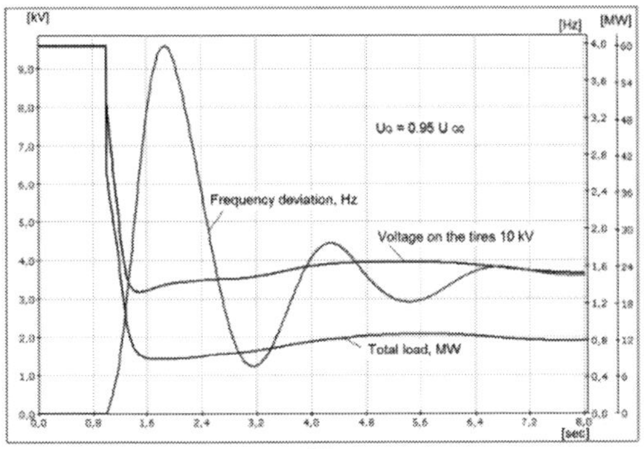

b)

Fig. 5. Reactive power regimes: a - zero reactive power flow from mains, b - non-zero reactive power flow from mains.

Quite opposite situation appears if generator excitation is relatively low. In pre-fault regime reactive power is taken from an external grid ($Q < 0$). At the moment of islanding the voltage drop takes place as low as preset U_{G0} value. Asynchronous motors' slip rises, leading to further voltage reduction and, finally, voltage collapse. The process is shown in Fig. 5b, where load outage happens initially due to voltage drop, then – due to voltage collapse on the time span of 0,1 s in overall. The frequency point reaches its peak at 54 Hz with steady-state value of 51,5 Hz. Eventually, all loads are inoperative, because of voltage value lower than 40% nominal. In the second case, voltage collapse and full outage take place.

Summing up, it is advisable to consider all the abovementioned processes while calculating U_{G0} threshold value.

5 Load stability margins influence toward emergency automatics selection

If islanding proceeds without load stability margins violation, lags in under-frequency load shedding (also, frequency thresholds lowering in comparison to maximum possible) do not cause influence toward post-emergency conditions, besides short-term frequency reduction is unacceptable for generators.

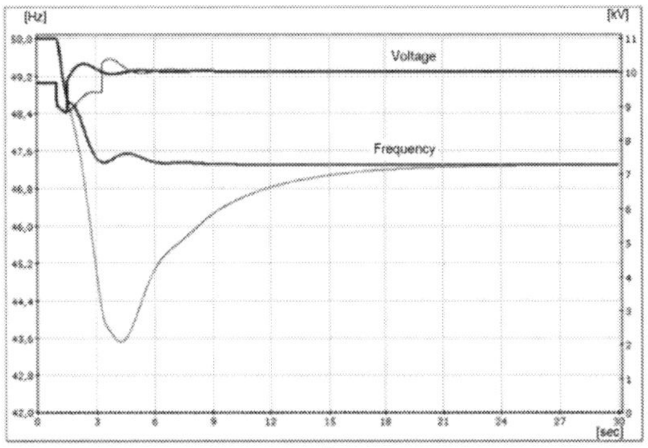

a)

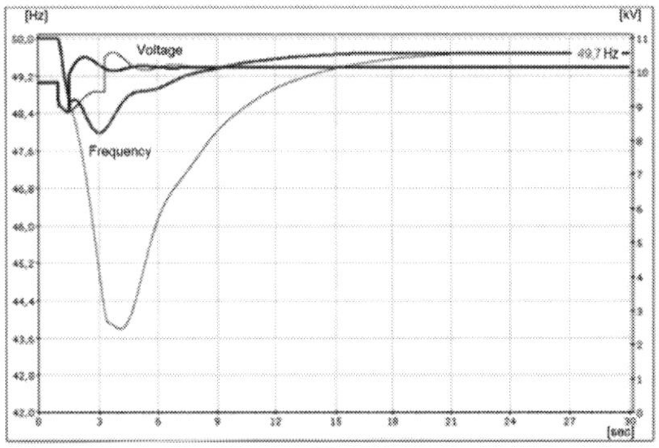

b)

Fig. 6. Islanding with initial active power deficit of 46%. Under-frequency load shedding trips at 49,2 Hz, the lag of load shedding are 0,2 s (bold curves) and 2,0 s (thin curves).

It is evident that lags do not affect the regime if transient frequency reduction satisfies relevant generator and load margins. Two processes distinctive in under-frequency load shedding volume and realization time are shown in Fig. 6 a), b). The assumptions made here are the following: shedding initialization at 49,2 Hz with 0,2 s lag for the first (bold) and 2,0 s for the second (thin), with 30% and 35% cut-off volume respectively.

Generally transient deep frequency reduction is allowed for vast majority of power receivers. Newer-the-less, selection of EAs, its thresholds and stimulus' volume should be accompanied by load stability analysis.

6 Islanding success under fault conditions

If micro-grid islanding is preformed under fault conditions (for instance, three-phase short-circuit in proximity of substation busbar is taken), motor transient stability upset takes place with extensive load outage and consequent frequency rise. Fig. 7 depicts characteristic curves both with short-circuit of 0,2 s duration and without short-circuit.

 DOI: 10.1051/e3sconf/20172502002

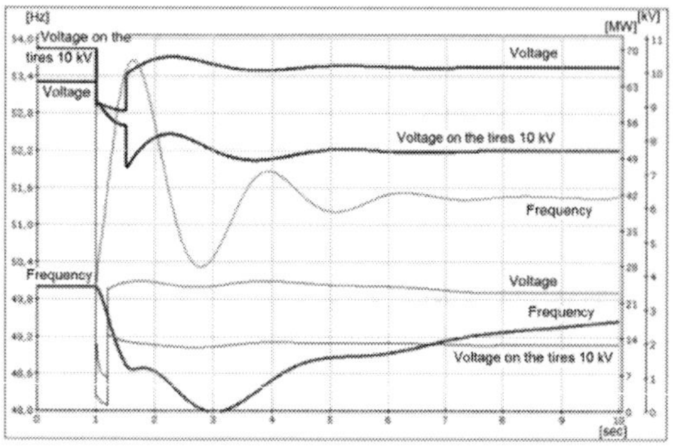

Fig. 7. Islanding under three-phase short-circuit conditions (thin curves) and without short-circuit (bold curves).

Drawn graphs depict about 30% load drop, voltage rise up to allowed boundaries after insignificant voltage drop and frequency rise up to 51,5 Hz, which in overall corresponds to successful islanding.

As a conclusion, short-circuit modeling should be conducted to assess the likelihood of successful islanding while planning micro-grids.

7 Isalding followed load shedding

Fast-operating load shedding is required in all of those cases characterized by significant voltage drop, load power drop and load transient stability upset.

As initial values for calculations the same cut-off volume (35%) is taken, but initializations are performed in 0,1 s after short-circuit inception and 0,3 s after clearance. Fig. 8 reveal self-acceleration of rotating load after fast shedding.

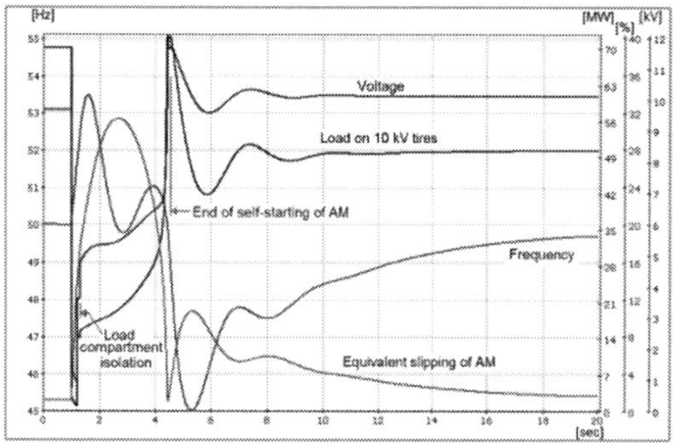

Fig. 8. Load shedding in 0,3 s after fault inception and 0,1 after clearance.

Implementation of load shedding, triggering in the moment of islanding, is justified if under-frequency load shedding operating speed is significant, this-type automatics failure due to deep voltage drops is inevitable (in most cases deep frequency reduction is accompanied by deep voltage decrease). Moreover, there is a probability to face lack in power available for shedding under high-power deficit conditions, which can end up with generation facilities' auxiliaries out of service.

8 Out-of-step protection

Inherent low inertia of reciprocating engines and two-/ three-shaft gas turbines (with free power turbine) possess considerable difficulties in transient stability ensuring of the abovementioned units. In addition, being the main reason of secondary dynamic stability upset, small electrical distances of MV/LV grids contribute to load stability loss in case of generator asynchronous regime takes place.

Fast clearance of generator asynchronous mode by its tripping off have the following consequences should be bear in mind while selecting out-of-step algorithm and corresponding thresholds:

- Microgrid-mains tie power flow increase in case of internal generation loss,
- The possibility of equipment overload (transformers, overhead lines) as vast majority of micro-grid loads is located at MV/LV levels,
- Voltage decrease at micro-grid nodes, where the most susceptible loads are synchronous and asynchronous motors of industrial facilities.

It is justified to allow asynchronous running of low-power generators if asynchronous satisfies the following:

- Short duration,
- Ending up with resynchronization of the generators,
- Not causing violation of normal regime bounds of other generators and power receivers (should be confirmed by calculations),
- The duration is controlled by out-of-step protection.

Tripping generators off by out-of-step protection is justified if asynchronous mode duration exceeds an allowed value, which, generally, occurs in maintenance schemes and emergency regimes with "weak" ties. The likelihood of spontaneous resynchronization of generators is high in micro-grids, making adequate to offset out-of-step protection tripping if it does not lead to multifrequency asynchronous running of the generators and motor stability loss.

Described above issue can be fully solved by conducting transients modeling, taking into account load transient parameters in adjacent nodes. It also allows to depict impedance hodograph diagram used for out-of-step protection parametrization.

9 Energy storage usage

Hereafter rapidly adjustable in active and reactive power energy storage is considered. The operating speed is assumed to be that high, so switching to controllable feeding/absorbing of active/reactive power within emergency regime is instantaneous in a first approximation.

The viability of energy storage devices is defined by economic factors. All expenses are counterbalanced and even outweighed for susceptible facilities.

Energy storages are applied within micro-grids under islanding mode in the following cases:

- Safety trips of power source, where energy storages are aimed to restore normal supply scheme and ensure failure-free manufacturing shutdown,

DOI: 10.1051/e3sconf/20172502002

- To cover local equipment overloads under extraordinary conditions (substantial weather changes, equipment emergency trips and ets.),
- Generator black-start source,
- Emergency source for reliable loads under autoreclosing and automatic transfer switch pauses,
- As a measure for voltage collapse tackling in grids with high share of rotating loads.

It worth to note, that voltage flattening out to pre-fault regime value cannot always be performed by energy storage effectively. For short-circuits in proximity of load terminals the energy storage power needed to lift voltage up to nominal can be significant. The fault clearance fastening can alleviate the situation.

Energy storage application within micro-grids provides several benefits, however the quantitative effect should always be assessed by calculations.

10 Conclusion

The analysis of successful islanding possibility and sustainable power supply in micro-grid island mode should always be conducted, taking into account all technical questions described in this paper.

It is justified to implement unified speed (frequency) control algorithm with frequency droop of 4-5%, depending on prime mover, irrespectively to projected generator operating regime.

The deep under-excitation of micro-grid generators can lead to significant voltage reduction, motor's slip rise and further process progress up to voltage collapse. Thus, the generator voltage preset value should be set relatively high.

To prevent voltage collapse of a micro-grid, islanding under fault conditions, the fast-operating load shedding, tripping by a signal formed at the moment of islanding, should be implemented, as under-frequency load shedding can operate ineffective.

As micro-grid generators spontaneous resynchronization is very likely, offsetting out-of-step protection by time is justified in all cases, but if leading to multifrequency asynchronous running of the generators or load transient stability upset.

Application of energy storage devices within micro-grids is economically effective for industrial facilities, characterized by high reliability requirements and financial losses for short-term power outages.

References

1. P. Ilyushin, S. Muzalev, Relay protection and automation **3**, 39-45 (2016)
2. N. Hatziargyriou, Wiley-IEEE Press **1**, 344 (2014)
3. P. Ilyushin, Electro **4**, 19-25 (2011)

Probabilistic assessment of power system mode with a varying degree of wind sources integration

Nariman Rahmanov[1], *Viktor* Kurbatskiy[2], *Huseyngulu* Guliyev[1*], *Nikita* Tomin[2], *Zaur* Mammadov[1]

[1] Azerbaijan Scientific–Research and Designed–Prospecting Institute of Energetics, Baku, Azerbaijan
[2] Melentiev Energy Systems Institute, 130 Lermontov str., Irkutsk, Russia

Abstract. At present among renewable sources the wind and solar plants have the most significant portion of power generation. Randomly changing and intermittent nature of this power leads to the stochasticity of the power grid mode, estimation of parameters of which requires application of probabilistic modeling. In the paper it is proposed an advanced algorithm of probabilistic load flow based on the development of two-point estimation method, the efficiency of which is confirmed on the basis of computational experiments and comparative analysis of the Monte Carlo simulation results. Calculations and analysis of the modeling results were carried out on standard 14-nodal scheme of IEEE and real electrical network of "Azerenerji" Grid.

1 Introduction

Calculation for the steady-state mode is a significant part of the total scope of researches on electric power systems (EPS) both at the design stage and during the operation of these systems [1, 2]. In recent years, the growth of power generation from renewable energy sources and mainly due to the large share of generation capacity from wind and solar energy converters with the stochastic nature increased the uncertainty in the power system state. This circumstance has led to larger need to solve the problems of planning, analysis and system management based on stochastic modeling techniques. For the first time the probabilistic modeling of the load flows in the power system was considered in [3], in which the stochastic input variable was taken to be the load power value in one of the nodes of the system.

At that the variability of the load power was reproduced by Monte Carlo method, which led to the necessity to consider the tens of thousands and more numbers of steady states. At the beginning of the 2000s a number of authors [4, 10, 11] proposed a simplified approach for probabilistic modeling of the load flows (PR), based on the analysis of initial data for load power values, power generation by wind turbines and relevant functions of their distribution in two given points.

Probabilistic methods of calculation and analysis of load flows (LF) in the system are based on two approaches: numerical simulation by Monte Carlo method [3-5,9,10], and analytical methods of the transformation in the form of cumulated representations [6], [7], or using of point values of probability distribution functions [8,9].

In a probabilistic formulation the problem of LF assumes the initial information assignment in the form of the probability distribution function of power at the nodes of generation and load, as well as the probability of random changes in the system diagram. Presence of description of random states of EPS with given probabilistic estimates of characteristics for the system input values allows for numerical or analytical determination of the probabilistic estimates of the variables at the system output - the voltage at the nodes, power flows and power losses in the Grid.

Modeling of the probabilistic state of the system, containing wind power plants (WPP), enables to solve the problem of planning and analysis of modes, taking into account their stochastic variation.

The paper presents the results of the modeling, calculation and analysis of EPS modes at stochastic variability of power values of load and sources, as well as considering the probability of emergency shutdown of transmission lines of backbone network.

To simulate such stochastic processes the combined approach of Monte Carlo method and two-point assessment method is used. It is known that the stochastic variability of the load power is fairly well described by a normal distribution law and the Monte-Carlo method is the most appropriate for its reproduction.

*Corresponding author: huseyngulu@mail.ru

DOI: 10.1051/e3sconf/20172502003

2 Probabilistic modeling of system state with wind farms

Probabilistic assessment of system state is caused by stochastic variability of generation in the nodes, containing sources with random-intermittent power output, fluctuation of power consumption in load nodes, as well as random changes in the topology (scheme) of the system. The values of the variables of the system state are determined based on recurring solutions of steady-state equations (SSE) for various combinations of values of variables in the input. The values of the variables at the output (the result of the SSE solution) are the values of flows of active and reactive power in the lines, the voltages at the subsystems. To solve the LF problem in deterministic formulation, various methods and algorithms are developed [11-15], which are widely used in practice.

Deterministic model of the SSE is described in the following form:

$$\sum_G P_i^G - \sum_D P_i^D = U_i \sum_{j=1}^{n} U_j \left(G_{ij}\cos\delta_{ij} + B_{ij}\sin\delta_{ij}\right) \quad (1)$$

$$\sum_G Q_i^G - \sum_D Q_i^D = U_i \sum_{j=1}^{n} U_j \left(G_{ij}\sin\delta_{ij} + B_{ij}\cos\delta_{ij}\right) \quad (2)$$

where P_i^G, P_i^D и Q_i^G, Q_i^D - active and reactive powers of generation and consumption respectively in $i-$ node;

U_i, U_j - voltage value in i and j nodes respectively;

G_{ij}, B_{ij} - active and reactive conductance;

δ_{ij} - phase shift between U_i and U_j vectors.

At random nature of variability of input variables, specified as a probability vector

$$X = \left[P_i^G, P_i^D, Q_i^G\ Q_i^D .....\right] \quad (3)$$

the SSE will be represented in the form of equation set combination (1), (2) with the values of the input variables defined as a series of random sample or distribution function of these variables. In this case, the stochastic system state is described in the form of relationship of randomly changing values of power overflows in the network and the voltages in the nodes from the random values of powers in the nodes of generation and load respectively.

$$Y = F(X) \quad (4)$$

where

$$Y = \left[P_{ij},\ Q_{ij},\ \delta_{ij},\ U_i......\right] \quad (5)$$

For wind turbine equipped with the asynchronous generator, the probabilistic model for generation of active and reactive power can be represented as the relationship from wind speed. At simplified representation of the asynchronous generator scheme, such relationship has the following form:

$$P_{wi}^G = \frac{-sR_2}{R_2^2 + s^2 X_i^2} \cdot U_i^2 = \\ = \begin{cases} 0, & v \le v_{ci} \\ P_{nm} \cdot \dfrac{v - v_{ci}}{v_H - v_{ci}}, & v_{ci} < v \le v_H \\ P_{nm}, & v_H < v \le v_{co} \\ 0, & v > v_{co} \end{cases} \quad (6)$$

$$Q_{mi}^G = \frac{sX}{R_2} P_{wi} \quad (7)$$

where P_{mi}^G, Q_{mi}^G - capacity of production of active and reactive power of wind farm connected to i - node of the system; S -sliding of wind turbine; R - active resistance of rotor of asynchronous generator; X - the sum of dissipation reactances of stator and rotor; v_{ci} - the wind speed at which the asynchronous generator starts; v_{co} - the wind speed at which the wind turbine is switched off ; v_H - the wind speed, which provides a nominal power output.

In equation (5) the influence of the stochastic variability of wind speed on the power generation value of wind turbine P_{wi} appears through the stochastic variability of the sliding value of the asynchronous generator. In this connection, during the probabilistic assessment of system state with stochastically varying power output of asynchronous wind turbines, one of the state parameters is the sliding value. At that it should be noted that the values of the generation of active and reactive power of the wind turbine will be random variables dependent on the stochastically changing wind speed.

Analysis of probabilistic LF in the electrical system will be performed based on modeling using the Monte Carlo method and analytical method of evaluation of the curve of the probability density function by two points (2TOM method).

The Monte Carlo method is the most commonly used due to the simplicity of the modeling implementation regardless of the number of random variables defining the system state, its complexity and the “output variable - input variable” nonlinear relationship. At the same time, the Monte Carlo method requires large expenditures for computing resources to conduct repetitive calculations carried out 10 thousand or more times. We will not consider in detail the algorithm of the Monte Carlo method because it is widely used in power industry [8,9].

2TOM method gives the opportunity for estimating the stochastic state of the system by determining the values of the variables or their distributions at the model output using the specified values of the variables or their distributions at the output at two points of the current state of the model.

For probabilistic models of LF, described in general form by equations (1), (2), (4), the state estimation by 2TOM method takes as input data the values of generation and load (variables x_i) that vary with the probability of distribution function chosen for two points, located on opposite sides of the mean value. At that, as a practice of this method application shows, for the function characterization it is enough to specify three moments - mean, standard deviation and asymmetry coefficient.

The algorithm for calculation of distribution probability functions and moments of random variables characterizing the state of load flows – voltages in nodes and power flows in the lines -consists of the following stages [11]:

1. The choice of the number of fuzzy variables n For 14-nodal IEEE test scheme n=3 is selected - one node of the load, which is taken into account by stochastic model of the power size; one node with connected wind farm and one linear element (transmission line) with a given probability of emergency shutdown.
2. It is assumed that the mathematical expectation and standard deviation of the state variables at the output are equal to zero

$$E(Y)=0; \qquad E(Y^2)=0$$

3. It is assumed that the number of input variables that are located above and below the mean value is $k=1$.
4. Determination of concentration point $\varepsilon_{k,1}$ and $\varepsilon_{k,2}$ and concentration probability $p_{k,1}$ and $p_{k,1}$ for the variables above and below their mean value

$$\varepsilon_{k,1}=\sqrt{n}$$

$$\varepsilon_{k,1}=-\sqrt{n}$$

$$p_{k,1}=p_{k,2}=\frac{1}{2n}$$

5. Determination of two values of input variable concentration

$$\chi_{k,1}=\mu_{x,k}+\varepsilon_{k,1}\cdot\sigma_{x,k}$$

$$\chi_{k,2}=\mu_{x,k}+\varepsilon_{k,2}\cdot\sigma_{x,k}$$

where $\mu_{x,k}$ и $\sigma_{x,k}$ – mean and standard deviation of variable x_k, respectively.

6. Calculation of LF at input variable values for both concentration positions $\chi_{k,i}$.
7. Calculation of new values of moments for random output variables Y.

$$E(Y)\cong\sum_{k=1}^{n}\sum_{i=1}^{2}\left(p_{k,i}h\left(\left[\mu_{x,1}\cdots x_{k,i}\cdots\mu_{x,n}\right]\right)\right)$$

$$E(Y^2)\cong\sum_{k=1}^{n}\sum_{i=1}^{2}\left(p_{k,i}h\left(\left[\mu_{x,1}\cdots x_{k,i}\cdots\mu_{x,n}\right]\right)^2\right)$$

8. Calculation of mean and standard deviation

$$\mu_Y=E(Y)$$

$$\sigma_Y=\sqrt{E(Y^2)-\mu_Y^2}$$

9. Calculations of probabilistic characteristics for the input variable $x_{2,1}$ ($k=2$).

3 Modeling results

Estimation experiments were conducted at 14- nodal test scheme of IEEE and real scheme of Azerenerji system with 219 nodes, in which the power values of load and generation nodes were considered as fuzzy random variables, as well as telemetry data, defining the probability of a normal (emergency) status of the monitored lines.

Probabilistic calculations for load flows were performed using ETAP-5 software complex.

Model of 14-nodal IEEE scheme. Single-line diagram of the system is shown in Fig.1, in which the 50 MW wind farm is connected to the system at node 8. It is assumed that the load powers at the nodes are stochastic with normal distribution function of probability density. The probability distribution of power generation of wind farm is determined based on the sample data of power values computed according to (5).

Let's perform a comparative analysis of results of modeling of electrical system modes for the following options of scheme state (Table1).

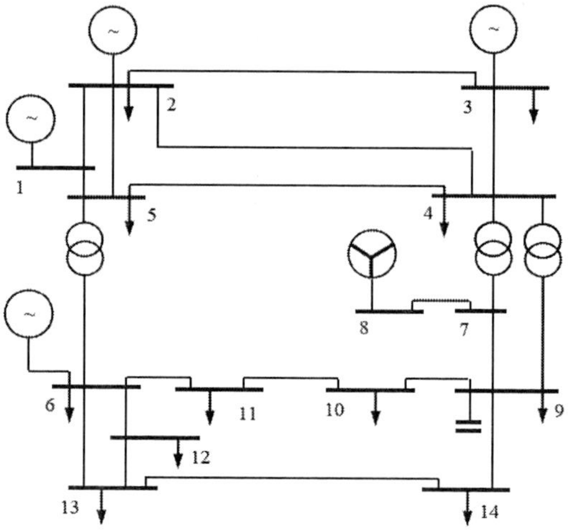

Fig. 1. Test IEEE scheme

Table 1. Options of stochastic states of power output of WPP, load capacities and the current scheme state probability

System state options	Model of load and wind farm		Network topology state
	Loads	Wind farms	
1	fixed	With changing generation (for periods with constant wind speeds)	Normal complete scheme
2	fixed	Stochastically changing	Normal complete scheme
3	Stochastically changing	Stochastically changing	Normal complete scheme
4	Stochastically changing	Stochastically changing	Probability of PTL outage

The calculations of the indicators, that determine the state of the system under study, are performed for each option. As it is seen from the analysis of the results, with the increase of stochasticity of power change, the greater variability is characteristic for the standard deviation values of the state variables. The increasing uncertainty of the input variables leads to an increase in stochasticity of state variables.

Fig. 2 and Fig. 3 show curves of integral functions of the probability distribution of voltages in the nodes 7 and 14 in the absence and connection of wind farm to the system (the second option of the scheme state).

From the comparison of the distribution characteristics one can see that the voltage fluctuation in the node 7 is the most significant compared to similar fluctuations in other nodes due to the remoteness of these nodes from the wind farm connection point.

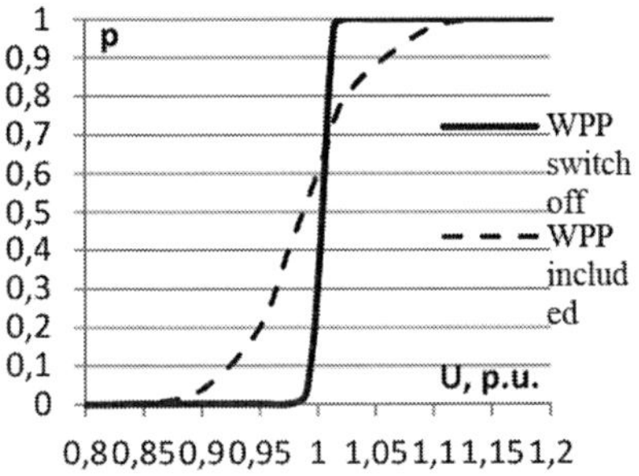

Fig.2. Cumulative distribution function of voltage probability at node 7

As can be seen from Fig. 2 and 3, when the WPP is connected, the voltage values in the nodes have random nature of the variability of probability distribution (shown in dash curve), which differs from the corresponding characteristic in the absence of the wind farm in the scheme.

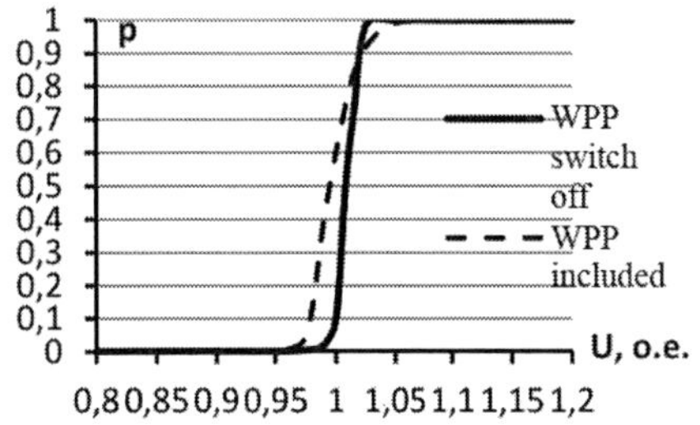

Fig.3. Cumulative distribution function of voltage probability at node 14

Model of "Azerenerji" real scheme. For the estimation of the influence of established power volume of WPP and the value of the random variation of power generation on the parameters of the electrical system state, the investigations were carried out on the model of real scheme of "Azerenerji". Estimation experiments were carried out on scheme models with connected WPP with capacity of 50 MW and 100 MW respectively. Fragment of the "Azerenerji" Grid with WPP connected to 330/220/110 kV "Yashma" substation is shown in the Fig.4. The system is presented in the form of two connected parts. To one of them a wind farm is connected, consisting of FL - 2500/90 type wind turbines with the capacity of 10x2,5 MW.

Technical parameters of FL -2500/90 DFİG type wind farm are as follows: rated power - 2500 кW; wind speed at WPP initiation (υ_{ci}) - 3,5 m/sec; rated wind speed (υ_r) -13 m/sec; maximum wind speed (υ_{co}) - 25 m/sec; output AC voltage – 690 V.

According to Table 1, the modeling of the probabilistic LF for all four options of the system state was performed twice - for the scheme with 50 MW and 100 MW WPP.

Analysis of modeling results shows that with the increase of generation power from the wind farm the dispersion of the voltage probability distribution in the nodes increases. The voltage distribution curves on 220/110 kV "Mushfig" SS with the capacity of wind power in the system 50 MW and 100 MW are illustrated as example in Fig. 5 and 6. As can be seen, the growth of wind power increases the stochasticity of the system state that indicates the difference of its parameters from the same parameters for the scheme without WPP.

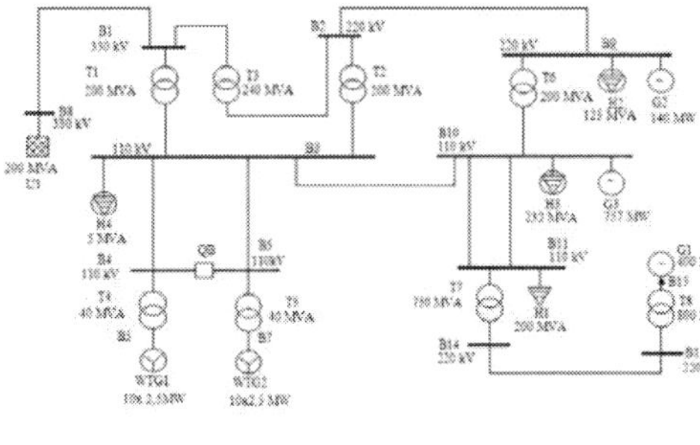

Fig.4. Simplified diagram of system under

 DOI: 10.1051/e3sconf/20172502003

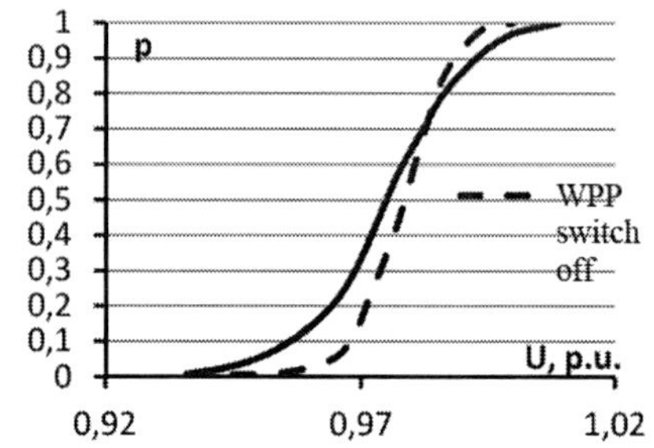

Fig.5. Cumulative distribution function of voltage probability at node B3 of "Azerenerji" scheme

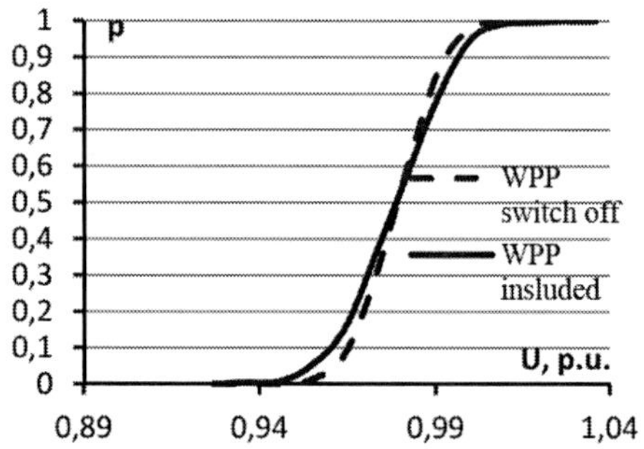

Fig.6. Cumulative distribution function of voltage probability at node B9 of "Azerenerji" scheme

4 Conclusion

Growth of capacity of WPP integrated to the grid power leads to the need for analysis and control of system mode based on probabilistic modeling that takes into account the stochasticity of the system state caused by WPP work. The paper presents the results of probabilistic LF modeling at different proportion of wind power integrated in the power system, assesses the probabilistic system state - voltage probability distribution in the nodes of the electrical network, power flows in the lines and the variability of the loss values. It is shown that the growth of the connected capacity of WPP can significantly change the probabilistic assessment of voltage in the network.

The model is proposed for calculation of power production by wind turbine equipped with a asynchronous generator. During the studies it is found out that when modeling the probability LF in the power system with integrated wind turbines, the stochastic variability of load, as well line failure flows in the calculated interval should be taken into account.

In the future it is necessary to develop and improve methods and tools for power grid management at stochastic changes of the state caused by renewable sources.

References

1. *Electrical systems. Mathematical problems of power engineering* (M: Higher school, 1981.- 288 p).
2. A. Gamm, *Probabilistic methods of calculation of modes of electric power systems* (Bratsk, BrRI. – 1990. – 91 p).
3. B. Borkowska, *IEEE Transactions*, Vol. PAS - 93, **No.3**, pp. 752-759. (1974).
4. P. Zang, S.T. Lee, *IEEE Trans. Power apparatus and Systems*. February 2004, **Vol. 19**.- pp. 676-682.
5. P. Jorgensen, J. Christensen, J. Tande, *Proc. 8th International Conf. Harmonics and Quality of Power*, **vol. 2**.- pp. 1146-1151, Athens, Greece. (1998).
6. N. Rahmanov, H. Guliyev, *International scientific seminar named after Rudenko Y. Methodical questions of research of reliability of large energy systems: Reliability of liberalized energy systems: Collection of research papers*. **Issue 65**, St. Petersburg, pp.- 511-516 (2015).
7. A. Hashimov, H. Rahmanov, H. Guliyev, R. Rahmanov, A. Mustafayev, *International scientific-technical journal*, Minsk, **№ 6**, pp.- 519-528. (2016).
8. El - Khattab, Y.G. Hegazy and M.M. Salama, *IEEE Transactions on Power Systems*, Vol. 21, **No. 2**.- pp. 524-532 (2006).
9. M. Basil, A. Jamieson, *16 th North Sea flow measurement workshop*, Gleneagles, -pp. 1-10. (1998).
10. C.L. Su, *IEEE Trans. Power Syst.*, Vol. 20, **No. 4**.- pp. 1843-1851. (2005).
11. G. Verbic, C.A. Canizures, *IEEE Trans. Power Syst.*, Vol. 21, **No. 4**.- pp. 1883-1893. (2006).
12. B. Stott, *Proceedings of the IEEE*. Vol. 2, **No. 7**,-pp. 916-929. (1974).
13. S. Yuferev, A.V.Pazderin, *Proceedings of the IASTED International Conference Power and Energy Systems (Asia PES 2013)*, April 10-12, 2013 Phuket, Thailand.
14. B. Ayuyev, V. Davydov, P. Yerokhin, V. Neuytin, *Computational models of load flow in electrical systems*. (M., Flint, Science, 2008).
15. O. Voytov, Y. Popova, L. Semenova, *Electricity*, **№3**, - pp. 19-26. (2013).
16. *Handbook on probability theory and mathematical statistics* (M.: Science. Chief editional board of physical and mathematical literature, 1985. — 640 p).

Method of adequacy optimization of electric power systems under market conditions

Natalia Aizenberg[1], *Sergey* Perzhabinsky[1]*

[1]Melentiev Energy Systems Institute, 130 Lermontov str., Irkutsk, Russia

Abstract. We suggest a new approach to optimization of the adequacy of electric power system under market conditions which involve competition of various interests of market agents. This feature makes unsuitable of traditional methods for optimization of the adequacy of electric power system. The reliability analysis of the variants of development of electric power system is carried out on the base of the Monte Carlo method. The profit of generating companies is calculated by the obtained reliability indexes which characterize some variant of electric power system development. The investigated variant of development of electric power system is effective if all generating companies choose a sufficient level of reliability that satisfies the consumer. The results of experimental research are presented in the paper.

1 Introduction

The reliability level of electricity supply depends of power reserves and power transmission lines capacity. These factors define additional costs and consequently electricity prices. If consumer is interested in high level of reliability, then he should agree to pay for this service.

The problem of electric power systems development taking into account adequacy is solved traditional on the basis of cost minimization for this development. There are following classical models of electric power systems adequacy optimization. The first one is problem of searching of ways for achieve of the given meaning of reliability index in nodes of electric power systems with minimal costs [1]. The second one is problem of minimization of costs and mathematical expectation of costs of undersupply of energy [2]. The problem of maximization of reliability index under the given cost level is third main classical problem. For solving these problems researchers take information of possible aggregate demand of power and available and planning generators in nodes, configuration and capacity of network. Today the basic method for realization of these models is interactive approach. The main idea of the method consists in adequacy analyze expert's variants of energy system development [3].

Today electric power systems are liberalized in many countries. Liberalized energy markets have different features [4] which we must take into account under planning of development of electric power systems. The type of market interaction is a reason of choice of behavior criteria of energy market agents such as profit maximization for generator and network companies or maximization of social welfare for energy system operator or utility maximization for customers. The result of interaction of agents is market equilibrium with formed price and quantity of power energy and consumption, quantity of used generators and others. This equilibrium is different of perfect competitive equilibrium. That's way we must use noncompetitive models for solving the problem of development of energy system.

Taking account of the reliability of energy supply under definition of plan on development of energy system in market conditions is seriously complicated of solving of the discussed problem. The model of regulated monopoly with traditional method of minimizing of costs could be expanded for analyze of perfect competitive market. Further development of model on the base of no perfect competitive is more complicated, for example, for oligopoly of energy market. The common way for solving of these problems is not exist. We'll describe some existing approaches and define the place of our model.

The authors in [5] offer to compute of efficiency of investment in electricity market with using reliability index LOLP (Loss of Load Probability) [6] which characterizes the probability of the load loss. Formed meaning of this index is took equal as one day of the load loss in ten years. In our model we use near idea and offer our approach of forming of consumer price.

Ensuring generation adequacy in electric power system in some market conditions such as conclusion of bilateral treaty is discussed in article [7]. There is the problem that the compute of reliability is very simplified. And the reliability index LOLP can't take into account more difficult events such as stimulus of keeping of generators which works in peak time. It can be reason of mismatch between prices on energy market and payments for power market.

*Corresponding author: sergperj@gmail.com

DOI: 10.1051/e3sconf/20172502004

There are articles in which possible development of electric power system is computed under competitive interactions at first. Then they estimate and compare reliability of the computed variants of the development of system [8, 9]. In our model we consider the set of variants of system development which be obtained by algorithm [1]. We analyze reliability indices and profit of companies for every electric power systems variant. After that the optimal variant for market agents is chose.

Choosing of the best variant of electric power systems development can be accepted on the government level or on level of generating and network companies. One of the way for union of the opposite interests is two stages approach. On a first stage investment decisions are accepted. Quantity of generated power and prices are computed on the second stage. Similar approach is presented in [10].

Some models of electric power systems development under regulated monopoly and oligopoly with taking account of quality of energy supply is discussed in [11]. Calculating of quality of energy supply is simplify in this model. The authors input additional condition for reliability index in the optimization problem.

We suggest of two-stage approach to optimization of adequacy of electric power systems under market conditions. At the first stage we consider and estimate of adequacy level of several expert variants of electric power system development. At the second stage parameters of market equilibrium are computed for all market agents such as consumers and generating or network companies. As a result, we choose of the most effective variant of development by maximization criteria of social welfare and high level of adequacy. The one of new ideas is presentation of price as function of reliability index. In addition, we can apply the developed method to estimate of effectiveness of different market types.

2 Method of reliability optimization

In market conditions the problem should be solved taking into account different interests of economic agents such as generation and network companies and consumers. This fact determines of changing of optimization criteria. Instead of minimization of system costs we should maximize payoff function of each agent [7, 12]. Decision making of economic agents is based on results of adequacy estimation of electric power system. Effective method for such problem is developed earlier [13].

We modified the method for adequacy analysis of electric power system in market conditions. The method consists of four main blocks.

1. *Modeling of random states of electric power system.* Each state is characterized of the set of random values such as load and available generator capacity in nodes, power line capacity.

2. *The model of power shortage estimation of electric power system.* Power shortage is estimated for each random state of electric power system. The model of power shortage estimation with quadratic power losses in power lines [13] is used in this stage. Taking account of quadratic losses guarantees uniqueness of deficit distribution over system nodes.

3. *Profit computation.* Profit of generation companies is computed with penalties for electricity sacrifice. There are two ways for price definition. The first one is considering of generation companies as price takers. The other one consists in taking into account service quality elasticity of demand.

4. *Computation of reliability indexes.* Reliability indices are probability of no shortage system work, mathematical expectation of power shortage and electricity sacrifice of consumers, coefficient of availability of power.

Optimal set of equipment in electric power system is defined as a result of comparison of variants of electric power systems development. That's way we are adding the next stage of computation.

5. *Comparison of reliability indices with companies' profit according to different configurations of electric power system.* On this stage it is choosing of optimal variant of system development which balances reliability level necessary for consumer and profit margin of generation and network companies. Maintaining the Integrity of the Specifications

3 The model of power shortage estimation

Let's consider the scheme of electric power system. The electric power systems scheme consists of n nodes and set of links between nodes. According to method for estimating of electric power systems adequacy it is necessary to simulate random states of electric power systems a many times. The simulations are occurred on a base of Monte Carlo method.

Let N is the given number of electric power systems states. Each state is characterized by set of means of random values such as generating capacity $\bar{x}_i^k$, load value $\bar{y}_i^k$ in the node i, line capacity $\bar{z}_{ij}^k$ between nodes i and j, $i = 1,...,n$, $j = 1,...,n$, $i \neq j$, $k = 1,...,N$.

We use following problem [13] for power shortage estimation of random states of electric power systems. The power x_i and the load y_i in node i, power flow z_{ij} from node i to node j, $i = 1,..., n$, $j = 1,...,n$ are variables of the problem. The considered problem for some k, $k = 1,...,N$, is

$$\sum_{i=1}^{n} y_i \to \max, \tag{1}$$

subject to the constraints

$$x_i - y_i + \sum_{j=1}^{n}(1 - a_{ji}z_{ji})z_{ji} - \sum_{j=1}^{n} z_{ij} = 0, \; i = 1,\dots,n, \tag{2}$$

$$0 \le y_i \le \bar{y}_i^k, \; i = 1,\dots,n, \tag{3}$$

$$0 \le x_i \le \bar{x}_i^k, \; i = 1,\dots,n, \tag{4}$$

$$0 \le z_{ij} \le \bar{z}_{ij}^k, \; i = 1,\dots,n, \; j = 1,\dots,n, \; i \neq j \tag{5}$$

where positive coefficients of power losses a_{ij} are given.

As a rule, the adequacy analysis of electric power system is realized for year. Every hour of work of electric power system is modeled. The failures of generators and power lines are used as random parameters. The repair time of equipment and fluctuations of load in the year are taking account in modeling. The rules of simulations of random values such as available capacity of generator and power line or load value are discussed in [14].

Let set of $\hat{x}_i^k$, $\hat{y}_i^k$, $\hat{z}_{ij}^k$ is optimal solution of the problem (1) – (5), $k=1,...,N$, $i=1,...,n$, $j=1,...,n$, $i \neq j$. The optimal value of power shortage in node i, $i=1,...,n$, is defined by the formula

$d_i^k = \bar{y}_i^k - \hat{y}_i^k$, $k=1,...,N$.

The state of electric power systems is deficit if the value

$$d^k = \sum_{i=1}^{n} d_i^k$$

is not equal to zero. Let is numerate all deficit states of electric power systems and define of the ircommon quantity as H. Then the probability of without shortage work of electric power systems are computed by next formula

$$P = 1 - \frac{H}{N}.$$

This index is corresponded to index LOLP.

Mathematical expectation of power shortage in nodes of energy system is computed by next rule

$$MD_i = \sum_{j=1}^{H} \frac{d_i^j}{N}, \quad i=1,...,n.$$

That's way mathematical expectation of power shortage in electric power systems is calculated by

$$MD = \sum_{i=1}^{n} MD_i.$$

4 Modeling of the economic agents behavior

As a result of the reliability assessment of each equipment we find the set of reliability indices in the electric power systems. From the value of reliability indices depends on the willingness of consumers to pay for electricity (demand for reliability). Let r is the some reliability index (for example, the probability of without shortage work P), characterizing the frequent, duration and depth of failures. In the general case, instead r, an n-dimensional vector can be used, the components of which are the values of some reliability index obtained from the nodes of the system. We assume that the values of the index r cannot be below a certain level $\underline{r}$ and not above the level $\bar{r}$. The decision of N tasks to assess the power shortage for each scenario of electric power systems development through simulation allows to estimate the satisfied demand, the electricity generation in each node and the volume of power flows in the system

$$\tilde{y}_i = \frac{1}{N}\sum_{k=1}^{N} \hat{y}_i^k, \quad \tilde{x}_i = \frac{1}{N}\sum_{k=1}^{N} \hat{x}_i^k, \quad \tilde{z}_{ij} = \frac{1}{N}\sum_{k=1}^{N} \hat{z}_{ij}^k,$$

$i=1,...,n$, $j=1,...,n$, $i \neq j$. Using these indexes, we evaluate the actions of agents interacting in the electricity market.

The payment for reliability is the price that the consumer agrees to pay. For definiteness, it can be determined in the form of an increasing function $g(r)$. Note that the choice of the reliability index depends on the form of the function $g(r)$, which is demand function of electricity of a certain quality (reliability).

4.1 Modeling the demand for electricity

Several options are possible.

The first complicated variant of the demand specification is accounting for different types of consumers. In this case, the price function is determined for each type of consumer, or, for each network node. This requires determination of the reliability indices r_i for the individual nodes $i=1,...,n$ or individual groups of nodes. Then the price that the consumer is willing to pay at node i can be represented in the form:

$$g_i(r_i) = (r_i)^m p_i(\tilde{y}_i). \tag{6}$$

where $p_i(\tilde{y}_i)$ is the maximum price that the consumer is willing to pay in the node i, $i=1,...,n$, without of power failures; m is the parameterthatdetermineshowmuchwetakeintoaccountth ereliabilityfactor, $m \in [0, 1]$ (the higher m, the more important for the consumer the qualitative power supply). The function $g(r)$ in the form (6) is concave, increasing with respect to r.

The value of $p_i(\tilde{y}_i)$ can be determined through the generally accepted inverse demand function: $p_i(\tilde{y}_i) = \beta_i - \alpha_i \tilde{y}_i$, where coefficients β_i, α_i are the parameters of inverse demand function that satisfy the standard requirements. In our study, we assume that demand is not elastic, so the price is identically equal to the constant part of the inverse demand function: $p_i = \beta_i$.

The second way, when consumers are homogeneous and have the same preferences (some aggregated consumer). Then $\beta_i = \beta$ and $\alpha_i = \alpha$ for all $i=1,...,n$ and the inverse demand function is the same in all nodes

$$p = \beta - \alpha \sum_{i=1}^{n} \tilde{y}_i. \tag{7}$$

The same variant of demand can be connected with the configuration of the wholesale electricity market organized as a bilateral auction. In this case, an aggregate demand function is formed to integrate all consumers.

For the inelastic variant, the inverse demand function is $p=\beta$, and the price in the model is defined as

$$g(r)=(r)^m\beta.$$

In this case the price is the same for all nodes and dependent on the resulting index of reliability r.

4.2 Modeling the behavior of generating companies

After selecting a scenario for the development of electric power systems and analyzing its reliability, it is necessary to assess the profitability of this development variant for generating companies. We take into account the fact that different generating companies may have different owners.

Generating companies maximize their profits depending on the emerging price, which in turn depends on the reliability index r_i, $i=1,...,n$, as well as on the parameters of the functions[a]. Suppose that a generating company $s=1,...,S$ operates in several nodes $l=1,...,L$. Then its profit is

$$\pi_s(r)=\sum_{l=1}^{L} g_l(r_l)\widetilde{x}_l - C_s\left(r,\sum_{l=1}^{L}\widetilde{x}_l\right), \qquad (8)$$

where $C_s\left(r,\sum_{l=1}^{L}\widetilde{x}_l\right)$ is costs of the generating company s, $s=1,...,S$ depending on the reliability index r_l in each node $l, l=1,...,L$, power generation and capital costs of electricity generation. The cost function is increasing and convex. In this case, the following relation must be carried out

$$C_s\left(\underline{r},\sum_{l=1}^{L}\widetilde{x}_l\right)<g(\underline{r}).$$

This property reflects the willingness of producers to develop a power system to increase their profits. As a cost function, it is possible to use a linear, quadratic and exponential function. These cases differ depending on the rate of increase in costs for an increasing level of reliability. Under the cost we mean the level of operational and capital costs, which determine the value of the index r. We form cost function on the basis of simulation modeling, comparing costs and balances reliability.

For the case of aggregate demand, the profit function for an individual company is as follows:

$$\pi_s(r)=\sum_{l=1}^{L} g(r)\widetilde{x}_l - C_s\left(r,\sum_{l=1}^{L}\widetilde{x}_l\right).$$

Each company s, $s=1,\ldots,S$, solves the problem of maximizing profit by the reliability index r

$$\pi_s(r)\to\max_r, \qquad (9)$$

$$\underline{r}\le r\le\bar{r}. \qquad (10)$$

It should be noted that the profit function is concave by virtue of convexity of costs and concavity of revenue for this index. The set of admissible solutions of the problem is compact, therefore, the problem has a solution. The solution of problem (9) - (10) is the value of reliability index that can provide this company to maximize its profits. At the same time, the profit of the company s, $s=1,\ldots,S$, depends on the actions of other generating and grid companies. Problem (9), (10) refers to the class of stochastic programming problems, for solutions to which are widely used approaches, based on the Monte Carlo method [15].

4.3 Modeling the behavior of grid companies

The network component influences the price of demand. For each case, we have the costs necessary for the transfer of electricity. In our model, we consider them aggregated without separately calculated profit network company. This is due to the usual market structure, where network companies are allocated to the natural monopoly segment and are regulated by the state. The profit of such a company can be considered zero (normal). Due to the network additive, electricity can be significantly more expensive, so this factor needs to be monitored.

5 The choice of the best case of the development of the electric power system

Thus, the reliability analysis of different variants for the development of electric power system can be compared with the level of balances reliability, with the profit generating companies $\pi_s(r)$ and with the transmission costs, which are reflected in the price level $g(r)$. In our model, the choice of the best equilibrium variant is based on the solution of the problem (9) - (10) by each company. The generating company maximizes profit, determining for itself the optimum level of reliability. The investigated variant of electric power systems development is considered effective if all generating companies choose a sufficient level of reliability or level of reliability falling into a predetermined interval that satisfies the consumer. If at least one company chooses low level of reliability of electricity supply, it then this indicates the ineffectiveness of installed capacity, and the scenario is recognized as not optimal.

6 Experimental researches

We carry out the verification of the developed approach by the test scheme of electric power systems consisting of 5 nodes and 5 links (Fig. 1). Table. 1 shows: characteristics of generating capacities (units with a capacity of 100 MW with an accident rate of 0.045); data on absolute load maxima; information on inter-node connections (in electric power systems power units are connected by a transmission line with a capacity of 135 MW with an integral accident rate of 0.011 per 100 km and a loss factor of 0.0000001661).

[a]In this formulation, the mechanism of competitive price formation in the market is not taken into account.

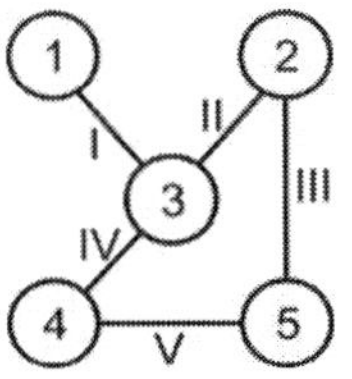

Fig. 1. The design scheme of the electric power system.

The given energy system is deficit in spite of excess of summary generation over absolute maximum of system load. The reliability index of system work without deficit P is 0.9629 under N=10000. The first node of these energy systems is deficit. The mathematical expectation of power shortage in this node is equal 1.4005 MW. The total mathematical expectation of power shortage is 1.4128 MW.

Table 1. Technical characteristics of the investigated EPS

Node number	Total power in the node, MW	Absolute maximum of load, MW	Link number	Number of transmission lines in the link	Length, Km
1	500	700	I	2	200
2	1000	700	II	2	100
3	1400	1300	III	1	300
4	1000	1000	IV	2	200
5	1200	1100	V	2	300
System	5100	4800			

For demonstration of developed method, we suggest to consider of four possible variants of development of EPS. The first one (case 1) is installation of additional generator with a power of 100 MW in node №1. The second one (case 2) is extension of capacity of link I by 135 MW with building corresponding transmission line. The third one (case 3) is installation of a two additional generators with a power of 100 MW in node №1. The fourth one (case 4) is installation of additional generator with a power of 100 MW in node №1 and extension of capacity of link I by 135 MW with building corresponding transmission line.

We consider the homogeneous consumers with similar preferences for modeling of economic agent's behavior. That is way demand function (7) is $p = \beta - \alpha\sum_{i=1}^{n}\tilde{y}_i$ with $\beta = 6845$, $\alpha = 0.81$. Payments for reliability are computed by formula (6). Here P is probability of system work with out power deficit. Reliability index P is computed for some variant of EPS development. m - parameter defining the type of consumer. If it is small, then reliability is not a priority for the consumer. If it is large, the consumer appraises very much the reliability. We consider $m \in \{0,2; 20\}$.

The first test was realized for one generating firm. The cost function of generating company is quadratic $C(\tilde{x}) = 0.08(\sum_{i=1}^{n}\tilde{x}_i)^2 + 50\sum_{i=1}^{n}\tilde{x}_i$. Generator profit is computed by

$$\pi = g\sum_{i=1}^{n}\tilde{x}_i - 0.08(\sum_{i=1}^{n}\tilde{x}_i)^2 - 5\sum_{i=1}^{n}\tilde{x}_i .$$

Table 2 shows the results of calculations for the 4-th cases of the development of EPS at N = 10000. The probability of failure, the price (7) and the total output are calculated.

Table 2. The characteristics of EPS

Case EPS	P	MD	$\sum_i y_i$	p
Initial	0.965	1.44	3560	3928.83
1	0.994	0.18	3563	3928.81
2	0.997	0.06	3558	3934,36
3	0.999	0.05	3566	3928.81
4	0.999	0.00	3560	3928.72

Specific capital cost for the installation of 1 kW of generating capacity is $1000. [16]. This level of cost corresponds to setting blocks of condensing power station running on gas. The unit capital cost for the installation of 1 kilometer of the electrical transmission line is 3.6 UE per 1 kW [16]. These costs correspond to 220 kW ETL. The coefficient of reduction is 0,065 (refund within 15 years).

We compare development cases for two kinds of consumers. The first one doesn't consider the energy reliability a priority. In this case $m = 0,2$ for realistic price g, which a consumer is willing to pay in accordance with reliability, decreasing insignificantly in comparison with the price without taking into account the reliability P. The table 3 shows the company profit (we are considering only one operating company here), the price taking into account the reliability, welfare for all four cases.

Table 3. Calculation results for $m = 0,2$

Variant of EPS	g	Profit	Welfare
Initial	3905	12673540	**12970983**
1	3925	6257528	6560656
2	3932	12763828	6740495
3	3928	-224835	81487
4	3933	6272965	252490

The original variant is advantageous for this kind of consumer. Costs for the installation of new generating capacity and electrical transmission lines don't pay off. We compare all cases by the criterion of the social welfare. The maximum is reached with the initial version. The third case is unprofitable for generating companies because of the negative benefit.

The results change if the consumer appreciates the reliability $m = 2$ (table. 4). Not the initial variant, but the variant with one additional power line (case 2) is a winning one. The others options are not the best. At the same time, if one chooses from them, then the case with the installation of the power transmission line (case 4) is more effective than the one with the installation of two additional capacities (case 2).

 DOI: 10.1051/e3sconf/20172502004

Table 4. Calculation results for $m = 2$

Variant of EPS	g	Profit	Welfare
Initial	1351	3580553	3877995
1	3281	3961147	4264275
2	3595	11565675	**5542342**
3	3928	-224835	81487
4	3933	6272965	252490

7 Conclusion

We suggest a new approach to optimization of reliability of electric power system under market conditions. In market conditions the problem should be solved taking into account a different interests of economic agents s. This fact determines of changing of traditional methods for solving the discussed problem.

The reliability analysis of variants of development of electric power system is based on method with using Monte Carlo method for simulations of random values. We use the model of power shortage estimation with quadratic power losses in power lines for analyzing each random state of energy system. The model guarantees uniqueness of deficit distribution over system nodes and single-valuedness of computed reliability indexes.

The reliability indexes are computed for each variant of development of electric power system. Profit of generating and network companies depends on meanings of reliability indexes. After analyzing all development variants companies will choose the reliability level which is optimal for them. They can provide this reliability level by inputting new generating and network equipment in the energy system. If all companies, choose the same development variant then this is Nash equilibrium which may be not the best solution for the system. If the companies, choose the different development variants then we accept a decision about an effective variant. Each company should have a positive profit and the reliability index should not be below some given level. The experimental research of the developed method is suggested on test scheme of electric power system which is constructed on the base of real technical data.

Further it is necessary concretization of network company's role to develop and verificate our model. Also we must will realize wide computing experiment and compare the different models of consumer behavior.

Acknowledgment

This work is partially supported Russian Foundation for Basic Research, № 16-06-00071a, № 16-37-00333mol-a.

References

1. D. Krupenev, S. Perzhabinsky, *Proc. of The 17th International Scientific Conference «Electric power engineering»*, 1-5 (2016)

2. D.S. Krupenev, S.M. Perzabinski, *4-th International Youth Conference on Energy. Proceeding*, 1-4 (2015)

3. H. Chen, *Generating System Reliability Optimization: Thesis of Doctor of Philosophy*, University of Saskatchewan (2000)

4. N. Aizenberg, *Lecture Notes in Computer Science*, **9869**, 469-479 (2016)

5. M. Assili, M. Hossein, D.B. Javidi, R. Ghazi,Ghazi, *Energy Policy*, **36**(10), 3703-3713 (2008)

6. R. Billinton, R.N. Allan, *Reliability Evaluation of Power Systems* (Plenum Publishing, New York, 540, 1996)

7. A.S. Chuang, F. Wu, P. Varaiya, *IEEE Trans. on Power Systems*, **16** (4), 885-891 (2001)

8. S.S. Oren, *Electricity deregulation: choices and challenges*, 388-414 (2005)

9. A. Motamedi, H. Zareipour, M. O. Buygi, W. D. Rosehart, *IEEE Trans. on Power Systems*, **25**(4), 1987-1995 (2010)

10. S. Rosekrans, D. Kirshner, Marnay, *Utilities Policy*; **7**(4), 210-219 (1998)

11. A.G. Kagiannas, D.Th. Askounis, J. Psarras, *Electrical Power and Energy Systems*, **26**, 413–421 (2004)

12. M. A. Ortega-Vazquez, D. S. Kirschen, *IEEE Trans. on Power Systems*, **22**(1), 24-33 (2007)

13. V.I. Zorkaltsev, L. M. Lebedeva, S. M. Perzhabinsky, *Numerical Analysis and Applications*, **3**(3), 231-240 (2010)

14. D. Krupenev, S. Perzhabinsky, *Izvestia Rossiskoi Academii Nauk. Energetica*, **2**, 39-47 (2017)

15. I.M. Sobol, *Monte Carlo method.* (Moscow: Nauka, 1968)

16. *Enlarged indicators of the cost of the construction (reconstruction) of substations 35-750 kV and transmission lines with voltage 6, 10-750 kV.* (Moscow: RAO EPS of Russia, 2007. – 13 p).

 DOI: 10.1051/e3sconf/20172502005

Method for analysing the adequacy of electric power systems with wind power plants and energy storages

Sergey Perzhabinsky[1*], *Dmitriy* Karamov[1]

[1]Melentiev Energy Systems Institute, 130 Lermontov str., Irkutsk, Russia

Abstract. Currently, renewable energy sources and energy storage devices are actively introduced into electric power systems. We developed method to analyze the adequacy of these electric power systems. The method takes into account the uncertainty of electricity generation by wind power plants and the processes of energy storage. The method is based on the Monte Carlo method and allowed to use of long-term meteorological data in open access. The performed experimental research of electrical power system is constructed on the basis of the real technical and meteorological data. The method allows to estimate of effectiveness of introducing generators based on renewable energy sources and energy storages in electric power systems.

1 Introduction

The modern development of electric power systems (EPS) is characterized by a significant increase in generating capacity with renewable energy sources (RES). Necessary highlights the use of RES in decentralized power systems of medium and low power levels, where both photovoltaics (PV) and wind turbines (WT) have become very popular. At the same time, Russian and foreign experience shows that the use of PV, WT and energy storages (ES) is an economically efficient way of energy supply to consumers. For example, in the Far East Federal District, there are fifteen small-scale solar power plants (from 10 to 100 kW) which are located in decentralized power systems. The larger projects carried out in the Far East Federal District use wind power. Starting from 2013 to the present time, three projects have been implemented, that using wind turbines (250 kW to 1100 kW).

It is worth noting that the generation of WT and PV has a stochastic, oscillating character. Of course, in such conditions, justified to use energy storages (ES) acting as a stabilizing element of a system operating in a buffer mode. In addition to a pronounced stabilizing effect, the presence of ES significantly improves the reliability of power supply, due to the availability of additional storage capacity, which allows to eliminate energy deficit caused by emergency shutdowns.

Possible approaches to the analysis of balance reliability of EPS are discussed in a number of works [1-3]. In [4], an effective methodology for estimating the balance reliability of EPS, based on the Monte-Carlo method, is presented. According to this methodology, at the first stage calculation of values of random variables (maximum possible generating capacity, power handling capacity of overhead line, electrical load) are simulated that characterize some operating state of the EPS. Then, ES power deficit is estimated, and reliability indicators are calculated (probability and mathematical expectation of undersupply of energy). This methodology has proven itself in the long-term planning of EPS and was implemented in the software. Based of this methodology have been developed algorithms to optimize reliability of EPS [6, 7].

To adapt the methodology to modern conditions, it is necessary to take into account the following provisions: accidental generation of energy of WT; modification the model to assess expectation of undersupply of energy with taking into consideration energy storage system in electrical power systems; take into consideration the charge-discharge processes of ES, with considering a number of operational and technical boundary conditions; to develop reliability indicators reflecting the availability of new types of equipment in electric power systems.

In [8], methods for simulating wind activity are presented and a modification of the methodology is given for analyzing the reliability of EPS with wind farms without energy storage system. In the literature, the task of analyzing the balance reliability of EPS with WT and energy storage system has been studied in detail [9-10]. It is worth noting that in [9] advantages are shown for using ES in electric power systems with WT. However, the reliability of the distribution network elements is not taken into account (overhead line, power electronics, etc.) of high and medium level of voltage, which undoubtedly affects the results. Also, in [9-10], clearly do not take into consideration mode indicators of ES, types of used ES, load factor in the moments short-term disturbances with the observance of a number of technical limitations and the period of their lifetime.

* Corresponding authors: sergperj@gmail.com, dmitriy.karamov@mail.ru

A number of approaches to modeling of wind activity and modeling of appropriate models of reliability assessment are suggested: analysis of the data of long-term observations of the wind speed of the territory under consideration and used of the mathematical expectation of the hourly wind speed [11]; construction of autoregressive models of wind speed [12]; application of the Monte Carlo method according to the scheme of Markov chains [13]; use of a hybrid algorithm of machine learning based on wavelet transform and artificial neural network [14], etc.

The goal of this article is to develop a methodology for analyzing the reliability of EPS with WT and ES. In the first section of this article, estimation model a modification of the undersupply of energy with quadratic power losses in overhead line is considered. Theoretical studies of the model with quadratic power losses, presented in [16], indicate the effectiveness of using this model in analyzing the reliability of EPS. The second section of this article discusses the features of modeling random variables and calculating reliability indicators. The third section is devoted to the description of the results of the computational experiment and the conclusions.

2 The model of power shortage estimation of electric power systems with energy storages

Let's consider a scheme of electric power system which contains n nodes and set of system links. The given power system contains energy storages. Let N is number of simulated states of electric power system. Each system state is characterized of set of random values such as of available generating capacity $\bar{x}_i^k$, value of load $\bar{y}_i^k$ in node i, power line capacity $\bar{z}_{ij}^k$ between nodes i and j, $i=1,\ldots,n$, $j=1,\ldots,n$, $i \neq j$, $k=1,\ldots,N$. Maximal $\overline{\Delta xi}^{\max}$ and minimal $\overline{\Delta xi}^{\min}$ capacities of storage are given for each node i. Current power $\underline{\Delta x}_i^k$ and available capacity of energy storage $\overline{\Delta xi}^k$ are computed on the base of characteristics of previous $k-1$ state of energy storage, $k=1,\ldots,N$.

Let x_i is power used at the node i, y_i is the power served at the node i, Δx_i is variation of battery power condition at i-th energy storage, z_{ij} is power flow from the node i to the node j, $i=1,\ldots,n$, $j=1,\ldots,n$, $i \neq j$. We propose following new problem for power shortage estimation of system state with the number k, $k=1,\ldots,N$.

$$G\sum_{i=1}^{n} y_i + \sum_{i=1}^{n} \Delta x_i \to \max, \quad (1)$$

subject to constraints

$$x_i - y_i + \sum_{j=1}^{n}(1-a_{ji}z_{ji})z_{ji} - \sum_{j=1}^{n} z_{ij} - \Delta x_i = 0,\ i=1,\ldots,n, \quad (2)$$

$$0 \le y_i \le \bar{y}_i^k,\ i=1,\ldots,n, \quad (3)$$

$$0 \le x_i \le \bar{x}_i^k,\ i=1,\ldots,n, \quad (4)$$

$$-\underline{\Delta x}_i^k \le \Delta x_i \le \overline{\Delta xi}^k,\ i=1,\ldots,n, \quad (5)$$

$$0 \le z_{ij} \le \bar{z}_{ij}^k,\ i=1,\ldots,n,\ j=1,\ldots,n,\ i \neq j. \quad (6)$$

Here a_{ij} are given positive coefficients of specific power losses during electric energy transfer from the node i to the node j, $i \neq j$, G is weight number for raising of priority of minimization of power deficit, $G \gg 1$, $i=1,\ldots,n$, $j=1,\ldots,n$. Variables Δx_i, $i=1,\ldots,n$, can be negative. It means that load in system is bigger than generating power and requisite power is consumed from energy storage.

Let Δx_i^{k-1} are optimal values that are got as a result of solving problem (1)-(6) for EPS state with number $k-1$, $i=1,\ldots,n$, $k=1,\ldots,N$. Values $\underline{\Delta x}_i^k$ and $\overline{\Delta xi}^k$, $i=1,\ldots,n$, $k=2,\ldots,N$, are recurrently computed by formulas

$$\underline{\Delta x}_i^k = \underline{\Delta x}_i^{k-1} + \Delta x_i^{k-1},\ \overline{\Delta xi}^k = \overline{\Delta xi}^{\max} - \underline{\Delta x}_i^k.$$

For the computing $\underline{\Delta x}_i^0$ you can use any meaning of random value ζ which is equally spaced on the interval $[\overline{\Delta xi}^{\min}, \overline{\Delta xi}^{\max}]$.

The process of charging and consumption of energy from energy storage can be presented in a view as Markov chain [15]. Applying of described approach is allowed to simplify a modelling of process of energy storage's work. We can consider a meaning of random values $\underline{\Delta x}_i^k$, $\overline{\Delta xi}^k$ as equally probable at the cost of big number of EPS states.

A solution of the problem (1) – (6) is exist because the feasible set of the problem is limited and not empty (solution $y_i=0$, $x_i=0$, $\Delta x_i=0$, $z_{ij}=0$, $i=1,\ldots,n$, $j=1,\ldots,n$, $i \neq j$ is a feasible) and objective function is linear. The uniqueness of optimal solution of the problem (1)-(6) by variables y_i, $i=1,\ldots,n$, follows from proved theorems in [1]. The next theorem is true.

Theorem. The solution of the problem (1)-(6) is uniqueness by variables y_i, $i=1,\ldots,n$.

The uniqueness of meanings of variables y_i, $i=1,\ldots,n$ in optimal solution of the problem (1)-(6) for all numbers k guaranties single-valuedness of computed reliability indexes.

3 Computation of reliability indexes

The modelling of random values is realized by next way. At first the number of working generators of every type is found. At next step the value of maximal generation is defined on the active generators. Additionally, random value of generated power is modelled for equipment with random generation. Procedure of random values modelling is detailed described in [8].

Adequacy analyzes is based on indexes which are computed as a result of many times estimation of random states of EPS. There are the most utilized indexes such as probability of non-deficit system work, mathematical expectation of power shortage and electricity supply coefficient [1]. The foreign indexes Loss of Load Expectation (LOLE) and Loss of Load Hours (LOLH) [17] are analogs of discussed indexes. For increasing quality of adequacy analysis it is possible to use special indexes which are took into account of peculiarities of researched energy system, for example, mathematical expectation of power generation on wind farms.

Let set of $\tilde{x}_i^k$, $\tilde{y}_i^k$, $\tilde{z}_{ij}^k$ is optimal solution of the problem (1)-(5), $k=1,\dots,N$, $i=1,\dots,n$, $j=1,\dots,n$, $i\neq j$. The optimal value of power shortage in node i, $i=1,\dots,n$, is defined by the formula

$$d_i^k = \bar{y}_i^k - \tilde{y}_i^k,\ k=1,\dots,N.$$

The state of EPS is deficit if the value

$$d^k = \sum_{i=1}^{n} d_i^k$$

is greater than zero. Let's numerate all deficit states of EPS and define of their common quantity as H. Then the probability of without shortage work of EPS are computed by next formula

$$P = 1 - \frac{H}{N}.$$

This index is corresponded to index LOLP.

Mathematical expectation of power shortage in nodes of energy system is computed by next rule

$$MD_i = \sum_{j=1}^{H} \frac{d_i^j}{N},\ i=1,\dots,n.$$

That's way mathematical expectation of power shortage in EPS is calculated by

$$MD = \sum_{i=1}^{n} MD_i.$$

For computing of mathematical expectation of power generation on a concrete type of equipment $l \in L$ we can use a next formula

$$MX_l = \sum_{i=1}^{n} \frac{\tilde{x}_{il}^k}{N}$$

where $\tilde{x}_{il}^k$ is optimal value of generation received on equipment type l in node i in k-state of EPS, $k=1,\dots,N$. The given index can be useful in estimation of potential of generated power on wind power farms.

4 Experimental research

To illustrate and verify the developed methodology, consider the test scheme of EPS, consisting of 5 node and 5 links (Fig. 1).

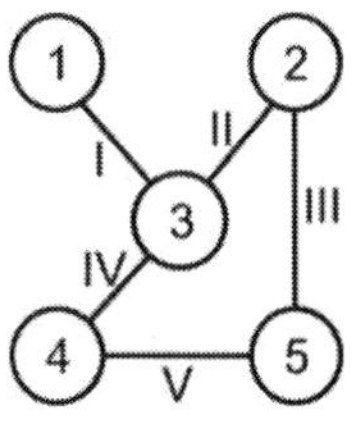

Fig. 1. The calculation scheme of the EPS

In Table. 1 are presented: characteristics of generating capacities (100 MW units with an accident rate of 0.045 p.u. were considered); the absolute maximum electrical load; in EPS nodes connected of 135-MW overhead lines with an integral accident rate of 0.011 p.u. / 100 km and a loss factor of 0.0000001661 MW / 100 km.

Table 1. Technical characteristics of the system.

Node Number	1	2	3	4	5	System
Total power in the node, MW	500	1000	1400	1000	1200	5100
Max load, MW	700	700	1300	1000	1100	4800
Overhead line number	I	II	III	IV	V	-
Number of overhead line	2	2	1	2	2	-
Length, km	200	100	300	200	300	-

In [8] the load graphs.

WT are located in node 1. Total 20 WT ENERCON E-82 E2 2 MW each. The accident rate of one WT is 0.9973 p.u. [18]. ES with a capacity of 2 MWh were also located in the node number 1. To simulate the wind speed, the statistics of meteorological observations obtained at station No. 20289 located on Russkiy Island were analyzed from 1967 to 1998. A total of 84,891 observations were processed. In Table. 2 shows the histogram of frequencies.

Table 2. The histogram of the frequencies of the average daily velocity of a vector.

Group interval, m/s	[1, 5)	[5, 10)	[10, 15)	[15, 20)	[20, 25]
Freq-cy	0,51152	0,40324	0,07523	0,00958	0,00001

DOI: 10.1051/e3sconf/20172502005

The treatment of long-term meteorological series in the meteorological station in question showed that during the autumn and winter seasons on the Russkiy Island, the direction of the wind predominates the northern direction (including the north-west and northeast), while in the spring-summer period the southern (including the southwestern and southeastern) - Fig. 2.

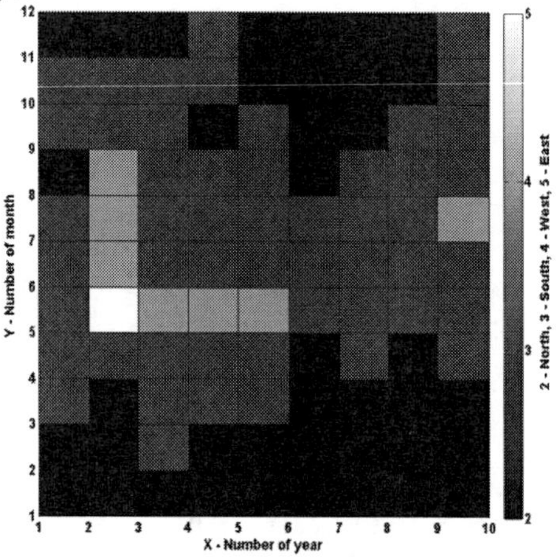

Fig. 2. The direction of the wind on the island Russkiy for 10 years the calculation period

At the same time, the average monthly wind speed exceeding 5 m / sec is observed in the autumn-winter period with the predominant northern wind direction. In Fig. 3 shows the average monthly wind speed relative to 10 years of the calculation period.

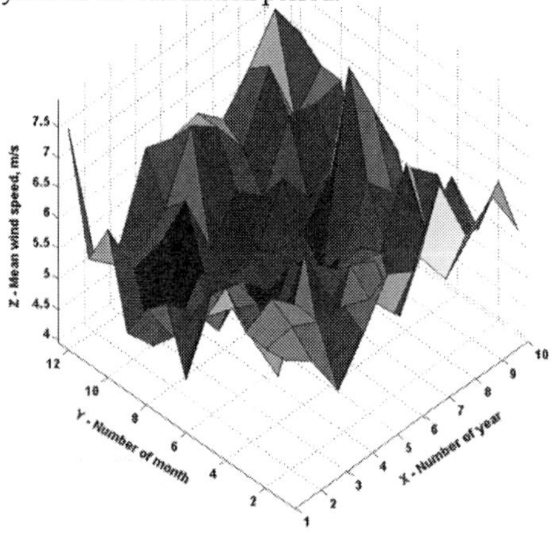

Fig. 3. Average monthly wind speed on the island Russkiy for 10 years of the calculation period

The simulated wind speed v_1 at a height of 10 m (h_1) to a height of 80 m (h_2) by formula [19]

$$v_2 = v_1 (h_1 / h_2)^{1/7}$$

In Table. 3. The results of the evaluation of options for the development of EPS with different composition of

Table 3. Indicators of balance reliability for different equipment.

Number of WT, ES, (installed capacity)	Survival rate of EPS	Power shortage, MW
10 WT (20 MW)	0,994920	0,196116
20 WT (40 MW)	0,996140	0,136028
30 WT (60 MW)	0,996460	0,119281
10 WT (20 MW) and ES (2 MWh)	0,996600	0,107055

The results show that the required reliability index (0.996) is achieved by installing in node 1 or 20 WT or 10 WT with ES with a total indicators obtained as a result of the input of 10 WT with ES are higher than when installing 30 WT.

5 Conclusion

The paper presents a new methodology that allows estimating the balance reliability of EPS with WT and ES. Simulation of wind activity is carried out on the basis of the Monte Carlo method using frequency histograms of observed wind speeds. To estimate the power deficit of the generated random states of EPS, the methodology uses a modification of the power shortage estimation model with quadratic power losses in the overhead lines (the model of power shortage estimation with quadratic power losses in overhead lines). This model guarantees the uniqueness of the distribution of the power deficit at the nodes of the system and the unambiguity of the calculated reliability indicators. A new method for organizing computations for modeling the operation of ES is proposed. This feature greatly simplifies the computational process. The developed methodology of reliability analysis is tested on the test scheme of EPS, built on the basis of real technical data. To model the wind speed, we used a sample of long-term meteorological series, consisting of 84,891 observations. The results of calculations showed that the construction of WT and ES in EPS increases the level of reliability to the required regulatory values. At the same time, the best reliability was achieved by installing a 10 WT and ES with a capacity of 2 MWh in a deficit node. Further, in order to assess the feasibility of building new power plants using renewable energy, it is necessary to develop a methodology that takes into account the environmental component, the damage from under-supply of electricity, the cost of consumed fuel for energy production on traditional generation sources, capital costs.

Acknowledgment

This work is partially supported Russian Foundation for Basic Research, № 16-37-00333mol-a.

References

1. G.F. Kovalev, L.M. Lebedeva, *Reliability of power systems* (Novosibirsk, Nauka, 2015, 224 p.)
2. R. Billinton, R.N. Allan, *Reliability Evaluation of Power Systems* (Plenum Publishing, New York, 1996)

3. Yu.Ya Chukreev, *Model of reliability of electric power systems* (Syktyvkar, Nauka, 1995, 176 p.)

4. Y.N. Rudenko, M.B. Chelcov *Reliability and redundancy in power systems* (Novosibirsk, Nauka, 1974, 263 p.)

5. G.F. Kovalev, L.M. Lebedeva, *Complex models of optimizing the design state regimes in assessing the reliability of electric power systems* (Irkutsk, ESI SB RAS, 2000, 73 p.)

6. D.S. Krupenev, S.M. Perzhabinskiy, *J. Proceedings of the Russian Academy of Sciences. Power Engineering*, **2**, (2014)

7. D.S. Krupenev, S.M. Perzhabinskiy, *J. Managing large systems,* **54**, (2015)

8. D.S. Krupenev, S.M. Perzhabinskiy, *J. Proceedings of the Russian Academy of Sciences. Power Engineering*, **2**, (2017)

9. Ungjin Oh, Jaeseok Choi, Hag-hyeon Kim, *IFAC-PapersOnLine* 49-27, (2016)

10. Nian Shi, and Yi Luo, *Energy Storage System Sizing Based on a Reliability Assessment of Power Systems Integrated with Wind Power, Sustainability*, (2017)

11. Mabel M. C, Raj R. E, Fernandez E, *Renewable and Sustainable Energy Reviews*, **15**, 2, (2011)

12. Billinton R, Huang D, *IEEE Transactions On Power Systems*, **26**, 4, (2011)

13. Almutairi A, Ahmed M. H, Salama M. M. A, *Electric Power Systems Research*, **133**, (2016)

14. Haquea A, Mandalb P, Mengc J, Negnevitsky M, *International Journal of Sustainable Energy*, **34** (2015)

15. Song J, Krishnamurthy V, Kwasinski A, Sharma R, *IEEE Transactions on Sustainable Energy*, **4**, 2, (2013)

16. V.I. Zorkaltsev, S.M. Perzhabinskiy, *J. Managing large systems*, **30**, 1, (2010)

17. Yu. YA. Chukreev, *J. Proceedings of the Russian Academy of Sciences. Power Engineering*, **6**, (2012)

18. Echavarria E, Hahn B, van Bussel G.J, Tomiyama T, *J. Sol. Energy Eng.*, **130**, 3, (2008)

19. Nikolaev V.G, Ganaga S.V, Kudryashov Yu.I, *National cadaster of Russia wind energy resources and methodology for their determination* (Moscow, Atmograph, 2008. 584 p)

 DOI: 10.1051/e3sconf/20172502006

Improving reliability of state estimation programming and computing suite based on analyzing a fault tree

Irina Kolosok[1*], *Elena* Korkina[1], and *Alexander* Tikhonov[2]

[1]Melentiev Energy Systems Institute, 130 Lermontov str., Irkutsk, Russia
[2] Joint Stock Company "Irkutsk Electronetwork company"

Abstract. Reliable information on the current state parameters obtained as a result of processing the measurements from systems of the SCADA and WAMS data acquisition and processing through methods of state estimation (SE) is a condition that enables to successfully manage an energy power system (EPS). SCADA and WAMS systems themselves, as any technical systems, are subject to failures and faults that lead to distortion and loss of information. The SE procedure enables to find erroneous measurements, therefore, it is a barrier for the distorted information to penetrate into control problems. At the same time, the programming and computing suite (PCS) implementing the SE functions may itself provide a wrong decision due to imperfection of the software algorithms and errors. In this study, we propose to use a fault tree to analyze consequences of failures and faults in SCADA and WAMS and in the very SE procedure. Based on the analysis of the obtained measurement information and on the SE results, we determine the state estimation PCS fault tolerance level featuring its reliability.

1 Introduction

At a supervisory control of a energy power system (EPS), the procedure of the state estimation (SE) is a basic program tool that provides the tasks of operational and anti-emergency control with reliable and quality data on the EPS current state. This information is based on telemeasurements (TMs), telesignals (TSs), and phasor measurements (PMs). The SE procedure enables to detect wrong measurements, and plays a role of a barrier for the distorted (by various reasons) information to penetrate into the EPS control subsystem. At the same time, the SE software itself is subject to technical failures, and the supervisor, making decisions, should be confident in correctness of an SE-obtained decision. In this report, we propose to use a fault tree to analyze the consequences of technical failures and faults in the data acquisition and data processing systems (SCADA and WAMS) and in the SE software. Based on analyzing the measurement information and on the SE results, we determine the reliability level for the programming and computing suite (PCS) that estimates the state, and propose some measures to increase reliability.

2 Energy Power System (EPS) State Estimation (SE) Programming and Computing Suite (PCS) computing environment

The EPS SE PCS is intended to obtain the EPS current state model from telemetery and telesignals arriving from SCADA and phasor measurements from WAMS.

2.1 Supervisory Control and Data Acquisition (SCADA) fault tolerance

A SCADA system includes: remote telemetry units (RTUs) installed at EPS substations to take telesignals on the switching equipment state and measurements of the state parameters, communication channels, database (DB), systems of on-line display of the state parameters, as well as the software (EMS-application suite) to process the measurement results and to form control commands for dispatching management objects. Figure 1 presents the structure of a SCADA installed at the control center of a regional network company.

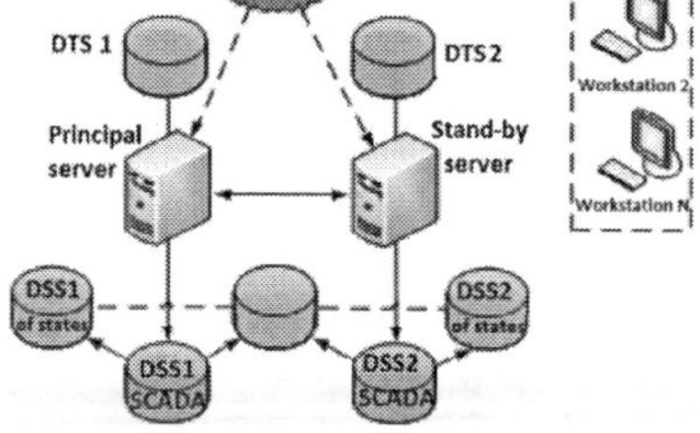

Fig. 1. SCADA structure

The SCADA operates at two independent servers, Principal (1) and Standby (2). The servers constantly

*Corresponding author: kolosok@isem.irk.ru

DOI: 10.1051/e3sconf/20172502006

exchange requests between each other. For each server, provided is its own system of data transmission (DTS) and of data storage (DSS). Herewith, DSS 1 and DSS 2 constantly synchronize their data. Once 30 minutes, the data are transmitted to the historic server, at which continuous reservation is provided.

A response absence from Principal Server 1 is considered as a fault, the system switches to Standby Server 2. This occurs at minimal delays not noticeable for the user.

There are two types of failures possible: 1) hardware failures, and 2) software failures. After a failure, there is software restart at the Principal Server (in case, it was a program failure), or switch to the doubling fitments that is in the server hot reserve. Client requests go to the Principal Server, and, in case of its fault, they are redirected to the Standby Server with no request latency change.

In case of the DSS1 failure, the Principal Server switches to DSS2 with minimal delays. In case of the DTS1 fault, the incoming data stream is directed to DTS2.

Like practice shows, failures seldom arise in such systems. Software failures occur once a month, on average, hardware failures occur once a year.

2.2 Wide-Area Measurement System (WAMS) Data Acquisition (DA) Automatic System (AS) fault tolerance [1]

To solve the tasks in acquiring and storing the WAMS information, in 2009-2011, a WAMS Data Acquisition Automatic System (WAMS DA AS), whose structure is shown in Figure 2, was created and put into industrial operation.

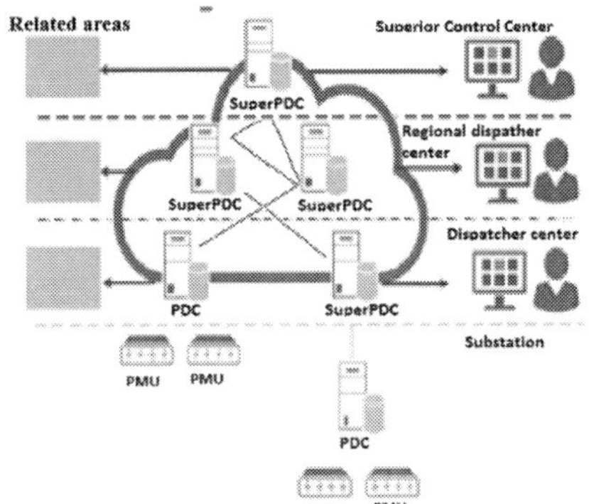

Fig. 2. Structure of WAMS Data Acquisition Automatic System [1]

The lowest hardware, PMU, transmits phasor measurements to the system under Protocol C37.118-2008/2011 into phasor data concentrators (PDCs) for further use in calculation tasks. PMD are relayed onto a higher level of dispatching control, into the super-PDC corresponding to the control hierarchy. This architecture is simple, reliable, and perfectly suitable to solve problems in the absence of restrictions for the computing and telecommunication infrastructure. The measurements are kept in the database of own design [1]. The system servers are united in a cluster operating synchronously: they interact with each other and exchange the information with data sources and clients. The databases included in the cluster are synchronized among themselves. The fault tolerance means within the system allow to create clusters of 2 and more servers. The storage is scaled over almost unlimited number of servers (up to 65535). Failure of any two servers will not lead to information loss. On the one hand, the DA AS architecture is hierarchical, because the servers are at all the levels of the dispatcher control; on the other hand, used is the cloudy technology, when the places of data storage and their traffic routes are not anchored rigidly. This increases the DA AS fault tolerance.

3 Block diagram of SE PCS for the reliability analysis

For a PCS to function necessary are measurements, EPS network, and software including the following algorithms:

to form the current (operational) network based on the basic scheme from telesignals;

to analyze the observability of the network;

to identify and detect gross errors in PMD and telemetry;

· to filter random telemetry errors, i.e., to receive their estimates and to finally calculate non-measured parameters.

For the SE PCS to operate under the WAMS conditions, used is a traditional algorithm of the linear state estimation through the state vector in rectangular coordinates [2]. Also, solving linear SE is successfully realized via the test equation method developed at the Melentiev Energy Systems Institute of the Siberian Branch of the Russian Academy of Sciences [3,4] that has a number of advantages over the traditional non-linear approach [5], the main of them being a possibility to a-priori detect bad data (BD).

In [6], to analyse the software reliability, there are the factors leading to software faults: errors in the program, use of non-optimal and imperfect algorithms (for example, heuristics use), restriction for real-time functioning (the system state changes faster, than the computing cycle lasts). An interaction of several factors and hardware problems in the computing system may also lead to software faults.

The software reliability degree features the probability of its fault-free operation over a certain time interval. In [6], the following is proposed as reliability indicators: the mean number of correctly solved tasks over a certain time interval Δt1, the mean number of errors for that interval, the probability of solving the set number of tasks for the time interval Δt2, the probability of emergence of the set number (probably, limit, being a fault fact) errors for that interval, etc. Such indicators may be used for any software. Conceiving the SE PCS object domain well, we will try to develop other reliability indicators that account for the peculiarities of the solved tasks and feature the fault tolerance of this

DOI: 10.1051/e3sconf/20172502006

very software.

Let us start with presenting the SE PCS, a program tool, as a technical system that is to operate safely and qualitatively. SE PCS will operate correctly, as long as all three elements are operable: measurements, the network, and SE algorithms. For the initial analysis, we make the block diagram (Fig. 3).

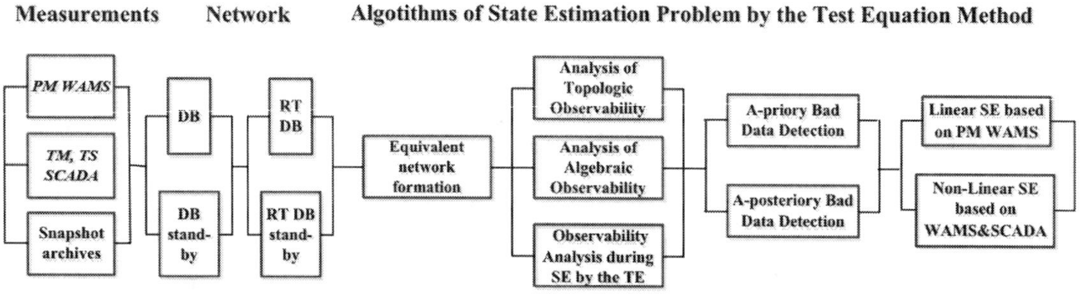

Fig. 3. SE PCS block diagram

From Figure 3, one can see that the bulk of the PCS components is reserved:

Measurements: when no PMs, then the SCADA telemetry is used; if there are no both PM frame and SCADA snapshot, PCS can operate with archival snapshot; besides, the incoming measurements are recorded in the real-time DB (RT DB) that is also reserved;

Network: the data on the diagram are stored at the DB server of constant information; at computer centers of major power facilities, a standby server and a standby DB are provided;

Algorithms: algorithms for the observability analysis, for the measurement validation, and for the state estimation are reserved by alternate blocks. The SE PCS produces the fastest solution, when the PMU LSE algorithm functions. Therefore, the Observability Analysis (OA) program should reply, whether the scheme plan is PMU-observable. If the answer is negative and LSE is impossible to run, OA is performed by a set of SCADA measurements (here is the example of the OA algorithm redundancy) for non-linear SE (here is the example of the SE redundancy). For the algorithm of the a-priori validation by control equations, there is a redundancy in the form of a-posteriori validation by the state estimation remainders.

However, when analysing the reliability of complex systems, it is necessary to find, which of the elements are critical, whose serious faults affect the system operability to a greater extent, in general. Typical criticality indicators [6] are the fault probability, the severity of consequences, the element tolerance to malicious activities, the risk value due to a fault, the possibility of fault localization, the controllability of the element state during the operation, reserving, etc. Ranging the elements by the criticality degree is possible at different levels of structuring the system objects. Critical elements may be visually provided by the Fault Tree technology.

4 Fault tree technology

For the first time, the term "fault tree" in Russian literature was mentioned in Yu. Guk's book [7]. Known since 1960s, the Fault Tree Analysis technology applied by expert systems in military aviation, then in nuclear power, and in some other industries [8], appeared a convenient means to analyse the operation capacity (fault tolerance) of any technical system or its separate complex nods.

The fault tree is presented in the form of a hierarchical structure:

top level - tree root - is the addressed technical system;

second level is the system indicators featuring this system;

third level - system elements - is the details of system indicators;

fourth level - tree leaves - is the events leading to a fault of the system operability (technological problems);

lowest level is the measures to suppress the fault causes.

Figure 4 presents the SE PCS fault tree to analyse the reliability of its operation.

At the top level, there is PSC itself. The system indicators (measurements, network, algorithms) is the second level. Those indicators contain basic elements (measurements types, databases, procedures etc.). Basic elements are exposed to these or those failures or technical faults which are on the forth level of Fault Tree. At the tree lowest level, there are counter-measures written in italics. The set of counter-measures enables to calculate the indicators for the EPS SE PCS algorithm operation efficiency and fault tolerance.

The figure 4 shows the block analyser, the program determining the most vulnerable PCS spots in terms of fault tolerance. This determination is based on the statistic block stored by the calculated indicators for a certain period of time (both blocks in bold).

DOI: 10.1051/e3sconf/20172502006

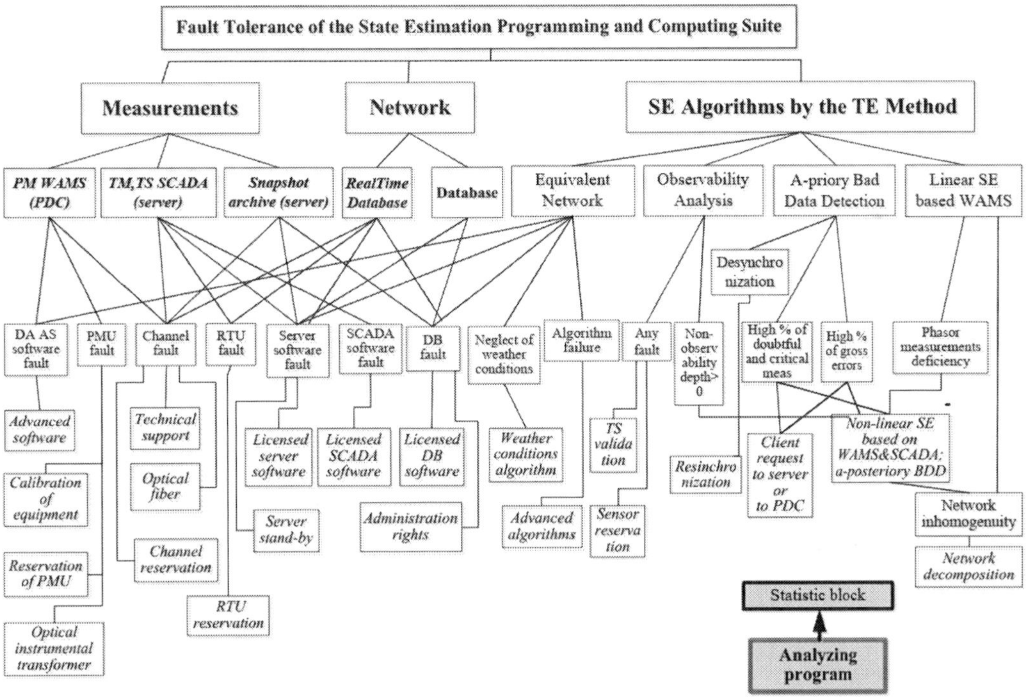

Fig. 4. EPS SE PCS Fault Tree

As compared with the known term "decision tree" [9], in a fault tree, the total of possible options to solve a problem decreases up to the number of options obviously threatening the operability of the addressed technical system. In fact, the fault tree is a diagnostic method offering a way out of a specific problem situation.

5 Fault probabilities for the Fault Tree elements

The analysing program is run several times per day after a certain time, in which N runs of the SE PCS are executed. The probability for the PCS fault tolerance is calculated like

$$P_{PCS} = P_{meas} * P_{network} * P_{a\lg},$$

in any other case, the PCS is non-operative, i.e.,

$$Q_{PCS} = 1 - P_{PCS}$$

The PCS operation fault is non-entry of measurements, or the absence of the network, or the software crash, i.e.,

$$Q_{PCS} = Q_{meas} \cup Q_{network} \cup Q_{a\lg}$$

where Q_{meas} is the fault probability of the suite measuring part, $Q_{network}$ is the fault of all components that are responsible for the assembly of the current network, $Q_{a\lg}$ is the fault of the suite computing part, i.e., non-entry of the calculation results. In turn,

$$Q_{meas} = 1 - (1 - Q_{PM})(1 - Q_{TM})(1 - Q_{archive})$$

where $Q_{изм} = 1$ means the total absence of measurements at the PCS input.

Equivalent scheme formation is switching of the scheme elements, which is imposing the telesignals on the basic network (scheme), combining the adjacent nods into one nod at the switched-on bus-tie switch, creating the common equivalent line for several parallel lines. Failure of forming the network at the PCS run readiness is possible at the errors downput in the algorithms for equalizing the network elements. The procedure provides an emergency shutdown, $Q_{network} = 1$ is the indicator of non-operability of forming the network:

Procedure for the observability analysis (OA) initially was in checking the correspondence of the number and structure of measurements in the SCADA-created snapshots to the network graph. To improve the network observability quality, a special algorithm for telemetry allocation enabled to indicate the spots poorly equipped with sensors on the network, provided recommendations on reserving sensors and measuring channels [10]. At present, along with telemetry, PMs are also used in EPS; therefore, now, developed are the algorithms for the WAMS sensor allocation accounting for possible shutdowns of individual connections, failures of separate sensors, loss of separate measurements [11].

OA enables to detect the observable and unobservable fragments in the network. It is known that, by the tension vector measurements in a node, and by the current vector measurements in the offline, one can obtain a PM calculated value in a node on other end of this line. Such sensor location on the network (area), when all the absent nodal measurements may be calculated by the available PMs, is called "non-observability depth equal to 0" [12]. If the available real and calculated PMs do not provide a completely PM-observable network(area), we consider the occurrence of the network(area), "unobservability depth" equal to 1 and more (which implies a measurement insufficiency for the linear SE operation) as an OA procedure failure for the linear SE (from the WAMS data).

$$Q_{OA} = \frac{n_{\mathrm{stop}\_unobserv=1}}{\mathrm{N}_{\mathrm{run}\_PCS}}$$

If there is no possibility to perform the linear SE, the procedure of nonlinear SE is run from the SCADA and WAMS data. An OA procedure failure for the nonlinear SE is practically impossible.

The procedure of the a-priori validation based on the test equation method is performed before the SE procedure run, and shows: the number of reliable and non-reliable measurements that arrived at the PCS input, the number of critical measurements, whose exclusion leads to presence of non-observable parameters whose errors cannot be detected; and the groups of doubtful measurements, where it is impossible to detect, which among them are erroneous. The measurements with the detected gross errors (bad data) are replaced with specified ones during the algorithm operation; thus, their dispersion values increase, showing a trust reduction concerning such measurements. It is much worse, if, among the measurements, there are critical ones that are not a part of test equations, and the doubtful ones, whose quality, at the given test equation set, is impossible to check. Therefore, we will consider a high percent of the doubtful and critical measurements present in the incoming snapshot a failure of the a-priori validation procedure:

$$Q_{apriri\_BDD} = \frac{n_{\mathrm{run}\_PCS\_with\_high\%\_doubt\&critical\_meas}}{\mathrm{N}_{\mathrm{run}\_PCS}}$$

Herewith, the PCS operation does not stop, because, further, the algorithms for the robust state estimation and for the a-posteriori validation are started. A correspondence of the SCADA snapshot time tags to those of WAMS is a strict requirement for the validation procedure.

The linear SE (LSE) procedure is solved non-iteratively, an indispensable condition for its start is the observability of the entire equivalent scheme through PMs. Therefore, the linear SE algorithm fault is:

$$Q_{LSE} = Q_{OA} = \frac{n_{\mathrm{stop}\_AO=1}}{\mathrm{N}_{\mathrm{run\_PCS}}}$$

In our Fault Tree, the algorithms for non-linear SE and for the a-posteriori validation are presented as the counter-measures against the PCS operability faults (see Figure 4). Therefore, we do not address their fault probability, but we surely account for the quality of their operation.

6 Case study

It is needed to calculate the SE PCS fault tree basic element fault probability values:

$$Q_{изм} = 1-(1-Q_{СВИ})(1-Q_{\mathrm{ТИ}})(1-Q_{архив})$$

$$Q_{расч\_сх} = 1-(1-Q_{ТИ})(1-Q_{архив})(1-Q_{БД})$$

$$Q_{алг} = 1-(1-Q_{форм\_сх})(1-Q_{АН})(1-\mathrm{Q}_{\mathrm{дост}})(1-Q_{ОС})\ 1.$$

1.The analyser is run 4 times per day, 1 time every 6 hours.

2. Software failures occur, on average, once a month, hardware failures occur ones a year

$$\mathrm{Q}_{\mathrm{server}} = \frac{1}{365*6*60} = 7.6*10^{-6}\ \mathrm{min}^{-1};$$

$$\mathrm{Q}_{\mathrm{SCADA\_software}} = \frac{1}{30*6*60} = 9.3*10^{-5}\ \mathrm{min}^{-1}$$

3. Let us take DA AS software failures (as newer, undergoing debugging) 3 times a month

$$\mathrm{Q}_{\mathrm{software\_DA\_AS}} = \frac{1}{10*6*60} = 28*10^{-5}\mathrm{min}^{-1};$$

$$\mathrm{Q}_{\mathrm{Database\_software}} = \mathrm{Q}_{\mathrm{SCADA\_software}}$$

4. PCS is run 3 times every 2 minutes, i.e., 540 runs in 6 hours.

5. In WAMS, there are 6000 frames over 2 minutes, but we also take 3 minutes (like in SCADA); i.e., over this interval, 540*6000=3240000 frames will arrive at PDC, of them 10 being faulty ones

$$Q_{PDC} = \frac{\mathrm{n}_{\mathrm{fault\_frames}}}{\mathrm{N}_{\mathrm{total\_PM\_frames}}} = \frac{10}{3240000} = 3*10^{-6}$$

6. Technical indicators for the PCS hardware reliability

$$Q_{PMU} = 0.0879\ [13];\quad Q_{channel} = 0.001\ [14];$$

$$Q_{RTU} = 0.0001\ ;$$

7. The number of PCS runs with a high % of gross errors

$$\mathrm{Q}_{\mathrm{gr\_err}} = \frac{33}{540} = 0{,}061$$

8. The number of PCS runs with a high % of critical measurements $\mathrm{Q}_{\mathrm{critical\_meas}} = \frac{15}{540} = 0{,}028$.

$$Q_{meas} = 1-(1-Q_{PM})(1-Q_{\mathrm{TM}})(1-Q_{archive})$$

$$Q_{network} = 1-(1-Q_{TM})(1-Q_{archive})(1-Q_{DB})$$

$$Q_{a\lg} = 1-(1-Q_{form\_net})(1-Q_{OA})(1-\mathrm{Q}_{\mathrm{BDD}})(1-Q_{SE})$$

The elements, whose operability may be improved through reservation, are presented with the n_i indicator:

$$Q_{PM} = 1-(1-Q_{PDC}^{\mathrm{n}_1})(1-Q_{channel}^{n_2})(1-Q_{PMU}^{n_3})(1-\mathrm{Q}_{\mathrm{DA\_AS}})$$

$$Q_{TM} = 1-(1-Q_{SCADA})(1-Q_{channel}^{n_4})(1-Q_{RTU}^{n_5})(1-\mathrm{Q}_{\mathrm{server}}^{\mathrm{n}_6})$$

$$Q_{a\lg} = 1-(1-Q_{network})(1-Q_{OA})(1-Q_{BDD}^{2})(1-Q_{\mathrm{SE}}^{2})$$

In Table 1, compared are the SE PCS fault tree basic element fault probability values and the PCS operability probability summary values. The calculation results corroborate that reserving individual elements demonstrates the PCS fault tolerance increase.

Table 1. Fault probabilities for the Fault Tree elements

	With no reserving the elements	With reserving the elements
Q_{PM}	0,0891	0,00802
Q_{TM}	0,0012	0,0000094
$Q_{archive}$	0,0001006	0,000093
Q_{meas}	0,0903	0,0081
$Q_{network}$	0,0781	0,0781
$Q_{a\lg}$	0,2245	0,0908
Q_{PCS}	0,3496	0,1686
P_{PCS}	0,65	0,83

One should note that the system element redundancy is not infinite, it should be guided by common sense: reservation of a system technical element is more preferable to implement through a functionally similar one, but made by another manufacturer to avoid possible recurrings of the discovered or (which is much more dangerous) undiscovered defects. Algorithm redundancy should be performed by producing new software written in such programming languages that do not enable to introduce malicious distortions into the program code during a cyberattack when operating the SE PCS.

7 Conclusions

There are universal indicators including those created based on the waiting theory for software reliability that enable to reveal the level of its working capacity in terms of functionality and delivery of result. Application packages, standard algorithm libraries, operating systems belong to this software type. SE PCS is specialized software, and it is intended to provide quality results (based on a deep analysis of the input data) affecting the control of a technological process. It is extremely important to know, what the SE PCS behaviour under real conditions is. Therefore, it is extremely important to develop such indicators for software fault tolerance that would enable to protect weak spots of program blocks, and, thereby, to increase the PCS operability in any hostile environment.

References

1. F. Gaidamakin, A.Kislovskiy. *Proc. Of the 5th Intern. Scientific and Technical Conf."Actual trends in development of power system relay protection and automation"*, Sochi, (2015).

2. Kolosok I., Korkina E., Buchinsky E. *Proc.of the 5th Intern. Conf. «Liberalization and Modernization of Power Systems: Smart Technologies for Joint Operation of Power Grid»*, Irkutsk, Russia, (2012).

3. A. Gamm, I. Kolosok. *Power and Electrical Engineering: Scientific Proc. of Riga Technical University*. Riga: RTU, (2002).

4. Kolosok I., Korkina E., Buchinsky E. *Proc. of the 18th Intern.Power System Computation Conference (PSCC-2014)*, Poland, Wroclaw, (2014).

5. Gamm A.Z. *Observability of power systems* (Moscow: Nauka. - 1990. - 220 p).

6. Polovko A.M., and S.V. Gurov, *Reliability theory grounds*. (SPb: BKhV-Petersburg. 2006. 704).

7. Guk Yu.B., *Computing the energy facility reliability*. (Leningrad.: EnergoAtomIzdat, 1988 – 224).

8. Ericson C.A. Fault tree analysis http://www.eecs.ucf.edu/~hlugo/cop4331/ericson-fta-tutorial.pdf

9. Quinlan J. R. Simplifying decision trees // *Intern. Journal of Man-Machine Studies*, V.**27** (3): 221 (1987). doi:10.1016/S0020-7373(87)80053-6.

10. Gamm A.Z., I.I. Golub *Observability of power systems*. (Moscow: Nauka (Science). - 1990. - 220 p).

11. Ya. Kuzkina. *Proc. of Intern. Conf. "Relay protection and automation for electric power system"*, 2017. S-Pb, S.1-8.

12. R. F. Nuqui. *State Estimation and Voltage Security Monitoring Using Synchronized Phasor Measurement, Ph.D.dissertation*, Virginia Polytechnic Institute & State University, Blacksburg, USA, 2001.

13. O. Gomez, C. Portilla, M.-A Rios. *IEEE Trans.on.Power Systems* ,**30** (2), (2015).

14. Aminifar F. *IEEE Transactions on power delivery*, V. **26** (2), (2011).

15. G.Kovalev. *Electrichestvo* **№4**. Pp.14-21. (2017).

Reliability and efficiency upgrades of power systems operation by implementing intelligent electronic devices with synchrophasor measurement technology support

Alexey Mokeev[1,2*]

[1]Northern (Arctic) Federal University, Electroenergetics and Electrotechnics Department, Severnaya Dvina emb.17, Arkhangelsk, Russia
[2]Engineering Centre "Energoservice", Kotlasskaya st. 26, Arkhangelsk, Russia

Abstract. This paper reviews issues of reliability and efficiency upgrades of power systems functions by means of a widespread implementation of intelligent electronic devices (IED) in various purposes supporting synchrophasor measurement technology. Thus, such issues as IED's operational analysis in the conditions of electromagnetic and electromechanical transient processes and synthesis of digital filters that improve static and dynamic responses of these devices play an important role in their development.

1 Introduction

Successful implementation of automated WAMS by applying special PMUs devices predetermined the potential use of synchrophasor measurement technology for solving other important tasks related to reliability and efficiency upgrades of power systems operation [1-4].

The use of IEDs with synchrophasor measurement technology support allows significant improving of power systems state estimation and creating new generation devices of relay protection and automatics [2, 5]. Such advanced areas of these IEDs application are DCS substations, WACS, WAPS, WAMPACS and SmartGrid. The need for fast synchronized measurements of power system condition parameters is also associated with the wide use of electrical network and substations automatic control systems in SmartGrid by means of various regulators, including reactive power compensation devices, network parameters control devices, etc.

About one hundred power system condition parameters for fundamental harmonic can be calculated on the basis of measurements of current and voltage synchrophasors at any level of control. In addition, parameters of the power system itself can be calculated as well. Various types of emergency and automatic control, relay protection devices of new generation can be also implemented on the basis of the information coming from different IEDs.

There are higher requirements specified to IEDs used in WAPS according to main performance criteria for signal processing (speed, accuracy of signal processing, synchronism of measurements and etc.) in terms of intense electromagnetic and electromechanical transient processes. It should be noted that an existing IEEE C37.118.1 standard is focused first of all on devices used in WAMS, and does not reflect a number of specific requirements to IEDs for WAPS and WAMPACS primarily related to the need for ensuring the reliability of measurements and the required quality of nonstationary input signal processing, providing a wider range of current and voltage phasor measurements, as well as the need for a higher data rate.

The IEEE C37.118.1 standard provides for PMUs testing, corresponding to the stationary and transient conditions of the power system [6]. With an expansion of the scope for PMUs application, it is necessary to increase the number of tests corresponding to electromagnetic and electromechanical transient processes. Moreover, mathematical modelling together with testing should be more actively used as a rather simple way of IEDs functional analysis in various conditions of power system operation. At the same time, the analysis of PMUs functioning in the conditions of transients processes listed above is associated with a large amount of computation. The report examines effective methods for PMUs functional analysis during electromagnetic and electromechanical transient processes, as well as methods of express analysis.

An expansion of PMUs application is directly related to an improvement of signal processing algorithms for better IEDs quality performance such as speed and accuracy of signal processing. Along with the development of effective methods of IEDs functional analysis it requires the development of new methods for digital filters synthesis that determine static and dynamic errors of current and voltage synchrophasors measurements in most devices.

2 PMU analysis

According to typical signal processing algorithm provided by IEEE C37.118.1 standard [6], two main

* Corresponding author: a.mokeev@narfu.ru

operations for PMU input signal are performed here. That is signal spectrum transformation due to multiplication by a complex exponential function with a following signal filtering with the help of averaging FIR-filter. PMU functional analysis in this case comes down to a filter analysis.

According to PMUs design for ensuring correct operation under stationary conditions, as well as in the conditions of electromagnetic and electromechanical transient processes, including IEEE C37.118.1 testing, a large number of calculations are necessary. In this regard, the development of effective express methods for frequency filters analysis is an actual task for today.

Currently, similar methods of PMUs functional analysis are widely used for input sine signals on the basis of filter frequency responses. In this case, PMUs input signals are completely described by a set of complex amplitudes and imaginary frequencies of current and voltage components. Thus, requirements to the filters at PMUs design are formed on the basis of these responses. Analysis express methods for input signals corresponding to electromagnetic and electromechanical transient processes are currently under development. The article summarizes author's works related to such express methods for PMUs analysis in the conditions of electromagnetic and electromechanical transient processes [7-12]. Given methods of analysis rely on generalizing the complex amplitude method in the case of signals with constant parameters in the form of a set of damped oscillatory components and signals with variable parameters.

The methods of filter analysis below are considered for the PMU analog prototype system in order not to take into account errors of microprocessor devices associated with time and level discretization of a signal. While it should be noted that these considered methods of analysis can easily be extended to the case of discrete signals and digital filters [8, 9].

2.1. Analysis for signals with constant parameters

PMUs input signals in many practical cases can be described by a set of damped oscillatory components [8]

$$x(t) = \mathrm{Re}\left[\dot{\mathbf{X}}^{\mathrm{T}} e^{\mathbf{p}t}\right] = 0.5\dot{\mathbf{X}}^{\mathrm{T}} e^{\mathbf{p}t} + 0.5\bar{\mathbf{X}}^{\mathrm{T}} e^{\bar{\mathbf{p}}t}, \qquad (1)$$

where $\dot{\mathbf{X}} = \left[\dot{X}_n\right]_N = \left[X_{m_n} e^{-j\varphi_n}\right]_N$ – is the vector of input impact complex amplitude and the vector of complex frequencies is $\mathbf{p} = \left[p_n\right]_N = \left[-\beta_n + j\omega_n\right]_N$, $\bar{\mathbf{X}}$, $\bar{\mathbf{p}}$ – is the complex-conjugate vectors.

Unlike the traditional method of complex amplitudes, signal (1) operates with complex frequencies. Signals (1) can be presented not only by signals of the normal mode of the power system as a set of harmonics, but signals at electromagnetic transient processes, for example, a signal in a set of a sine signal and free components of transient processes (exponential and/or damped oscillatory components), and in other cases PMUs signals at some electromechanical transient processes, for example, in the case of PMU tests according to IEEE C37.118.1 for amplitude and phase modulation.

The input signal FIR-filter accounting multiplication of signal (1) by complex exponential function $2e^{-j\omega_0 t}$

$$\dot{z}(t) = \dot{\mathbf{X}}^{\mathrm{T}} e^{(\mathbf{p}-\mathbf{C}j\omega_0)t} + \bar{\mathbf{X}}^{\mathrm{T}} e^{(\bar{\mathbf{p}}-\mathbf{C}j\omega_0)t} = \dot{\mathbf{Z}}^{\mathrm{T}} e^{\mathbf{p}_1 t}, \qquad (2)$$

where ω_0 – nominal frequency of power system, $\mathbf{C} = \left[1\right]_N$, $\dot{\mathbf{Z}} = \left[\dot{\mathbf{X}} \;\; \bar{\mathbf{X}}\right]$, $\mathbf{p}_1 = \left[\mathbf{p}-\mathbf{C}j\omega_0 \;\; \bar{\mathbf{p}}-\mathbf{C}j\omega_0\right]$.

According to expression (2) the input signal of FIR-filter contains 2 times more components than PMU input signal (1). In this case, from one useful component of PMU input signal (1) of industrial frequency ω_1 with complex amplitude $\dot{X}_1$ at the input of the filter, we have two components; the low-frequency component (constant component at $\omega_1 = \omega_0$) and the component with frequency $\omega_1 + \omega_0$ ($2\omega_0$ at $\omega_1 = \omega_0$). The first component should pass through the FIR-filter without any distortions, and another one should be suppressed.

It should be noted that under the conditions of stationary mode of the power system and electromagnetic transient processes such PMU input signal (1) contains one useful component of fundamental harmonic ($\dot{X}_1$, ω_1). As for modulated PMU input signals (amplitude and phase modulation according to IEEE C37.118.1 tests the useful signal is described as a set of complex amplitudes and imaginary frequencies. Further we are going to consider the case with one useful component of input signal (1).

As for express analysis of PMU filters at input signals (1) there are several methods [8] developed on the basis of generalizing the complex amplitude method and corresponding extended frequency responses in coordinates of the complex frequency. Let's study the analysis method based on the use of the time-dependent transfer function

$$K(p,t) = \int_0^t g(\tau) e^{-p\tau} d\tau, \qquad (3)$$

where $g(t)$ – impulse function of the filter.

The output signal of the filter can be fully described by a set of time-dependent complex amplitudes

$$\dot{\mathbf{Y}}(t) = \dot{\mathbf{Z}} \circ K\left(\mathbf{p}_1, t\right), \qquad (4)$$

where $\circ$ – operation of vector multiplication.

Each time-dependent complex amplitude of vector $\dot{\mathbf{Y}}(t)$ contains information both on a forced and free component of the own transition process in the filter. At $t \to \infty$ the expression is converted to the form

$$\dot{\mathbf{Y}} = \dot{\mathbf{Z}} \circ K\left(\mathbf{p}_1\right) \qquad (5)$$

and instead of a set of time-dependent complex amplitudes $\dot{\mathbf{Y}}(t)$, we get a set of complex amplitudes $\dot{\mathbf{Y}}$ corresponding to those forced components of filter reaction.

The output signal of the FIR-filter, which is synchrophasor estimation, is determined as follows

$$\hat{X}_1(t) = \mathrm{Re}\left[\dot{\mathbf{Y}}(t)^{\mathrm{T}} e^{\mathbf{p}_1 t}\right].$$

It should be noted that this method of analysis can be extended to cases where PMU input signal is described

DOI: 10.1051/e3sconf/20172502007

by a set of signals in the form of finite damped oscillatory components [8, 9].

On the basis of above expressions (4) and (5), we obtain a simple analytic method for filter analysis using only algebraic operations on complex amplitudes and values of extended frequency responses of the filter at complex frequencies of the input signal. Thus, the express analysis of filters is carried out on the basis of extended frequency responses [7, 8].

Express analysis, if necessary to take into account the quality of the own transient processes in the filter, is performed on the basis of the time-and-frequency approach based on the construction of frequency responses $K(p,t)$ in coordinates of the complex frequency. In most simply cases, it is preferable to use express analysis on the basis of frequency responses $K(p)$ when using spectral representations of the Laplace transform.

In the first case we are dealing with 4D MFR and FFR of the filter, and in the second case with 3D MFR and FFR. Magnitude frequency and phase-frequency function of the filter in the first case will be functions of three variables, i.e. responses must be displayed in four-dimensional space. For practical visualization of frequency responses, we can use an approach based on a three-dimensional frequency responses with a fixed value of the real or imaginary part of the complex frequency.

Let's consider the express analysis of the simplest averaging FIR-filter with a rectangular time window length T_0 = 20 ms with an input signal as a set of a sine component of industrial frequency and a free exponential component of the transients process in the power system. PMU input signal in this case can be described by the set of the following complex amplitudes and frequencies $\dot{\mathbf{X}} = \begin{bmatrix} \dot{X}_1 & -X_2 \end{bmatrix}$, $\mathbf{p} = \begin{matrix} j\omega_1 & -\beta_2 \end{matrix}$. Let's set the ranges of parameter variations of the useful signal and the exponential disturbance: $\omega_1 = 2\pi\ 45 \div 55$ rad/s, $\beta_2 = 2 \div 200\,\text{s}^{-1}$, $X_{2max} = |\dot{X}_1|$.

The input signal of the filter is completely determined by the following set of complex frequencies and amplitudes

$$\mathbf{p}_1 = \begin{bmatrix} j\ \omega_1 - \omega_0 & j\ \omega_1 + \omega_0 & -\beta_2 - j\omega_0 \end{bmatrix},$$

$$\dot{\mathbf{Z}} = \begin{bmatrix} \dot{X}_1 & \bar{X}_1 & -2X_2 \end{bmatrix}.$$

Fig. 1 shows 3D MFR of the filter when using the spectral representations of the Laplace transform.

As follows from the above dependence (5), the values of the transfer function $K(p)$ of the filter at complex frequencies of the components of the filter input signal determine the variation law of the corresponding complex amplitudes. Therefore, it is quite enough to consider two cross-sections of 3D MFR here: $p = j\omega$ (curve 1) and $p = -\beta + j\omega_0$ (curve 2).

These two cross-sections determine variation laws of three complex amplitudes of filter input signals: the first two complex amplitudes are determined by the cross-section $p = j\omega$, and the third complex amplitude is determined by the cross-section $p = -\beta + j\omega_0$. To simplify the analysis with exponential disturbance at PMU input, MFR is multiplied by function $e^{-\beta T_0}$. The factor $e^{-\beta T_0}$ at $\beta \neq 0$ takes into account the damping of the aperiodic forced components by the end of the transient process in the filter. The FIR-filter speed is determined by its length T_0 only if damped components of the disturbance are suppressed to the level of the permissible error for a time equal to the length of the filter.

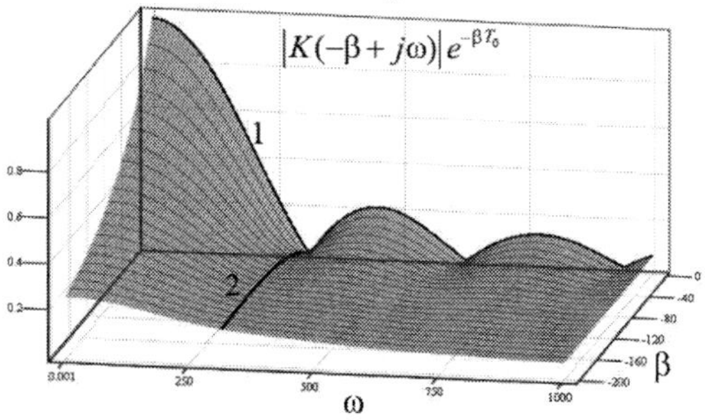

Fig. 1. 3D MFR in coordinates of complex frequency

For a more detailed analysis of filter suppressed component from the exponential disturbance we plot the function graph $2|K(-\beta + j\omega_0)|\,e^{-\beta t}$ (Fig. 2).

As follows from the graphs (Fig.1-Fig.3), the analysed filter has large errors in the estimation of the synchrophasor at the deviation of power system frequency from the nominal value, as well as has insufficient suppression of the component from exponential disturbance, especially at damping decrement $\beta_2 = 20 \div 60\,\text{s}^{-1}$. So, accounting an allowable error of 5%, the filter suppression time of the component from exponential disturbance is more than 50 ms.

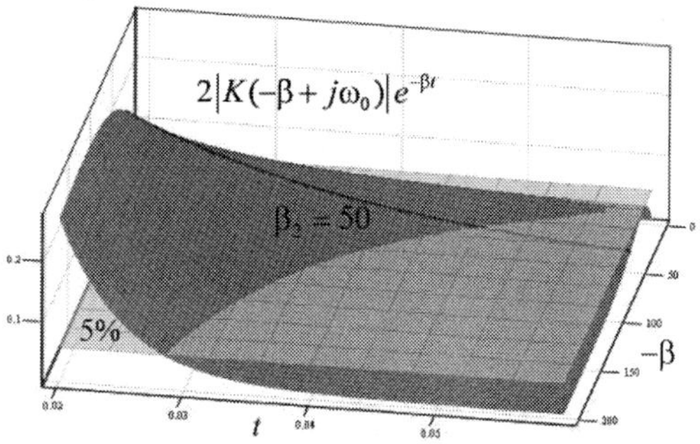

Fig.2. Exponential disturbance suppression

To analyse the filter taking into account the quality of the own transient processes in the filter, one should use extended time-and-frequency responses of the filter based on the time-dependent transfer function (3).

Fig. 3 and Fig. 4 show 3D MFR of the filter: $p = j\omega$ (Fig. 3) and $p = -\beta + j\omega_0$ (Fig. 4), respectively. Curves 1 and 2 in Fig. 3 correspond to the envelope of the transient process in the filter for the useful component for the synchrophasor estimation at $\omega_1 = \omega_0$ and disturbance caused by multiplying PMU input signal by a complex exponential function. Curve 1 in Fig. 4

 DOI: 10.1051/e3sconf/20172502007

corresponds to the transient process from the exponential component at the input of the PMU.

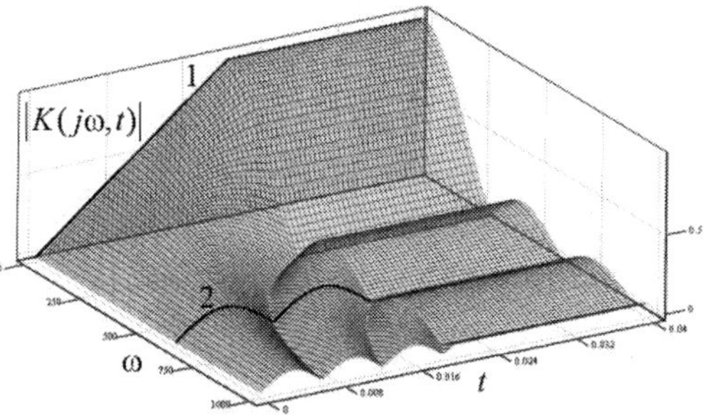

Fig. 3. Extended MFR filter (section $p = j\omega$)

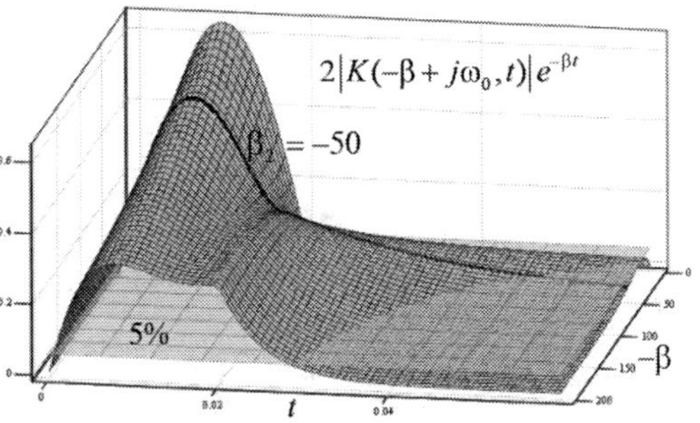

Fig.4. Extended MFR filter (section $p = -\beta + j\omega_0$)

These graphs clearly demonstrate the efficiency of the proposed methods for express analysis of filters for PMU signals, which can be represented by a set of damped oscillatory components.

2.2 Analysis for signals with variable parameters

For the sake of simplicity, we shall restrict our analysis to the analysis of FIR-filters with only one component of PMU input signal with variable parameters, as given in IEEE C37.118.1 standard.

$$x(t) = \mathrm{Re}\ \dot{X}_1(t)e^{j\omega_0 t}\ = 0.5\dot{X}_1(t)e^{j\omega_0 t} + 0.5\overline{X}_1(t)e^{-j\omega_0 t}.$$

Then the signal at the input of the FIR-filter

$$\dot{z}(t) = \dot{X}_1(t) + \overline{X}_1(t)e^{-j2\omega_0 t} = \dot{\mathbf{Z}}(t)^{\mathrm{T}} e^{\mathbf{p}_1 t},$$

where $\dot{\mathbf{Z}}(t) = \left[\dot{X}_1\ t \quad \overline{X}_1\ t\ \right]$, $\mathbf{p}_1 = \ 0 \quad -j2\omega_0\ $.

For the analysis of PMU functioning, the definitions of complex amplitudes as components of synchrophasor estimation are actual, i.e. the use of generalizing complex amplitudes and methods for filter analysis in the case of signals with variable parameters corresponding to electromechanical processes in the power system.

The determination of complex amplitudes for two components of the filter output signal can be made on the basis of the following proposed dependence [11]

$$\dot{\mathbf{Y}}(t) = \int_0^t \dot{\mathbf{Z}}\ \tau\ \circ K'\ \mathbf{p}_1, t-\tau\ d\tau,$$

where $K'\ p,t\ = \dfrac{d}{dt}K\ p,t\ $.

At $\dot{\mathbf{Z}}\ t\ = \dot{\mathbf{Z}}$ the expression for complex amplitudes calculation generates dependence (4).

Under certain assumptions (signal modulation depth, rate of change of frequency), complex amplitudes can be approximately estimated using the following dependence [11]

$$\dot{\mathbf{Y}}(t) \approx \dot{\mathbf{Z}}\ t\ \circ K\ \mathbf{p}_1\ t\ , t\ , \quad (6)$$

where $\mathbf{p}_1\ t\ = \left[\, j\ \omega_1\ t\ - \omega_0 \quad j\ \omega_1\ t\ + \omega_0\ \right]$,

$$\omega_1(t) = \omega_1 + \frac{d\arg(\dot{X}_1(t))}{dt}.$$

The use of simplified dependence (6) for determining complex amplitudes of the output signal of the filter makes it possible to use previously discussed methods for express analysis of filters at PMU input signals with variable parameters that correspond to electromechanical processes in the power system. It is necessary to note the important advantage of the proposed methods of analysis, which is associated with the possibility of formalizing the requirements for frequency responses of filters in coordinates of the complex frequency that makes it possible to simplify the solution of the task of filters synthesis for PMU.

3 Filters synthesis

The method of FIR-filters synthesis [8, 9, 12] proposed by the author is based on the following basic concepts: the use of extended spectral representations of signals and filters based on the Laplace transform [7] and methods of nonlinear optimization with constraints [13]. In this case, the restrictions are imposed on 4D or 3D frequency responses of the filters.

If it is necessary to provide only the specified quality of signal processing in the filter in a steady-state condition, the constraints are imposed on 3D frequency responses. If it is necessary to additionally ensure the necessary quality of the own transient processes in the filter, then it is necessary to impose constraints on 4D frequency responses of the filters. During the synthesis of frequency filters, taking into account the nature of the input signals of the IEDs, it is advisable to use the impulse responses of filters in a set of semi-infinite (IIR-filters) or finite (FIR-filters) damped oscillatory components [8].

The efficiency of the proposed synthesis method for PMU is discussed in detail in [12]. Thus, for PMU of class M, it was possible to double the speed compared to the filter proposed in IEEE C37.118.1 standard with the same accuracy requirements for the deviation of the frequency of the power system from the nominal value, with the same requirements for suppressing the sine component of the double frequency resulting from the operation of the spectrum shift of PMU input signal, as well as the suppression of higher harmonics and interharmonics. With a slight modernization in the signal processing algorithm for PMUs, the speed has been increased by 2 times more.

Let's consider the synthesis of FIR-filters for PMU of class P, as well as multifunctional IEDs, performing

DOI: 10.1051/e3sconf/20172502007

the functions of PMU of class P and devices of relay protection and automatics.

Fig. 5 and 6 show the impulse and magnitude frequency responses of FIR-filters for the following cases, respectively: newly synthesized filters (curves 3, 4, 5, 6) recommended by IEEE C37.118.1 for PMU of class P (curve 1), widely used in microprocessor-based relay protection and automatics devices (curve 2).

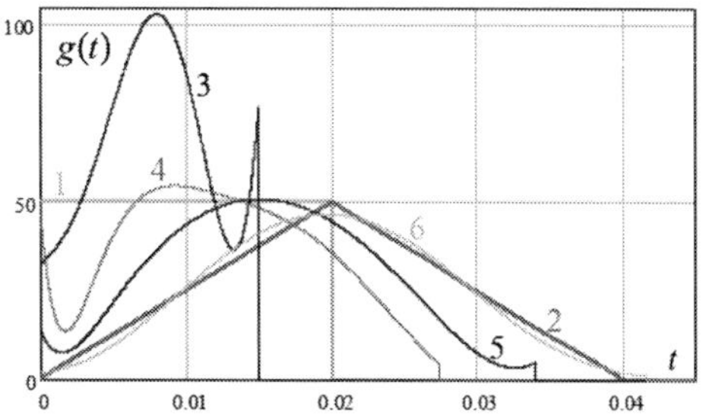

Fig. 5. Impulse responses filters

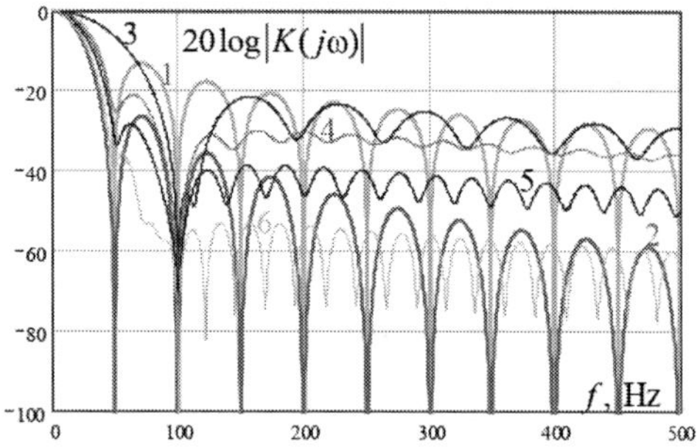

Fig. 6. MFR filters (section $p = j\omega$)

All synthesized filters, except for the first one (curve 3 in Fig. 5 and Fig. 6), are developed for PMU modernization algorithm [12]. Here, the requirements for the permissible range of frequency variation are substantially tightened: ± 5 Hz instead of ± 2 Hz in accordance with IEEE C37.118.1 standard.

The first of the synthesized filters (curve 3) has an accuracy class 0.5 and the shortest length of filter impulse response. This filter can be used only in devices for synchronized phasor measurements of voltage in a homogeneous electrical network, when exponential components are practically absent in a voltage.

Three other synthesized filters provide suppression of the component from the exponential disturbance at the input of the device, respectively, by 20, 50 and 100 times. The last filter (curve 6 in Fig. 5 and Fig. 6) provides high accuracy of measurements (accuracy class 0.2).

Fig. 7 shows MFR of the fourth synthesized filter (curve 5 in Fig. 5 and Fig. 6) in cross-section $p = -\beta + j\omega_0$, where this filter provides a given level of the suppressed component from exponential disturbance. In view of this circumstance, these filters have better speed than the widely used in relay protection averaging FIR-filter with rectangular impulse response of 20 ms.

We would like to note the following peculiarity of the impulse responses of synthesized filters, consisting in the asymmetrical form of these responses. At the same time, we would like to draw your attention to the fact that the impulse responses inverted with respect to the axis of ordinates will have exactly the same traditional MFR as the original ones. But such two filters with mirror-symmetrical finite impulse responses will have different properties for suppressed exponential and damped oscillatory components. This indicates the constraints of usual frequency responses based on the Fourier transform and the appropriateness of using spectral representations on the basis of the Laplace transform.

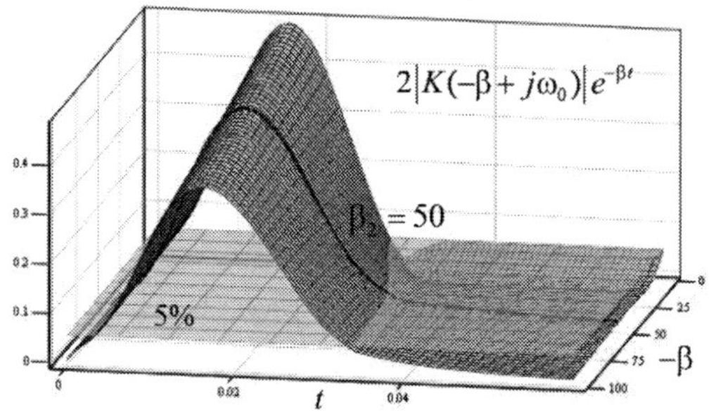

Fig. 7. MFR filter (section $p = -\beta + j\omega_0$)

The conversion of analog filter prototypes to digital filters is performed using known approximation methods. Direct realization of synthesis for digital filters is also possible [8, 9].

4 Implementation

The specialists of "Engineering Centre "Energoservice" LLC developed a number of intelligent electronic devices supporting synchrophasor measurement technology, in which frequency filters synthesized using the synthesis method discussed above were implemented [14, 15].

Thus, the multifunctionaldeviceENIP2-PMU has been developed for use in WAMS. ENIP2-2-PMU devices can be configured for phasor measurements under class P or M. Optionally, ENIP2-2-PMU contains inboard GPS-GLONASS receiver device for time sync, as well as TFT LCD touch screen and inboard phasor data concentrator. The presence of a full set of options allows to install one PMU on the site and reduce the requirements for data transfer channels with a minimal cost.

It should be noted that ENIP2-PMU devices have very small overall dimensions in comparison with other PMUs. At present, preparations for the serial production of the multifunctional intelligent ESM device, the dimensions and cost of which are comparable with those of the multifunction measuring transducers, are being completed [15]. This makes it possible to apply them widely in the future both in SCADA and in WACS.

For use in WAMPACS and in the structure of digital substations, a multifunctional merging unit with ENMU process bus (Fig. 8) has been developed. The main purpose of ENMU is stand-alone merging unit (SAMU). ENMUs are connected to the relay and measuring windings of electromagnetic current transformers, as well as to measuring voltage transformers, and provide the generation of sampled values of currents and voltages in accordance with IEC 61850-9-2.

In addition, the ENMU device along with the implementation of SAMU functions performs the functions of several intelligent electronic devices, including a multifunctional measuring transducer, a synchronized phasor measurements unit, an oscillographic fault data recorder and a phasor data concentrator [15]. Phasor measurements can be considered as an addition or as an alternative to sampled values, particularly for relay protection and automatics devices, especially at implementing longitudinal differential protection and differentiation busbar protection.

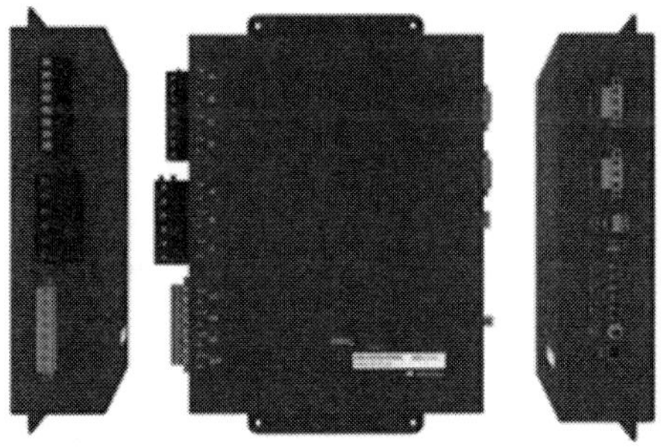

Fig. 8. Stand-Alone merging unit ENMU

The ENMU merging unit contains three optical 100BASE-FX Ethernet ports (two main and one service) and ENMU has the following protocols: IEC 61850-9-2LE for transmitting sampled current and voltage values (80 and 256 samples per cycle), IEEE C37. 118.2 for transmission of synchronized current and voltage phasor values, frequency and rate of change of frequency, IEC 60870-5-104 for transmission to SCADA data by measuring the parameters of the fundamental harmonic and taking into account higher harmonics. For reservation of data transfer channels, PRP backup protocol (IEC 62439-3) is implemented. The real-time clock is synchronized via a special optical port with signals PPS, IRIG-A or IRIG-B.

The ENMU device is constantly being improved. Currently, the work is underway to expand the functionality of the ENMU by adding discrete inputs and outputs and supporting the IEC 61850-8-1 protocols (MMS and GOOSE). It greatly enhances the ENMU's capabilities from the point of view of WAMPACS use.

The most important advantages of the current and voltage synchrophasors are associated with the possibility of calculating the parameters of the electrical network condition, as well as the parameters of the network itself, at any level of control. For the purpose of measuring the parameters of the network condition, taking into account the higher harmonics, work is underway to implement ENMU measurements of equivalent current and voltage synchrophasors taking into account the higher harmonics.

A prototype of the digital combined current and voltage sensor (DSCV) TECV.P1-10 with inboard merging unit has been developed for use at indoor switchgears 6-10 kV [14].

TECV.P1-10 is intended for joint use with devices of relay protection and automatics and measurements IEDs. DSCV contains a low-level electromagnetic current transformer with an inboard shunt (Low Power Current Transformer, LPCT), Rogowski coil and a capacitive voltage sensor (Fig. 9). The inboard merging unit is located in a limited space at the base of DSCV.

Fig. 9. Digital sensors of combined current and voltage

Measurements IEDs use digitized current data from LPCT in the range from 1 to 200% of the nominal value with accuracy class not worse than 0,2S. Relay protection and automatics devices use digitized current data from Rogowski coil and LPCT.

The future use of DSCV with inboard MU in comparison with similar sensors with analog low-level output signals is associated with better metrological performance due to the use of two current sensors, increased noise immunity, the ability to monitor DSCV and digital communication channels. The inboard AMU provides storage of data on the characteristics of current and voltage sensors and calibration coefficients[14].

DSCV is not only a source of information for the IEDs in various functional purposes, but it is also capable to perform a number of IEDs functions by implementing additional functions: telemetry, measurements of synchronized current and voltage phasors, etc. The use of synchronized current and voltage phasors measurements will significantly reduce the requirements for microcontrollers or digital signal processors of relay protection and automatics devices in 6-10 kV switchgears.

5 Conclusions

The increase of the efficiency of power system control is largely determined by the quality of information coming from the information control objects. Best implementation of modern requirements for the quality of IEDs data with the support of synchrophasor measurement technology. At IED design of new generation, the task of analysing the quality of the current and voltage estimation can be effectively solved by using the above methods of express analysis of filters that determine the dynamic responses of IEDs with the support of synchrophasor measurement technology.

DOI: 10.1051/e3sconf/20172502007

The use of extended spectral representations allows us to formalize the requirements for frequency responses of filters and simplify the task of synthesizing filters for IEDs in various purposes. This, in turn, allows the implementation of more advanced IEDs supporting by synchrophasor measurement technology with extended functionality.

References

1. A.G. Phadke, J.S. Thorp, *Synchronized Phasor Measurements and Their Applications* (Springer, NY, 2008).
2. H.J.A. Ferrer, E.O. Schweitzer III (eds.), *Modern Solutions for Protection, Control, and Monitoring of Electric Power Systems* (SEL, Pullman, 2010).
3. M. Kezunovic, S. Meliopoulos et al., *Application of Time-Synchronized Measurements in Power System Transmission Networks* (Springer, NY, 2014).
4. Integrating Synchrophasor Technology into Power System Protection Applications (NASPI, 2016).
5. A.Z. Gamm, Y.A. Grishin, A.M. Glazunova, I.N. Kolosok, E.S. Korkina, *Proceedings International Conference PowerTech* (2007).
6. Standard IEEE C37.118.1-2011.
7. A.V. Mokeev, *Proceedings International TICSP Workshop Spectral Methods and Multirate Signal Processing*, 159-167 (TICSP, Tampere, 2007).
8. A.V. Mokeev, F.P.G. Márquez (ed.), *Digital Filters*, 27-52 (InTech, Rijeka, 2011).
9. A.V. Mokeev, F.P.G. Márquez and N. Zaman (ed.) *Digital Filters and Signal Processing*, 81-107 (InTech, Rijeka, 2013).
10. A.V. Mokeev, *Proceedings IEEE International Conference SIBCON* (2013).
11. A.V. Mokeev, *Proceedings IEEE International Conference SIBCON* (2015).
12. A.V. Mokeev, *Proceedings IEEE International Conference SIBCON* (2017).
13. F. Wilhelm, D. Hoffmann, *Optimization - Theory and Practice* (Springer, 2010).
14. A.V. Mokeev, V.N. Bovykin, L.V. Perelygin, Relay Protection and Automation, **4**, 18-24 (2016).
15. A.V. Mokeev, V.N. Bovykin, E.I. Khromtsov, Relay Protection and Automation, **4**, 58-63 (2015).

Methods and models of optimal managing of district heating systems with prosumers

Valery Stennikov[1], *Ivan* Postnikov[1,*], *Andrey* Penkovskii[1]

[1]Melentiev Energy Systems Institute, 130 Lermontov str., Irkutsk, Russia

Abstract. The paper presents a statement of the problem on load distribution in the district heating systems with prosumers. The research suggests a bi-level model for solving this problem to determine an optimal balance between the load of district heat sources and prosumer-owned heat sources. The case study results obtained by using the developed methodology are presented. The potential economic effect of the prosumer adoption under specified conditions is demonstrated. The conclusions and directions for further research are formulated.

1 Introduction

Intensive development of energy and information technologies affected greatly the theory of energy system design. The key principles of this theory are integration of different generators, transition to intelligent systems (i.e. capability of a system to generate and implement decisions based on the forecast and analysis in combination with self-learning), and an increase in the participant of the consumer in energy supply.

The latter principle is implemented within the prosumer (producer-consumer) concept. The functions of prosumers are regulation and optimization of their demand to enhance the efficiency and reliability of both the consumer itself and the entire energy system. Prosumers have their sources and energy storage devices that allow them based on the balance between their needs and capabilities to vary the amount and properties (reliability, losses, quality, etc.) of the power received from the system.

Theoretical and practical studies on the prosumer are conducted by researchers in many countries. The investigations to be emphasized are presented in [1–7]. They are focused on various aspects of prosumer operation and control within power supply systems. In [8, 9] the authors consider some issues of prosumer operation in the district heating systems (DHS).

Based on the analysis of publications on the prosumer topic, which is part of a more general theory of the establishment of intelligent and integrated energy systems (Smart Grid), we can conclude that almost all these investigations concern electric power systems. At the same time, these technologies are also relevant for DHS which are the largest fuel consumers, especially in Russia, where fuel consumption for heating needs exceeds 45% of the total fossil fuel consumed in the country.

One of the main goals of involving the prosumer (hereinafter we mean heat prosumer) in the DHS is to enhance the efficiency and cost-effectiveness of system operation by managing optimal distribution of sources supplying heat loads. The methodological problems which need to be solved to achieve this goal are various and traditional for energy systems on the whole and DHS in particular. The involvement of prosumer in the DHS, however, brings about new aspects in these problems and calls for new methodological approaches to solve them. The development of these methods is a subject of the presented paper.

2 Bi-level model of DHS managing with prosumer

Management of DHS with prosumer implies distribution of heat load among system district heat sources (HS) and local HS that belong to prosumers, according to some criteria providing the required (anticipated) parameters of system operation. The inclusion of prosumers with their HS into the DHS changes the organizational model of the system operation. Thus, the need arises to consider the control at two levels: from the viewpoint of district DHS (system level) and from the standpoint of the prosumer with their generation (prosumer level).

Mathematically, the problem of the management of DHS with the prosumer is solved by the bi-level programming methods [10–12].

The management structure of DHS with the prosumer is based on the hierarchical two-level approach and is as follows. At the first level, we solve the problem of cost minimization for the prosumer. This problem represents a sum of heat purchase costs and heat production costs of the prosumer-owned HS. The second level of management corresponds to the system for which the problem of profit maximization is posed. The profit is

* Corresponding author: postnikov@isem.irk.ru

 DOI: 10.1051/e3sconf/20172502008

defined as a revenue from selling the heat produced at district HS less its production costs.

The DHS and prosumer interact as follows:

1) The prosumer submits a bid for the amount of heat it needs;

2) The system submits a price offer obtained according to its purpose of profit maximization;

3) The prosumer chooses an optimal distribution of loads for their supply from the system (by district HS) and by its own HS according to its purpose of cost minimization and makes a bid for the load again.

In this cycle, an equilibrium that satisfies both participants of the interaction is determined.

The bi-level optimization model of managing the DHS with the prosumer is formulated as follows:

1) The objective functions are:

$$F_{\text{obj}}^{(1)} = \sum_{j\in J}\sum_{\tau\in T}(c_{j\tau}^{h}q_{j\tau}^{\text{sys}} + \alpha_j q_{j\tau}^2 + \beta_j q_{j\tau} + \gamma_j) \rightarrow \min; \quad (1)$$

$$F_{\text{obj}}^{(2)} = \sum_{j\in J}\sum_{\tau\in T} c_{j\tau}^{h}q_{j\tau}^{\text{sys}} - \sum_{i\in I}\sum_{\tau\in T}(\alpha_i q_{i\tau}^2 + \beta_i q_{i\tau} + \gamma_i) \rightarrow$$
$$\rightarrow \max; \quad (2)$$

2) Load curve of the consumer is:

$$q_{oj\tau} = q_{oj}[1-(1-\omega_j)(\tau/\tau_o)^{\sigma_j}], j\in J, \tau\in T; \quad (3)$$

3) The flow distribution model is:

$$\mathbf{A}\mathbf{x}_\tau = \mathbf{q}_\tau, \tau\in T; \quad (4)$$

$$\overline{\mathbf{A}}^{\text{T}}\mathbf{p}_\tau = \mathbf{h}_\tau - \mathbf{H}_\tau, \tau\in T; \quad (5)$$

$$\mathbf{S}\mathbf{X}_\tau\mathbf{x}_\tau = \mathbf{h}_\tau, \tau\in T; \quad (6)$$

4) Energy flow balances are:

$$\sum_{i\in I} q_{i\tau} - \sum_{j\in J} q_{oj\tau} = 0, \tau\in T; \quad (7)$$

$$q_{oj\tau} = q_{j\tau}^{\text{sys}} + q_{j\tau}, j\in J, \tau\in T; \quad (8)$$

5) Constraints on variables and parameters are:

$$q_{oj\tau} > 0, q_{j\tau}^{\text{sys}} \geq 0, c_{j\tau}^{h} > 0, j\in J, \tau\in T; \quad (9)$$

$$0 \leq q_{j\tau} \leq q_{j\max}, j\in J, \tau\in T; \quad (10)$$

$$q_{i\min} \leq q_{i\tau} \leq q_{i\max}, i\in I, \tau\in T; \quad (11)$$

$$p_{j\min} \leq p_{j\tau} \leq p_{j\max}, j\in J, \tau\in T. \quad (12)$$

Here: j – heat consumer; J – set of heat consumers (including prosumers); i – district HS; I – is a set of district HS; τ – time instant corresponding to the number of hours with a specified load of consumers, h; T – set of time instants corresponding to the number of hours with a specified load; $F_{\text{obj}}^{(1)}$ – objective function of the first level (consumer), rub; $F_{\text{obj}}^{(2)}$ – objective function of the second level (system), rub; $c_{j\tau}^{h}$ – heat price for consumer j at time instant τ, rub/GJ; $q_{j\tau}^{\text{sys}}$ – part of heat load of consumer j, supplied from the system (by district HS) at time instant τ, GJ/h (only for prosumer); $q_{j\tau}$ – part of heat load of consumer j, supplied by their own HS (only for prosumer) at time instant τ, GJ/h; α_j, β_j, γ_j –coefficients of cost function of the HS prosumer j; $q_{i\tau}$ –performance of i-th district HS at time instant τ, GJ/h; α_i, β_i, γ_i – coefficient of cost functions of the i-th district HS; q_{oj}, $q_{oj\tau}$ – design (maximum) heat load and the heat load of consumer j that corresponds to time instant τ, GJ/h; ω_j, σ_j – heat load irregularity factors of consumer j; $\mathbf{A}$ – incidence matrix of linearly independent nodes of heat network (HN); $\overline{\mathbf{A}}^{\text{T}}$ – complete transpose of node-branch incidence matrix for HN; $\mathbf{x}_\tau$ –vector of heat carrier flow rates in the branches of the HN at time instant τ, t/h; $\mathbf{q}_\tau$ – vector of flow rates at network nodes at time instant τ, t/h; $\mathbf{p}_\tau$ – vector of nodal pressure of HN at time instant τ, mm wc; $\mathbf{h}_\tau$ – vector of head losses in the network branches at time instant τ, mm wc; $\mathbf{H}_\tau$ – vector of operating heads at the sources at time instant τ, mm wc; $\mathbf{S}, \mathbf{X}_\tau$ – diagonal matrices of coefficients of hydraulic resistances of branches, m/(h^{2}t^2), and absolute values of flow rates in them, t/h, at time instant τ; $q_{i\min}, q_{i\max}$ – minimum and maximum value of the i-th district HS capacity, GJ/h; $q_{j\max}$ – maximum value of the HS capacity of consumer j (for prosumer), GJ/h; $p_{j\min}, p_{j\max}$ – minimum and maximum value of the heat carrier pressure at consumer nodes j, mm wc.

Expression (3) is used to specify the load of consumer at each instant of the considered period (a heating season) based on the Rossander equation [13]. Expressions (4)–(6) represent a model of flow distribution in HN written in a matrix nodal form which is traditional for the theory of hydraulic circuits [14]. Parts of expressions in (1) and (2) that define the heat production costs at HS are presented as quadratic dependences derived by approximating the actual data on heat production and fossil HS costs corresponding to it, by using the method of least squares.

With the methods presented in [15–17], the considered problem (1)–(12) can be transformed as follows:

Find:

$$F_{\text{obj}\tau}^{(1)} = c_{i\tau}^{h}q_{i\tau} + \sum_{j\in J_i}(\alpha_j q_{j\tau}^2 + \beta_j q_{j\tau} + \gamma_j) \rightarrow \min; \quad (13)$$

subject to:

$$q_{i\tau} = (c_{i\tau}^{h} - \beta_i)/2\alpha_i, i\in I, \tau\in T; \quad (14)$$

$$q_{oj\tau} = q_{oj}[1-(1-\omega_j)(\tau/\tau_o)^{\sigma_j}], j\in J, \tau\in T; \quad (15)$$

$$\mathbf{A}\mathbf{x}_\tau = \mathbf{q}_\tau, \tau\in T; \quad (16)$$

$$\overline{\mathbf{A}}^{\text{T}}\mathbf{p}_\tau = \mathbf{h}_\tau - \mathbf{H}_\tau, \tau\in T; \quad (17)$$

$$\mathbf{S}\mathbf{X}_\tau\mathbf{x}_\tau = \mathbf{h}_\tau, \tau\in T; \quad (18)$$

$$\sum_{i\in I} q_{i\tau} - \sum_{j\in J} q_{oj\tau} = 0, \tau \in T; \quad (19)$$

and constraints:

$$q_{oj\tau} > 0, c_{i\tau}^h > 0, j \in J, i \in I, \tau \in T; \quad (20)$$

$$0 \le q_{j\tau} \le q_{j\max}, j \in J, \tau \in T; \quad (21)$$

$$q_{i\min} \le q_{i\tau} \le q_{i\max}, i \in I, \tau \in T; \quad (22)$$

$$p_{j\min} \le p_{j\tau} \le p_{j\max}, j \in J, \tau \in T. \quad (23)$$

where: $F_{\text{obj}\tau}^{(1)}$ – objective function of the first level, determined at time instant τ, rub/h; $c_{i\tau}^h$ – price of heat generated at the i-th district HS at time instant τ, rub/GJ.

Thus, after the transformations, the management problem of DHS with the prosumer (13)–(23) represents, unlike the initial form (1)–(12), a traditional problem of mathematical programming. Fig. 1 demonstrates a graphical illustration of the computational process of search for an optimal solution to the management problem of DHS with the prosumer.

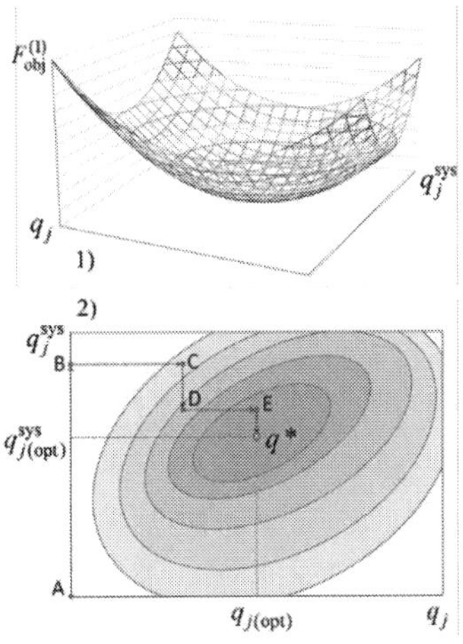

Fig.1. A graphical illustration of a computational process of search for an optimal solution to the management problem of DHS with the prosumer: 1) the objective function form in a three-dimensional system of coordinates; 2) a projection of the objective function plot to the plane of coordinates of heat loadings of sources and the solution search procedure.

The spatial view of the objective function is presented in Fig. 1-1 and its projection on the system of coordinates of heat loads of the sources – in Fig. 1-2. Elliptic lines in this diagram are isocost lines, i.e. the lines of equal costs, the dotted line A-B-C-D-E-$q*$ corresponds to a step-by-step procedure of search for a solution. The formulated problem is solved by the coordinate descent method with a simple iteration within the cycle to reduce the multidimensional optimization problem to one-dimensional one with a step-by-step procedure for improvement in solutions according to the heat production output by all the sources.

3 Case study and an analysis of results

Consideration is given to DHS scheme presented in Fig. 2-1. The scheme consists of two district HS (HS1 and HS2), seven consumers (nodes 1–7) and a looped HN consisting of 18 sections. One of the consumers is a prosumer (P6) with a load of 400 GJ/h, and a HS with a capacity of 200 GJ/h. The technical characteristics of the HN components, loads of consumers and capacities of HS are presented in Table 1.

HS1 supplies heat to P6 according to the hydraulic calculation of the system. Therefore, the problem of optimal management of heat source load distribution is solved at the level of interaction between these entities. Fig. 2-2 demonstrates a scheme of their interaction.

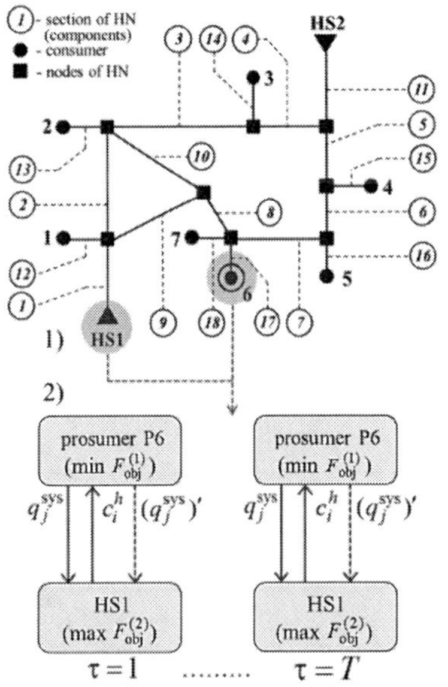

Fig. 2. Test study of the heating system: 1) an aggregate scheme of the heating system; 2) a two-level scheme of interaction between the prosumer (P6) and heat source 1 (HS1).

According to the management model presented in the methodological part of the research, search for an optimal (corresponding to the minimum costs) choice of the sources supplying heat to the node P6 is performed for each time instant τ within the heating period that lasts $T = 6000$ h.

The HS of prosumer P6 is assumed to operate on fossil fuel. For this HS, we specify a quadratic cost function which is part of the objective function (1) and has the following values of approximation coefficients: $\alpha_j = 0{,}108$ rub/(GJ)^2; $\beta_j = 131$ rub/GJ; $\gamma_j = 10240$ thousand rub. The production costs of HS1 are also defined by the quadratic function that is included in expression (2) and has the following values of approximation coefficients: $\alpha_i = 0{,}077$ rub/(GJ)^2; $\beta_i = 76$ rub/GJ; $\gamma_i = 12800$ thousand rub.

Table 1. Technical characteristics of components of considered heating system scheme (corresponding to scheme on Fig. 1)

Heat network parameters		
Section	Diameter,m	Length, m
1	0,65	1200
2	0,6	2000
3	0,55	2600
4	0,5	1300
5	0,6	1000
6	0,55	900
7	0,5	1700
8	0,55	900
9	0,55	2000
10	0,5	1900
11	0,7	1500
12	0,35	800
13	0,4	800
14	0,35	800
15	0,5	800
16	0,4	600
17	0,45	600
18	0,35	600

Consumer load and HS capacity			
Consumer	Load/ HS capacity of prosumer, GJ/h	HS	Capacitiy, GJ/h
1	280	HS1	1254
2	343	HS2	1463
3	268	-	-
4	502	-	-
5	351	-	-
P6	400/200	-	-
7	196	-	-
Total	2340	Total	2717

Table 2 presents the optimal loads of HS in the considered DHS scheme for prosumer P6 during the entire heating period with a step of 1000 h. These data are illustrated in Fig. 3 where diagram 1 shows the heat load curves of P6 with highlighted amounts of heat generated by HS1 and HS of P6. Diagram 2 presents the graph of costs of heat supply to P6, that are calculated for the cases with and without involving the HS of P6 in its heat supply.

Table 2. Optimal distribution of sources to supply heat to P6 during heating period

Heating period hours, h	Load of P6	Loading of HS1, GJ/h	Loading of P6 HS, GJ/h	Costs of heat supply to P6, disregarding its HS, thousand rub/h	Costs of heat supply to P6, considering its HS, thousand rub/h	Cost cutting due to the use of HS of P6, thousand rub/h
1	400	219	181	153	116	37
1000	244	187	57	71	64	7.2
2000	190	187	2.9	49.2	49	0.3
3000	150	150	0	35.4	0	0
4000	117	117	0	25.5	0	0
5000	89.4	89.4	0	17.9	0	0
6000	64.4	64.4	0	12	0	0

According to Fig. 3-1 and Table 2, the HS of P6 operates during the time of consumer peak loads and supplies heat in the amount of about 180,6 GJ/h (which corresponds to the maximum load of the consumer).

Further, its load is reduced, and by the time corresponding to 2000 h, becomes equal to zero. The total heat consumption of P6 during the heating period is 1026.5 thousand GJ, including 890,7 thousand GJ (86,8%) from the district HS1, and 135.9 thousand GJ (13,2%) – from its own HS.

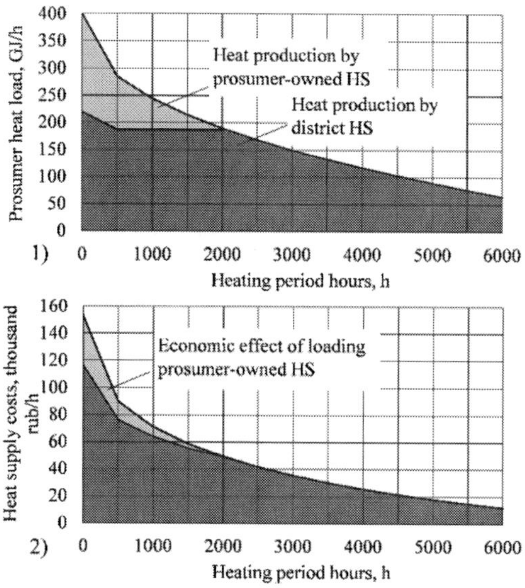

Fig. 3. Solution to the load management problem for the test scheme of heating system with prosumer: 1) optimal load of P6-owned HS and HS 1 during the heating period, 2) a graph of heat production costs for P6.

The economic effect due to the operation of P6 HS can be seen in Fig. 3-2. Its value corresponds to the area of a figure limited by two graphs of heat supply costs of prosumer P6: 1) – disregarding its heat source (heat is fully supplied from HS1), and 2) – considering it. Thus, the total costs over the heating period without the HS of P6 made up 283.5 million rub, and with it – 262 million rub. Therefore, the economic effect due to the use of the prosumer HS made up 21.6 million rub or 7.6%.

4 Conclusion

The relevance of studies on heat prosumer is explained by the objective problems in heat supply, that are related to low cost-effectiveness of operating DHS, insufficient quality and reliability of heat supply to consumers, particularly now when the technologies develop and loads increase.

This research states the load management problem of DHS with prosumer. Solving this problem is aimed at cost-effective distribution of HS to supply heat to consumers both from district HS and from prosumer-owned HS. A mathematical tool of bi-level programing is used to solve this problem because it defines most accurately the organizational model of interaction between the system and prosumer. Additionally, the methods of the theory of hydraulic circuits were applied.

The results obtained in the calculation of DHS scheme show the operability of the developed mathematical model and possibility of gaining the economic effect owing to the use of a prosumer-owned HS. The case study demonstrates that the application of the prosumer-owned HS is most effective during peak demand. The economic effect obtained in the test calculations is not final to substantiate the prosumer involvement in the operation of the existing DHS. Further research in this area will provide a more informed assessment of the effectiveness of prosumer adoption in district heating.

The research was performed at Melentiev Energy Systems Institute SB RAS under the support of Russian Science Foundation (Grant №17-19-01209).

References

1. Ni Zhang, Yu Yan, Wencong Su. *Applied Energy*, **vol. 154**, p. 471–479. (2015).
2. Luka Perkovic, Hrvoje Mikulcic, Neven Duic. *Journal of Cleaner Production*, **xxx**, p. 1–12. (2016).
3. Hongming Yang, Tonglin Xiong, Jing Qiu et al. *Applied Energy*, **vol. 167**, p. 353–365, (2016).
4. Rehman Zafar, Anzar Mahmood, Sohail Razzaq et al. Prosumer based energy management and sharing in smart grid // *Renewable and Sustainable Energy Reviews,* in press, available online 11 July 2017.
5. Stig Ødegaard Ottesen, Asgeir Tomasgard, Stein-Erik Fleten. *Energy*, **vol. 94**, p. 828–843, (2016).
6. Dimitrios J. Vergados, Ioannis Mamounakis, Prodromos Makris, Emmanouel Varvarigos. *Sustainable Energy, Grids and Networks*, **vol. 7**, p. 90–103, (2016).
7. Lakshmi Prakash, Sugatha Kumari P.R, Sharanya Chandran et al. *Procedia Technology*, **vol. 21**, p. 338–344, (2015).
8. Lisa Brange, Jessica Englund, Patrick Lauenburg. *Applied Energy*, **vol. 164**, p. 492–500, (2016).
9. Lisa Brand, Alexandra Calven, Jessica Englund et al. *Applied Energy*, **vol. 129**, p. 39–48, (2014).
10. M.S. Ershova. *Introduction into a bi-level programming*/ Manual. – Irkutsk: ISU, 2006.– 76 p.
11. J.F. Bard *Practical Bilevel Optimization* (Dordrecht, Netherlands: Kluwer Academic Publishers, 1998. – 488 p).
12. S. Dempe *Foundations of Bilevel Programming*. (Dordrecht, Netherlands: Kluwer Academic Publishers, 2002. – 320 p).
13. E.V. Sennova, V.G. Sidler *Mathematical modeling and optimization of developing heat supply systems*. (Novosibirsk, 1985. – 222 p).
14. A.P. Merenkov, V.Y. Khasilev *Theory of hydraulic circuits*. (M.: Nauka, 1985. – 278 p).
15. V.A. Stennikov, O.V. Khamisov, A.V. Pen`kovskii. *Thermal Engineering*, **Vol. 58**, No.12. pp. 1043–1048, (2011).
16. V.A. Stennikov, O.V. Khamisov, A.V. Penkovskii *Izv.RAN, Energetika*, **No.3**. – P. 27–36.
17. A.V. Penkovskii, V.A. Stennikov, O.V. Khamisov, E.E. Mednikova, I.V. Postnikov. *Energy Procedia* **105**, – P. 3158–3163, (2017).

 DOI: 10.1051/e3sconf/20172502009

Methods and models of ensuring reliability of district heating systems with prosumers

Ivan Postnikov[1,*], *Valery* Stennikov[1]

[1]Melentiev Energy Systems Institute, 130 Lermontov str., Irkutsk, Russia

Abstract. The paper is concerned with the statement of a district heating systems reliability problem. The reliability is planned to be ensured through the optimal distribution of reliability parameters among components of system, considering the capabilities of the prosumer to supply part of load by their heat sources. The methods and models are suggested to solve this problem. They are based on the theory of random processes, theory of hydraulic circuits and basic laws of cogeneration. The case study results obtained using the developed technique are presented. A potential economic benefit of involving the prosumer is demonstrated in the case study. The conclusions and directions for further research are formulated.

1 Introduction

One of the major principles of a contemporary energy systems paradigm is an increase in the role of consumer in energy supply. This principle is implemented in the framework of a prosumer (producer-consumer, consumer that have self-generation) concept which implies control and optimization of the consumer load curve by using the consumer-owned sources and storage devices with a view to enhance the efficiency and reliability of both the consumer and the entire energy system.

The authors of [1–7] present the studies on various aspects of operation and control of prosumer within the electric power systems. Although, the issues of prosumer operation in district heating systems (DHS) are raised much more rarely [8, 9], they are relevant and important since DHS are the largest fuel consumers in many countries.

Nowadays, modern DHS are experiencing an increase in heat loads, which causes some problems related to the fulfilment of volume and reliability requirements in heat supply. The creation of the prosumers with their heat sources (HS) will provide additional capacity and time reserve which will reduce the load of district HS and thus improve the quality and reliability of both heat supply to consumers and operation of the system as a whole.

Thus, one of the key prosumer goals within the DHS is to increase reliability of the system operation. The methodological problems to be solved to achieve this goal are mainly traditional for the analysis and synthesis of reliability of energy system as a whole and DHS, in particular. At the same time, the involvement of prosumer in the DHS operation introduces new aspects in these problems and requires new methodological approaches to solve them.

2 Methods and models to ensure reliability of heating systems with prosumers

2.1 Statement of the problem

The development of methods for reliability analysis and optimization in the DHS with prosumer generates two main lines of investigations. The first line is related to an analysis and optimization of heat supply reliability, considering the prosumer functions in the system, provided the quantity, nodes and capacity of the prosumer sources are specified. The second line represents the problem aimed at determining the amount of prosumers, their connection points and capacities.

This paper is focused on the first reliability problem which implies the determination of the values of reliability parameters of the DHS components (failure and restoration rates) that ensure the required reliability level of heat supply to consumers, including prosumers (considering their additional reserve), at minimum cost of reaching these indices within their feasible range.

Two main reliability indices are assumed to assess the level of heat supply reliability [10]. These are failure free operation probability R_j (FOP) and availability factor K_j (AF). These indices are determined for each consumer $j \in J$, where J is a set of consumers in the system (including prosumers). The standard values of these indices are denoted by R_{oj} and K_{oj}.

The stated problem is solved with respect to these reliability indices according to a technique consisting of 4 stages: 1) determine relationships between the average reliability parameters of DHS components; 2) model post-emergency conditions of the DHS; 3) model the

* Corresponding author: postnikov@isem.irk.ru

DOI: 10.1051/e3sconf/20172502009

functions of the prosumers; 4) formalize the problem of determine optimal reliability parameters of the DHS components.

2.2 The relationship between the average reliability parameters of DHS components

Average reliability parameter of components is taken to mean their failure or restoration rate that preliminarily has the same value for these components, which provides the required level of reliability indices. These parameters are determined by the formulas intended for the calculation of nodal reliability indices of DHS [10], Rossander equation that determines annual heat load curves of consumers [11], and some basic laws of district heating and thermal physical processes involved in DHS [12].

Based on the theoretical studies we determined the following relationship between the average reliability parameters of DHS components (failure rate and restoration rate), considering meeting the conditions of FOP and AF:

$$\bar{\lambda}_j = \frac{1}{\tau_o}\left(\ln\frac{1}{R_{oj}}\right)\times\left(1-N_s(1-K_{oj})\Big/\sum_{s\in E}L_s^{1/\sigma_j}\right)^{-1}\times \left(\sum_{s\in E}M_s^{1/\sigma_j}\right)^{-1}; \quad (1)$$

where

$$L_s = \frac{1}{1-\omega_j}\times \left[1-\frac{1}{q_{oj}}\left(q_{sj}+\varphi_j\left(t_{sj}-\frac{C_1-C_2\exp B_j}{C_3(1-\exp B_j)}\right)\right)\right]; \quad (2)$$

$$M_s = \frac{1}{1-\omega_j}\left(1-\bar{q}_{sj}+\frac{\varphi_j t_{sj}}{q_{oj}}-\varphi_j\frac{C_1-C_2\exp B_j}{C_3 q_{oj}(1-\exp B_j)}\right); \quad (3)$$

$$C_1 = t_{oj}(1-\bar{q}_{sj}), C_2 = t_{j\min}-t_{oj}\bar{q}_{sj}, C_3 = 1-\bar{q}_{sj}; \quad (4)$$

$$\bar{q}_{sj} = q_{oj}/q_{sj}, s\in E, j\in J; \quad (5)$$

$$B_j = 1/(\varepsilon_j\bar{\mu}_j), j\in J. \quad (6)$$

Here: $\bar{\lambda}_j$ and $\bar{\mu}_j$ – average failure and restoration rates for consumer j, respectively, 1/h; τ_o – time instant corresponding to a total number of hours of the considered (heating) period, h; N_s – the quantity of system states; s – number of system state; E – a set of system states; ω_j, σ_j – irregularity factors of heat load curve of consumer j [11]; q_{oj} – design (maximum) heat load of consumer j, GJ/h; q_{sj} – level of heat supply to consumer j in system state s, GJ/h; φ_j – coefficient of specific heat losses for consumer j, GJ/(h°C); t_{sj} – current (actual) internal air temperature for consumer j in system state s, °C; $\bar{q}_{sj}$ – relative heat supply to consumer j in system state s, GJ/h; t_{oj} – design temperature of internal air for consumer j, °C; $t_{j\min}$ – minimum admissible temperature of internal air for consumer j, °C; ε_j – coefficient of thermal energy storage for consumer j, h; $L_s, M_s, C_1, C_2, C_3, B_j$ – assumed abbreviations of expressions.

2.3 Modeling of prosumer functions

The prosumer functions in the emergency conditions leading to limitation or total termination of heat supply are aimed at eliminating heat undersupply by their HS, and increasing time reserve. These properties of the prosumer are considered in the calculation of indices $\bar{q}_{sj}$ and B_j. To this end, the following expressions are added to equations (5) and (6):

$$\bar{q}_{sj} = q_{oj}/(q_{sj}^{\mathrm{sys}}+q'_{sj}), s\in E, j\in J; \quad (7)$$

$$B_j = 1/[(\varepsilon_j+\Delta\varepsilon_j)\bar{\mu}_j],\ j\in J; \quad (8)$$

where q_{sj}^{sys} – part of heat load of consumer j, supplied from the system (by district HS) in system state s, GJ/h; q'_{sj} – capacity of HS for consumer j (for prosumer) in system state s, GJ/h; $\Delta\varepsilon_j$ – additional passive time reserve of prosumer (caused by the use of their heat generators and/or heat storage devices), h.

Index q'_{sj} can be fixed and correspond to a rated (required) value of capacity of the prosumer HS. This index can be variable and take into account the components failures decreasing the performance of the HS. Here, the corresponding failures should be added to a set of the system states. In this case, decision should be made based on an analysis of initial characteristics of equipment reliability: if the values of failure flow for the HS equipment of the prosumer are much lower than for the DHS equipment, then the failures at prosumer HS can be neglected, and q'_{sj} can be assumed to be fixed according to their required performance under specified emergency heat supply to consumers.

2.4 Modeling of post-emergency conditions in heating systems

Post-emergency hydraulic conditions are determined by steady flow distribution in the heat network (HN) after disconnection of an emergency component. The hydraulic conditions are calculated by the methods of theory of hydraulic circuits [13]. The nodal form of writing the model of flow distribution (hydraulic conditions) in the network is represented by a matrix and has the form [13]:

$$\mathbf{A}\mathbf{x}_s = \mathbf{q}_s, s\in S; \quad (9)$$

$$\overline{\mathbf{A}}_s^{\mathrm{T}} \mathbf{p}_s = \mathbf{h}_s - \mathbf{H}_s, s \in S; \quad (10)$$

$$\mathbf{S}\mathbf{X}_s \mathbf{x}_s = \mathbf{h}_s, s \in S; \quad (11)$$

where: $\mathbf{A}_s$ – incidence matrix of linearly independent nodes in the network under emergency system state s (considering failure of some component); $\overline{\mathbf{A}}_s^{\mathrm{T}}$ – full transposed node-branch incidence matrix; $\mathbf{x}_s$ – vector of heat carrier flow rates in the network sections (branches) under emergency system state s, t/h; $\mathbf{q}_s$ – vector of flow rates at network nodes under emergency system state s, t/h; $\mathbf{p}_s$ – vector of nodal pressures of HN under emergency state s, mm wc; $\mathbf{h}_s$ – vector of heat losses in the sections under the emergency system state s, mm wc; $\mathbf{H}_s$ – vector of operating heads at sources in the emergency system state s, mm wc; $\mathbf{S}, \mathbf{X}_s$ – diagonal matrices of coefficients of hydraulic resistance of sections, m/(h^{2}t^2), made up from the values of hydraulic resistances of sections and absolute values of flow rates in them, t/h.

Modeling of an emergency situation in some state s of the system is performed by excluding a component whose failure corresponds to this state from the calculation DHS scheme.

2.5 Formalization of a problem of determination of optimal reliability parameters of DHS components

The values of average reliability parameters determined by expressions (1)–(6) are distributed among the system components according to the following equations of conservation of the sum of system state probabilities:

$$\bar{\lambda}_j \sum_{s \in E} p_s = \sum_{n \in N} \sum_{s \in E(n)} \lambda_n p_s, j \in J; \quad (12)$$

$$\bar{\mu}_j \sum_{s \in E} p_s = \sum_{n \in N} \sum_{s \in E(n)} \mu_n p_s, j \in J; \quad (13)$$

where: p_s – probability of system state s; n – number of system component; N – a set of system components; $E(n)$ – is a subset of system states for which the system can transition because of failure or restoration of component n; λ_n, μ_n – failure or restoration rates of component n, 1/h.

The probabilities of system states are determined by solving the equations of Markov process that describes a sequence and structure of events that characterize the DHS operation. The Markov models are a justified and universal tool for the assessment of reliability of restorable systems [14–17].

A stationary Markov model of DHS operation can be represented by the following system of linear equations:

$$p_s \left(\sum_{n \in N(s)} \lambda_n + \sum_{n \in N(s)} \mu_n \right) = \sum_{z \in E(s)} \left(\sum_{n \in N(z)} p_z \lambda_n + \sum_{n \in N(z)} p_z \mu_n \right), s \in E; \quad (14)$$

where: p_z – probability of the system state z (division of state into s and z is necessary to write the system of equations of random process); $N(s)$ – subset of system components whose failure or restoration corresponds to a direct transition of the system from state s to some other state z; $N(z)$ – subset of system components whose failure or repair corresponds to a direct transition of the systems from state z to some other state s; $E(s)$ – subset of the system states from which the system can transition to state s.

The rationale of using the stationary Markov model as well as the other aspects of using the tool of Markov random processes in the problems of DHS reliability are considered in more detail in [14, 18–22].

According to the problem stated in 2.1, the objective function of DHS reliability optimization is expressed as a sum of costs of ensuring the values of reliability parameters of the DHS components:

$$F_{\mathrm{obj}} = \sum_{n \in N} f_{n\lambda}(\lambda_n) + \sum_{n \in N} f_{n\mu}(\mu_n), \quad (15)$$

where $f_{n\lambda}(\lambda_n), f_{n\mu}(\mu_n)$ – cost functions of ensuring reliability parameters of components, i.e. their failure and restoration rates, respectively, rub. The form of functions (15) and their quantitative parameters are determined by the methods of approximation on the basis of an analysis of the actual data on the cost of equipment with different reliability characteristics, costs of installing the backup components; costs of establishing and maintaining the emergency and restoration services (ERS) and other measures capable to improve the reliability.

Technically possible values of reliability parameters of the components are specified by the following constraints:

$$\lambda_n^{\min} \le \lambda_n \le \lambda_n^{\max}, n \in N; \quad (16)$$

$$\mu_n^{\min} \le \mu_n \le \mu_n^{\max}, n \in N. \quad (17)$$

Thus, the optimization problem of component reliability of DHS considering prosumers lies in the following:

– conditions (1)–(6), that determine the relations between average reliability parameters of DHS components considering the prosumer functions;

– conditions (9)–(11) according to which the levels of heat supply under different system states are calculated;

– conditions (12) and (13) that determine the principles of allocating the average reliability parameters among the system components;

– constraints (16) and (17) on the reliability parameters of components;

Nonlinearity of considered optimization problem is connected firstly with function (15), which, as a rule, is a power function. Solution of such problems is based on method of coordinate-wise relaxation (or method of coordinate-wise descent) with use of simple iteration method within solve cycle. The GAMS software package is used as a solver.

3 Case study and analysis of results

Consideration is given to DHS scheme presented in Fig. 1-1. The scheme consists of two district heat sources (HS1 and HS2), seven consumers (nodes 1–7) and a looped HN consisting of 18 sections (components). One of the consumers is a prosumer (P6) with a load of 400 GJ/h, and a HS with a capacity of 200 GJ/h. An aggregate scheme of components for both district HS is demonstrated in Fig. 1-2 and consists of: a boiler 19, a turbine 20, network heaters 21 and 23 GJ/h and network pumps 22 and 24. The indicated numbers of components correspond to HS1, the numbers of components for HS2 are assigned in the order from 25 to 30. Technical characteristics of the HN components, consumer loads and capacities of HS are given in Table 1.

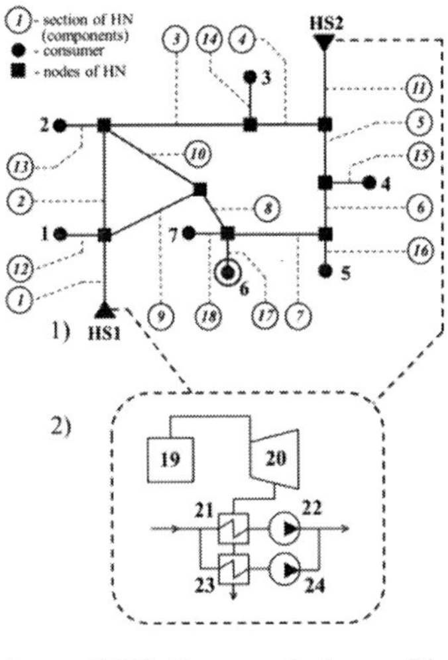

Fig. 1. A scheme of DHS: 1) a general scheme of the DHS; 2) a simplified scheme of HS

The calculation scheme of a considered DHS is constructed according to a comprehensive approach to the research into the reliability of heat supply [18–20] as a joint structure integrating the components of the schemes of HS and HN and is given in Fig. 2. The components 1–18 are network sections, components 19–30 correspond to HS1 and HS2.

The random process of DHS operation is modeled for the following conditions: each component can be in two states – operable and failed, and the flow of events within one subsystem (HN, HS1 and HS2) is the simplest. The latter condition suggests a simultaneous failure of components only from different subsystems of DHS. Here we will confine ourselves to the consideration of the state when no more than 2 components fail at once. Thus, the set of states is formed by individual states of HN, HS1, HS2 and combinations of these states for HN+HS1, HN+HS2 and HS1+HS2. The graph of DHS states in a reduced form is presented in Fig. 3. The number of component of the graph presented in Fig. 3 corresponds to the number of a failed component (components 1–18 are network sections, components 19–30 correspond to the sources). Markov random process that meets the specified conditions is described by the system consisting of 283 stationary linear equations of form (14).

Table 1. Technical characteristics of components of considered DHS (corresponding to scheme on Fig. 1)

Heat network parameters			
Section	Diameter,m	Length, m	
1	0,65	1200	
2	0,6	2000	
3	0,55	2600	
4	0,5	1300	
5	0,6	1000	
6	0,55	900	
7	0,5	1700	
8	0,55	900	
9	0,55	2000	
10	0,5	1900	
11	0,7	1500	
12	0,35	800	
13	0,4	800	
14	0,35	800	
15	0,5	800	
16	0,4	600	
17	0,45	600	
18	0,35	600	
Consumer load and HS capacity			
Consumer	Load/ HS capacity of prosumer, GJ/h	HS	Capacitiy, GJ/h
1	280	HS1	1254
2	343	HS2	1463
3	268	-	-
4	502	-	-
5	351	-	-
P6	400/200	-	-
7	196	-	-
Total	2340	Total	2717

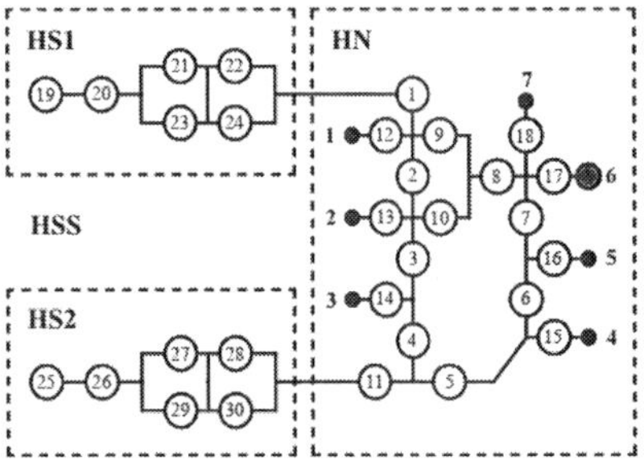

Fig. 2. Calculation scheme of HSS for reliability analysis (integrating component of schemes of HS and HN)

The level of heat carrier delivery to consumers in different states of the DHS is determined by multi-variant calculations of flow distribution in the HN, considering the failures of HS equipment, with the model (9)–(11). The results of the calculation of probabilities of system states and post-emergency hydraulic conditions are not presented due to a large data array.

The reliability parameters of the DHS are optimized provided the following standard values of nodal reliability parameters [10] are met: $K_{oj} = 0,97$ and $R_{oj} = 0,905$. The following ranges of possible values of optimized reliability parameters of the DHS components are assumed: 0,0002–0,0025 1/h – for failure rate; 0,007–0,09 1/h – for restoration rate.

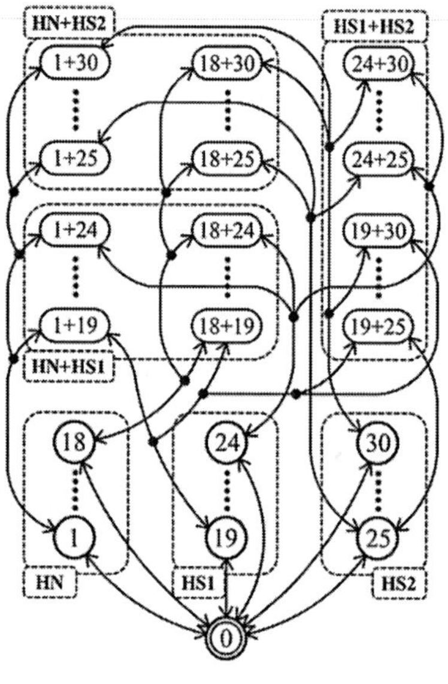

Fig. 3. Graph of HSS states and relation between them

Analytical functions of the costs required to restore the components of heat supply system (15) are assumed in an aggregate form, using the data on composition and specific costs of backup components of DHS, as well as the data on the establishment and maintenance of the ERS [10]. Analysis of these data made it possible using the approximation methods to obtain the required relationships between the costs and reliability parameters in the form of power objective functions.

Table 2 presents the calculated optimal average reliability parameters of DHS components and corresponding total costs of ensuring reliability of heat supply to consumers. There are two results for the costs: without prosumer P6 and with prosumer P6.

Fig. 4 illustrates a search for an optimal relationship between the average reliability parameters of DHS components that correspond to the minimum costs of ensuring reliability in the system when the required reliability indices (FOP and AF) are met. The presented diagram demonstrates three graphs: line 1 represents a relationship between the average reliability parameters of components, and graphs 2 and 3 show a change in the costs of ensuring the required reliability level in the DHS depending on the relationship between the average reliability parameters of components with and without HS of prosumer (P6), respectively.

Table 2. Result of optimization reliability parameters of HSS components (corresponding to diagram on Fig. 4)

Average reliability parameters of DHS components, 1/h		Total costs of ensuring required reliability of HSS, million rub	
failure rate	restoration rate	without prosumer P6	with prosumer P6
0.0007	0.01	30.4	30.34
0.0008	0.02	28.17	27.64
0.0009	0.03	26.82	25.83
0.0011	0.04	25.92	24.74
0.0013	0.05	25.35	24.19
0.0015	0.06	25.25	24.1
0.0019	0.07	25.31	24.38
0.0024	0.08	25.7	24.98
0.0031	0.09	26.38	25.89
0.0044	0.1	27.28	27.02

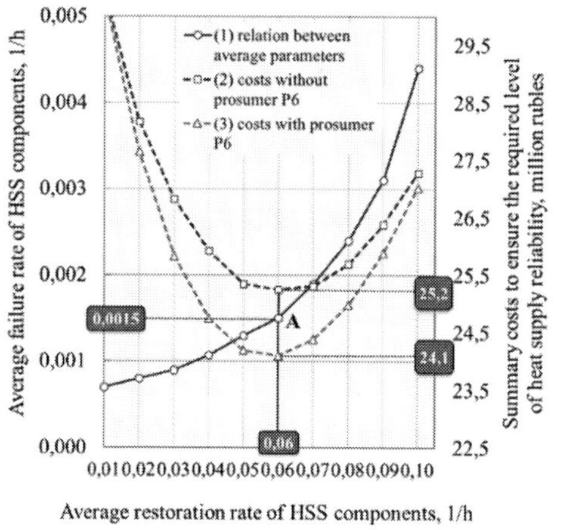

Fig. 4. Search for an optimal relationship between average reliability parameters of HSS components

The obtained solution (point A) corresponds to the costs necessary to ensure reliability in the amount of 25,25 million rub without participation of HS of prosumer P6 and 24,1 million rub with the prosumer participation; i.e. an economic effect from the involvement of one prosumer in the considered system makes up 1.15 million rub, or 4.5%. Thereby, the values of the average reliability parameters are: failure rate is 0.0015 1/h, and restoration rate equals 0,06 1/h.

4 Conclusion

The relevance of the research into heat prosumer is explained by objective heat supply problems related to low cost-effectiveness of operating systems, insufficient quality and reliability of heat supply to consumers, particularly due to technological development and growing loads.

The research formulates the statement of DHS reliability problem considering the prosumer functions as a way to provide additional capacity and time reserve owing to the prosumer HS. To solve this problem, we applied nodal reliability indices, Markov random process, models of the theory of hydraulic circuits and some basic laws of cogeneration and thermal-physical processes in heat supply.

The main advantage of the proposed technique lies in the integration of measures aimed at reducing the failures and enhancing restoration of the components in a joint procedure of search for reliability parameters, which makes it possible to most rationally distribute the total potential enabling the increase in reliability of DHS components.

The practical applicability of the developed methods for optimization of reliability parameters of DHS components is confirmed by calculations carried out on a test scheme. The calculation for the case with prosumer participation shows some economic effect and fulfillment of system reliability requirements.

The result obtained in the test calculations is not final to provide the rationale for the involvement of prosumer in ensuring the reliability of existing DHS. Further research aimed at a comprehensive analysis of the aspects of prosumer operation in the DHS (operation costs of prosumer sources, analysis of their operation, heat production cost cutting in the system, and other aspects) make it possible to formulate all the necessary criteria to substantiate the involvement of prosumers in DHS.

The research was performed at Melentiev Energy Systems Institute SB RAS under the support of Russian Science Foundation (Grant №17-19-01209).

References

1. Ni Zhang, Yu Yan, Wencong Su. *Applied Energy*, **vol. 154**, p. 471–479, (2015).
2. Luka Perkovic, Hrvoje Mikulcic, Neven Duic. *Journal of Cleaner Production*, **xxx**, p. 1–12, (2016).
3. Hongming Yang, Tonglin Xiong, Jing Qiu et al. *Applied Energy*, **vol. 167**, p. 353–365, (2016).
4. Rehman Zafar, Anzar Mahmood, Sohail Razzaq et al. *Prosumer based energy management and sharing in smart grid // Renewable and Sustainable Energy Reviews*, in press, available online 11 July 2017.
5. Stig Ødegaard Ottesen, Asgeir Tomasgard, Stein-Erik Fleten. *Energy*, **vol. 94**, p. 828–843, (2016).
6. Dimitrios J. Vergados, Ioannis Mamounakis, Prodromos Makris, Emmanouel Varvarigos. *Sustainable Energy, Grids and Networks*, **vol. 7**, p. 90–103, (2016).
7. Lakshmi Prakash, Sugatha Kumari P.R, Sharanya Chandran et al. *Procedia Technology*, **vol. 21**, p. 338–344, (2015).
8. Lisa Brange, Jessica Englund, Patrick Lauenburg. *Applied Energy*, **vol. 164**, p. 492–500, (2016).
9. Lisa Brand, Alexandra Calvén, Jessica Englund et al. *Applied Energy*, **vol. 129**, p. 39–48, (2014).
10. *Reliability of heat supply systems* (Novosibirsk: Nauka, 2000. – 360 p).
11. E.V. Sennova, V.G. Sidler. *Mathematical modeling and optimization of developing heat supply systems* (Novosibirsk, 1985. – 222 p).
12. V.Y. Sokolov *District heating and heat networks* (M.: MEI Publishing House, 1999. – 472 p).
13. A.P. Merenkov V.Y. Khasilev *Theory of hydraulic circuits.* (M.: Nauka, 1985. – 278 p).
14. *Reference book on general models for analysis and synthesis of energy system reliability/ Ed. by Y.N.Rudenko. – M.: Energoatomizdat, 1994. – 474 p.*
15. Chervonny A.A., Lukiyashchenko V.I., Kotin L.V. Reliability of complex systems. 2nd revised and corrected edition. (Mashinostroenie, 1976. – 288 p).
16. A.M. Plovko, S.V. Gurov *Foundations of the reliability theory.* (St. P.: BHV-Peterburg, 2006. – 702 p).
17. V.A. Kashtanov, A.I. Medvedev *Theory of reliability of complex systems.* (M.:Fizmatlit, 2010. – 608 p).
18. V.A. Stennikov, I.V. Postnikov *Izv. RAN, Energetika.* **No.2**, M.: Nauka, p.107-121, (2011).
19. V.A. Stennikov, I.V. Postnikov *Power Technology and Engineering.* **Vol. 47**, No. 6, March, p. 446–453, (2014).
20. V.A. Stennikov, I.V. Postnikov *Sustaining power resources through energy optimization and engineering, Vasant P., Voropai N.I., editors. – Hershey PA: Engineering Science Reference* (an imprint of IGI Global), p. 102–126, (2016).
21. Ivan V. Postnikov, Valery A. Stennikov, Ekaterina E. Mednikova, Andrey V. Penkovskii. *Energy Procedia* **105**, P. 3083–3088. (2017).
22. Ekaterina E. Mednikova, Valery A. Stennikov, Ivan V. Postnikov. *Energy Procedia* **105**, P. 3152–3157, (2017).

 DOI: 10.1051/e3sconf/20172502010

Identification of lines of electric lines of three-phase distribution networks in the composition of ASMAE

Turatbek Omorov[1,*], *Beishenaly* Takyrbashev[2], and *Kubanychbek* Zakiriaev[1]

[1] National Academy of Sciences, Bishkek, Kyrgyz Republic
[2] PC "National electric network of Kyrgyzstan", Chui enterprise of high-voltage electrical networks, Bishkek, Kyrgyz Republic

Abstract. The problem of protection of a three-phase four-wire distribution network (DEN) with a voltage of 0.4 kV from the interruption of electric power lines is considered. On the basis of the analysis of the values of complex resistances of interpersonal sections of the main line of the network, criteria are proposed that allow detecting critical situations associated with discontinuities in power transmission lines. Using these criteria makes it possible to localize the places of breaks of phase and neutral wires. Mathematical models and methods that are used to evaluate the complex resistances of interpersonal sections of a three-phase network are briefly considered. The obtained results are oriented for use in the automated system of electricity control and accounting (ASMAE).

1 Introduction

At present, automated systems for monitoring and accounting of electric power (ASMAE) are widely implemented in automation and informatization of power consumption processes in distribution electrical networks (DEN) with a voltage of 0.4 kV [1-4]. The main structural and functional elements of these systems are data concentrators (DC), a set of electricity meters (EM) installed at the subscribers of the network. For the exchange of technological and service information between the subsystems of the ASMAE in the DC and EM there are telecommunication modules (TCM) built using modern data transmission technologies (GSM, Zigbee, PLC, etc.). The main functions of the DC: operational data collection from a group of electricity meters (EM); storage of received data; digital processing of information to solve relevant functional problems; information exchange with the upper level of management. In general, it should be noted, that in the framework of ASMAE, the tasks of commercial metering of electricity are being solved. At the same time, in order to increase the efficiency and intellectual level of automated systems, it is advisable to solve such important functional tasks as operational diagnostics of states [5-7] and optimization of operation modes of DEN [8-10]. Among the diagnostic tasks, it is especially important to identify the problem of identification and localization of breakage places of electric lines of a three-phase distribution network. To date, a number of approaches and tools for its solution have been developed [7].

The paper proposes one of the possible methods for solving it as part of ASMAE using real-time data from electricity meters installed at the network subscribers.

2 Method of solution

Three-phase DEN with a voltage of 0.4 kV is considered, the design scheme of which is shown in the fig.1. For convenience in the scheme, the phases A, B, C are numbered respectively through the index variable ($k = \overline{1,3}$). The remaining notations have the following meaning: E_A, E_B, E_C – instantaneous phase EMF; $Z_{\nu k}$ - is the designation of the ν - th load (electric receiver) ($\nu = \overline{1,n}$), connected to the phase with the number k; $I_{\nu k}$, $U_{\nu k}$- instantaneous current and voltage on the load $Z_{\nu k}$; $\iota_{\nu k}$, $z_{\nu k}$ – instantaneous current and resistance of the ν-th interpersonal section (IS) of the k-th phase; $u_{\nu k}$, u_ν - instantaneous voltages respectively on the ν- th IS of the k-th phase and the neutral wire; J_ν, z_ν- instantaneous current and resistance of the ν-th section of the neutral wire; U_{0k}, ι_{1k} – are instantaneous voltages and currents, respectively, at the inputs of the respective phases.

Further, it is assumed that the following conditions are satisfied:

1) DEN operates in an asymmetric mode;

2) the phase and neutral wires of the network can have different cross sections, i.e. the resistances $z_{\nu k} \neq z_\nu$ ($k = \overline{1,3}$, $\nu = 1, n$), which are a priori unknown;

3) the system uses technical means to suppress the higher harmonic components of currents and voltages in the network;

* Corresponding author: omorovtt@mail.ru

DOI: 10.1051/e3sconf/20172502010

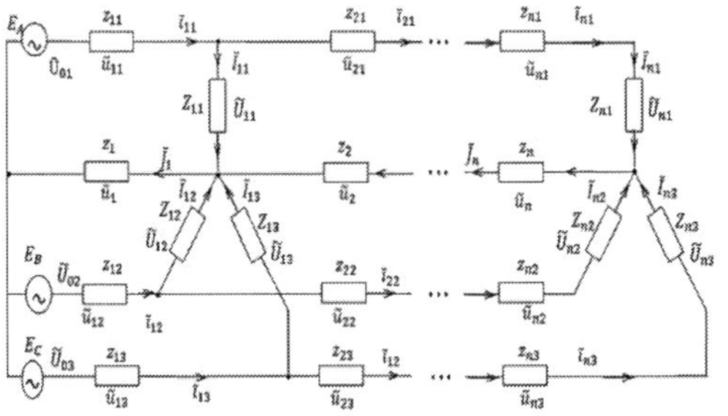

Fig.1. The design scheme of a three-phase DEN.

4) from electricity meters ($Em_{\nu k}$) to the concentrator (DC) via communication channels at discrete instants $t \in [t_\xi, t_{\xi+1}]$ with a sampling rate $\Delta t_\xi = t_{\xi+1} - t_\xi$ ($\xi = 1,2, \dots$) the following data arrive:

- current values of currents $I_{\nu k}$ and voltages $U_{\nu k}$ on loads $Z_{\nu k}$;
- power factors $cos\, \varphi_{\nu k}$, o determined by the phase shifts $\varphi_{\nu k}$ between the corresponding voltages $U_{\nu k}$ and currents $I_{\nu k}$.

Note that the electrical state of the network at time $t \in [t_\xi, t_{\xi+1}]$ is described by the following vectors:

$$I_k = \; I_{1k}, I_{2k}, \dots, I_{nk} \;, \qquad U_k = \; U_{1k}, U_{2k}, \dots, U_{nk} \;,$$
$$u'_k = \; u_{1k}, u_{2k}, \dots, u_{nk} \;, \; \imath_k = \; \imath_{1k}, \imath_{2k}, \dots, \imath_{nk} \;,$$
$$u = \; u_1, u_2, \dots, u_n \;, \qquad J = \; J_1, J_2, \dots, J_n \;,$$
$$Z_k = \; z_{1k}, z_{2k}, \dots, z_{nk} \;, \qquad Z_N = \; z_1, z_2, \dots, z_n \;, k = \overline{1,3},$$

where $I_{\nu k}$, $U_{\nu k}$, $u_{\nu k}$, $\imath_{\nu k}$, u_ν , J_ν – are complex representations of the corresponding electrical variables indicated on the three-phase DEN circuit. The task is as follows:

1) define the vectors I_k, U_k, u'_k, $\imath_k$, u, J, Z_k, Z_N, describing the electrical state of the network;

2) to identify the coordinates of the breaks of the phase and neutral wires of DEN.

Under the condition that the DEN operates in the normal (state) mode, the of determining the problem of the network is solved, in particular, the interpersonal complex currents $\imath_{\nu k}$ for k-th phase, as well as the resistance values z^*_ν of neutral wire segments. Further on the basis of these data we make the following vectors:

$$i^*_k = \; i^*_{1k}, i^*_{2k}, \dots, i^*_{nk} \;, \qquad Z^*_N = [z^*_1,\; z^*_2, \dots ,\; z^*_n].$$

We note that the values of the base resistances z^*_ν, $\nu = 1, n$, very slowly depending on the current climatic conditions and for a sufficiently long period of time remain practically constant, i.e $z^*_\nu = const$.

As is known, there are two types of breaks in power electric lines in the DEN:

1) breaks of phase lines;

2) a neutral wire breaks.

In the first case, when the wire of the q-th section of the k-th phase is broken, the vector i_k has the following form:

$$i_k = \; i_{1k}, i_{2k}, \dots i_{q-1,k}, 0, 0, \dots, 0 \;. \qquad (1)$$

In the second case, if the wire of the η-th section of the neutral wire is broken, the vector Z_N has the following form:

$$Z_N = [z^*_1,\; z^*_2, \dots, z^*_{\eta-1}, z_\eta, z^*_{\eta+1}, \dots, z^*_n].$$

The analysis shows that in this case the following relations hold for the components of the vector Z_N:

$$z_\eta \neq z^*_\eta, \qquad (2)$$
$$z_\nu = z^*_\nu, \qquad \nu = 1, 2, \dots, \eta - 1, \eta + 1, \dots, n,$$

where z_η – is the current resistance value determined for the considered discrete time moment t. Thus, second case, when condition (2) is satisfied, the vectors Z^*_N и Z_N are not equal, i.e.

$$Z_N \neq Z^*_N. \qquad (3)$$

Obviously, in the normal operation mode of the DEN (in the absence of line breaks), the condition $Z_N = Z^*_N$ is satisfied with a certain accuracy.

As a result, to localize breaks of phase and neutral network wires, the following criteria can be formulated:

- **Criterion 1:** when the wire of the q-th section of the k-th phase breaks off for the vector $\imath_{\nu k}$ the condition (1).
- **Criterion 2:** if the neutral wire η-th section is cut off, condition (3) is fulfilled.

It should be noted that in order to identify and localize breaks in electrical lines of DEN on the basis of the formulated criteria, it is necessary to solve the problem of determining the electrical state of a DEN. Because of its complexity, the use of known methods for calculating the three-phase network [11-14] presents certain difficulties. In this connection, in [4, 6, 15], the methodological and algorithmic bases of estimating the variables and network parameters, oriented for application in the ASMAE structure, were proposed. The basis of this research area is the mathematical model [15], which allows currents and voltages on the network loads to be presented in a complex form:

$$I_{\nu k} = I^{в}_{\nu k} + jI^{м}_{\nu k} = I_{\nu k} e^{j\alpha_{\nu k}}, \qquad (4)$$

$$U_{\nu k} = U^{в}_{\nu k} + jU^{м}_{\nu k} = U_{\nu k} e^{j\psi_{\nu k}},\; \nu = 1, n,\; k = 1,3, \;\; (5)$$

where the symbols "b" and "m" denote the real and imaginary parts of the corresponding complex variables; $I_{\nu k}, U_{\nu k}, \alpha_{\nu k}, \psi_{\nu k}$ – are the moduli (acting currents and voltages) of the corresponding complex variables and their phase shifts, respectively; $j = \overline{-1}$ – is an imaginary number.

The representations (4) and (5) make it possible to determine the complex currents $i_{\nu k}$ и J_ν, occurring in the interpersonal sections of the network, as well as the corresponding voltages $u_{\nu k}$ и u_ν. The corresponding models and algorithms are described in [4, 6, 15]. Knowledge of these data allows us to determine with some accuracy the estimates of the current values of the

DOI: 10.1051/e3sconf/20172502010

components of the vector parameters $Z_k = z_{1k}, z_{2k}, \dots, z_{nk}$ and $Z_N = z_1, z_2, \dots, z_n$ of the three-phase network:

$$z_{\nu k} = u_{\nu k} \ \iota_{\nu k}, \qquad \nu = \overline{1,n}, \quad k = 1,3,$$
$$z_\nu = u_\nu \ J_\nu \qquad \nu = \overline{1,n}.$$

3 Conclusion

Implementation of the above procedures (algorithms) for estimating the electrical state of the network is carried out in a data concentrator (DC) in real time.

The proposed models and computational algorithms for localizing faults of phase and neutral wires of a three-phase distribution network based on the formulated criteria are oriented to create special software for the corresponding subsystem within the ASMAE system.

References

1. A. Ozhegov. ASMAE systems (Kirov, VyatSU, 2006)
2. A. Sapronov, S. Kuzhekov, V. Tyniansky. *Izv.vuzov. Electromechanics*, **1**, 55 (2004)
3. A. Soldatov, U. Evdokimov. *Devices and systems. Management, control, diagnostics*, **3**, 1 (2017)
4. T. Omorov, B. Takyrbashev. *Electricity*, **11**, 4 (2016)
5. A. Soldatov. *Devices and systems. Management, control, diagnostics*, **6**, 1 (2016)
6. T. Omorov, B. Takyrbashev. *Control. Diagnostics*, **5** (2017)
7. A. Ershov, O. Filatov, A. Mlotok etc. *Electric station*, **5**, 28 (2016)
8. V. Idelchik. *Calculations and optimization of modes of electrical networks and systems* (M. Energoatomizdat, 1988)
9. T. Omorov, B. Takyrbashev. *Instruments and systems, Management, control, diagnostics*, **6**, 11 (2016)
10. F.D. Kosoukhov, N.V. Vasilev, A.O. Filippov *Electrical engineering*, **6**, 8 (2014)
11. K. Demirchyan, L. Neiman, A. Korovkin. *Theoretical bases of electrical engineering* (Spb. Peter, 2009)
12. U. Zhelezko. *Loss of electricity. Reactive power. The quality of electricity* (M. EHAS, 2009)
13. A. Harutyunyan. *Electricity*,**10**, 55 (2015)
14. Kochergin, A. Kobelev, N. Khrebtov, P. Kitashin, K. Terekhov. *Fractal simulation*. **1**, 5 (2013)
15. T. Omorov, B. Takyrbashev, R. Osmonova. *Energetik*, **4**, 28 (2017)

E3S Web of Conferences **25**, 03001 (2017) DOI: 10.1051/e3sconf/20172503001
RSES 2017

Using of sensitivity models of the steady-state of electric power systems for account of failure of elements

Dmitry Krupenev[1*]

[1]Melentiev Energy Systems Institute, 130 Lermontov str., Irkutsk, Russia

Abstract. The article presents an algorithm for obtaining sensitivity models of the first and second orders of the steady-state regime of electric power systems (EPS). The sensitivity models are intended for express calculations of steady-state when estimating the static security of EPS. The use of sensitivity models allows one to simulate failures of EPS elements without calculating new steady-state. To verify the reliability of the sensitivity models obtained, the results of an experiment performed on a 3-node test pattern are presented.

1 Introduction

For minimize of the consequences of electrical equipment failures in electric power systems (EPS), it is necessary to made a set of specific actions, one of which is an operational estimation of the security of EPS. The security estimation of EPS is determine of the indicators that most fully characterize the current steady-state from the standpoint of opposing possible disturbances in EPS. Estimation of the security of EPS is technologically capacious and computationally expensive. There are the static and the dynamic security of EPS. In this paper, we describe static security. The estimation of a static security of EPS is based on the assumption that the transition of EPS to a new state due to an element failure is not accompanied by a violation of dynamic stability.

When we estimate of security of EPS very important parameters are accuracy and time of the estimation, since the estimation should be carried out in real time. Various simplifications when we solve this problem can lead to distortion of the result and, ultimately, acceptance of a non-optimal or incorrect set of control actions.

There are some approaches to the estimation of security of EPS. We can divide such approaches as the approach based on the application of the criterion n-i [1-3], the machine-based approach [4], the Jacobi matrix based approach [5], In [6] to estimate the security of EPS using the Monte Carlo method. In most of the present approaches (methods), one of the main stages of estimation, and the most time-consuming expense, is the stage of modeling the effect of failures of power equipment on the EPS parameters or the calculation of steady-state of EPS at failures of its elements. When we estimate of security of EPS, it is necessary to analyze single and group failures of EPS elements in a minimal time. EPS consists of a many of elements, the number of possible states is large. In this paper, we study the problems of accelerating the calculation of steady-state of such steady-states of EPS. We propose to use steady-state sensitivity models or to replace the exact model of calculating of steady-state of EPS by its approximation.

Models of first-order sensitivity of steady-state of EPS are used in solve deferent electric power problems, among them one can single out the problem of analysis of voltage stability of EPS [7]. In [8] the first-order sensitivity models (matrices) are used to calculate the steady-state of EPS when to change the reactive power at the nodes of the EPS. So, in [7], [8], using the models of the first-order sensitivities, linearly approximate the steady-state and search for the steady-state parameters of the EPS during the introduction of the perturbation. The aim of this paper is to obtain sensitivity models of not only the first, but also the second order for the rapid calculation of the steady-state of the EPS. Models of second-order sensitivity give a quadratic approximation to the new steady-state point, which increases the accuracy of calculating the steady-state and, accordingly, the accuracy of the estimation of the security of EPS.

2 Formation of sensitivity models of the steady-state of electric power systems

The steady-state of EPS can be described by a vector equation of the form [9]:

$$W(X,V)=0, \quad (1)$$

where: W – vector-function, $W : G \to \mathbb{C}^{n+m-1}$, $G \subseteq \mathbb{C}^{n+m-1}$; X – vector of input parameters of the EPS steady-state, $X \in \mathbb{C}^{n}$; V – vector of output parameters of the EPS steady-state, $V \in \mathbb{C}^{m-1}$, – number of power lines of EPS; n – number of nodes of EPS.

When we deal estimate of security of EPS, we should to find the dependencies of some parameters on others,

---

* Corresponding author: krupenev@isem.irk.ru

DOI: 10.1051/e3sconf/20172503001

and the set of input and output parameters is not known in advance.

In accordance with the implicit function theorem [10] for Banach spaces [11], to which the space of complex numbers $\mathbb{C}$, which characterize the parameters of steady-state operation of EPS, if the mapping $W : O \rightarrow \mathbb{C}^{n+m-1}$, defined in the neighborhood of the O point $(x_0, v_0) \in \mathbb{C}^{n+m-1}$, $x \in \mathbb{C}^n$, $v \in \mathbb{C}^{m-1}$, such is that

1. $W \in \mathbb{C}^{(1)}(O, \mathbb{C}^n)$ (vector function is continuously differentiable),
2. $W(x_0, v_0) = 0$ (the equation has a solution at the point (x_0, v_0)),
3. $W'_v(x_0, v_0)$ – reversible matrix,

then, for any point from the neighborhood (x_0, v_0)

$$W(x,v) = 0 \Leftrightarrow v = f(x)$$

$$f'(x) = -[W'_v(x, f(x))]^{-1}[W'_x(x, f(x))], \quad (2)$$

The expression (2) is first-order sensitivity matrix of steady-state EPS [12-14]. To determine the change in the vector of output variables v relative to the vector of the input variables x, it is necessary to find the corresponding sensitivity matrix or differential in accordance with relation (2).

Let's write the system of steady-state equations in the vector form to obtain the sensitivity models.

$$\overline{A}I = (\widetilde{U})^{-1}S, \quad (3.a)$$

$$A^T U = \widetilde{Y}^{-1}I, \quad (3.b)$$

where: A is the incidence matrix of an EPS graph with size of $m \times n$; $\overline{A}$ is the truncated incidence matrix by the basis node; I is a vector that composed of the components which are the power flows along the power transmission lines of EPS, $I \in \mathbb{C}^n$; S is the vector whose components are the power in the nodes of EPS (the residual between generation and demands) $S \in \mathbb{C}^{m-1}$; U is a vector that composed of the components which are the nodes voltage of EPS, $U \in \mathbb{C}^{m-1}$; $\widetilde{U}$ is a diagonal matrix with $m-1$ order of the EPS nodes voltage; $\widetilde{Y}$ is a diagonal matrix with n order of EPS lines conductivity.

The system (3.a), (3.b) will be called as the initial model. This record of the vector equations system of EPS steady-states is not complete, because of these equations are nonlinear and there are solution set that appear in case of the solution this system. For example in determining the voltages in EPS nodes, the number of solutions is 2^{m-1}. Therefore, to make the system with a single solution, it is necessary to supplement this system of equations by an additional condition:

$$Re\ U > 0. \quad (4)$$

Moreover, to obtain a sensitivity model of 1st dimension the vector equations system (3) should be differentiated by variable that change own value during the process of reliability assessment. This variable depend on current problem. It is worth to be noting that the EPS elements failures and random deviations of consumers demand are simulating during the process of the system reliability assessment. Hence, it is required to differentiate the system (3.a), (3.b) by the Y (vector of lines conductivity) for the EPS lines failure simulating (modeling) and by S for the simulating (modeling) of random deviations of consumers demand and failures of generating blocks. In this work we will focused on the differentiation by Y. With the aim of uniquely defined the differentials $d_Y I$ (differential of I vector by Y variable) and $d_Y U$ (differential of U vector by Y variable) it is necessary to fix the diagonal matrix $\widetilde{U}$ at the solution point, before differentiation of the vector equations system (3.a), (3.b). The values of voltages in nodes will be located on the diagonal of this matrix.
on the right-hand side.

$$\overline{A}d_Y I = 0, \quad (5.a)$$

$$A^T d_Y U = -\widetilde{Y}^{-1}(d\widetilde{Y})\widetilde{Y}^{-1}I + \widetilde{Y}^{-1}d_Y I, \quad (5.b)$$

where: $d_Y I \in \mathbb{C}^n$; $d\widetilde{Y} \in \mathbb{C}^n$; $d_Y U \in \mathbb{C}^{m-1}$.

The differentiation of the inverse matrix function was carried out according to [15]: $d(\widetilde{Y}^{-1}) = -\widetilde{Y}^{-1}d\widetilde{Y}\widetilde{Y}^{-1}$.

Next, let us convey from equation (5.b):
on the right-hand side.

$$d_Y I = \widetilde{Y}A^T d_Y U + (d\widetilde{Y})\widetilde{Y}^{-1}I. \quad (6)$$

Replace the $d_Y I$ from (6) to (5.a):

$$\overline{A}\widetilde{Y}A^T d_Y U = -\overline{A}(d\widetilde{Y})\widetilde{Y}^{-1}I, \quad (7)$$

as follows from (7) equation we will take:

$$d_Y U = -(\overline{A}\widetilde{Y}A^T)^{-1}\overline{A}(d\widetilde{Y})\widetilde{Y}^{-1}I = -(\overline{A}\widetilde{Y}A^T)^{-1}\overline{A}\widetilde{Y}^{-1}\widetilde{I}dY, (8)$$

Where $\widetilde{I}$ is the diagonal matrix, where the power flows along the power transmission lines of EPS located on the diagonal, $I \in \mathbb{C}^n$.

The matrix permutation was carried out in the mathematical expression (8) that is possible for diagonal matrices.

The linearized dependence of the voltages in the nodes of the EPS on the conductivities change in the elements of the EPS can be represented in the form $U \approx U_0 + d_Y U$, where U_0 is a column vector whose components are values of the voltages in the nodes of the EPS at the solution point $U_0 \in C^{m-1}$. The column vector ΔY will be used instead dY for defining the differential $d_Y U$. That column vector composed of the components which are increments of the conductivities on the corresponding lines $\Delta Y \in C^n$.

Let's substitute $d_Y U$ (8) into (6) for to get $d_Y I$ .Then

$$d_Y I = \tilde{Y} A^T (\bar{A} \tilde{Y} A^T)^{-1} \bar{A} (d\tilde{Y}) \tilde{Y}^{-1} I + d\tilde{Y} \tilde{Y}^{-1} I. \quad (9)$$

The linearized dependence of the change of power flows along the EPS power transmission lines with depending on conductivities change can be represented and look like $I \approx I_0 + d_Y I$, where I_0 is column vector, whose components are of power flow values along the power transmission lines in the solving point, $I_0 \in C^n$.

Besides, to get the sensitivity models of 2nd order it is necessary to get the second differentials. Hence, the first differentials (8) and (9) must be differentiated. In summary, the second order differentials will have the next form:

$$d^2_{YY} U = 2(\bar{A} \tilde{Y} A^T)^{-1} \bar{A} d\tilde{Y} A^T (\bar{A} \tilde{Y} A^T)^{-1} \bar{A} \tilde{I} \tilde{Y}^{-1} dY, \quad (10)$$

$$d^2_{YY} I = d\tilde{Y} A^T d_Y U + \tilde{Y} A^T d^2_{YY} U - \tilde{Y}^{-2} \tilde{I} d\tilde{Y} dY. \quad (11)$$

In this case the 2nd order sensitivity models, namely the dependence of voltage in EPS nodes on conductivity change in EPS elements can be represent as: $U \approx U_0 + d_Y U + \frac{1}{2} d^2_{YY} U$, the dependence of power flows along the EPS power transmission lines on EPS elements conductivity change could be can be represent as: $I \approx I_0 + d_Y I + \frac{1}{2} d^2_{YY} I$.

In event of EPS lines failure the disturbance of voltages and streams along the power transmission lines of EPS will be determined as follows:

$$\Delta_{YY} U \approx d_Y U + \frac{1}{2} d^2_{YY} U, \quad (12)$$

$$\Delta_{YY} I \approx d_Y I + \frac{1}{2} d^2_{YY} I \quad (13)$$

where: $\Delta_{YY} U$ is a vector of voltages change in EPS nodes, $\Delta U \in C^{m-1}$, $\Delta_{YY} I$ is a vector of EPS power flows change, $\Delta I \in C^n$.

Thus, vector ΔY that characterizes the values of conductivities that equivalent to switching off the transmission line in equations (8-13) remains undefined. However, the conductivity Y_i power transmitting line i has a definite value during the normal functioning. In case of power transmission line i failure the value of its conductivity become equal 0, $0 = Y_i + \Delta Y_i$. In summary the value of conductivity that equivalent of power transmission line i disconnections will be defined as:

$$\Delta Y_i = 0 - Y_i = -Y_i \quad (14)$$

During the use sensitive models for the steady-states of EPS calculation by one iteration it is possible to get the vector of voltage changes in EPS nodes in case of single and non-single disconnections of power transmission lines. It could be done if represent the ΔY as a matrix $\Delta \tilde{Y}$ (instead vector) with size nЧn and values of $-Y_i$ by the main diagonal, of course if it is necessary to get the assessment of single failures of EPS power transmission lines, and other zero elements. If the assessment of multiple failures required then it is necessary to fill matrix $\Delta \tilde{Y}$ in a certain way.
preparing them.

3 Case study

The sensitivity models of the first and second orders applications are demonstrated for steady state calculation of the three-node EPS test scheme shown in fig.1. The experiment includes power line I capacity changing, with its final shutdown.

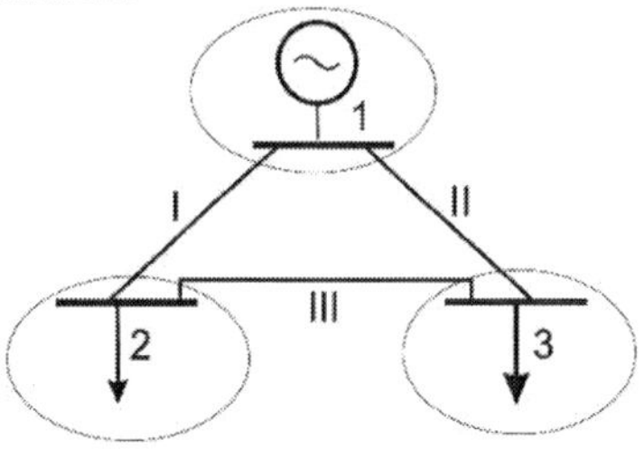

Fig. 1. EPS test scheme.

The main characteristics of EPS necessary for steady state calculation are shown in table 1.

Table 1. The initial data of the investigated EPS.

Power line	Impedance, Ω	Node	Load, MVA	Voltage, kV
I	12,1+ i43,5	1	0	220,0+ i0
II	12,1+ i43,5	2	329,0-i114,0	-
III	12,1+ i43,5	3	669,0-i173,0	-

As a result of the steady-state calculation of the initial model (3.a), (3.b), the following voltages: U_2 = 234,94 - i7,73 kV; U_3 = 213,67 - i71,69 kV, and power lines flows: I_{12} = 0,08 + i0,36 kA, I_{13} = 1,57 + i0,29 kA, I_{23} = 1,49 - i0,07 kA were obtained at the nodes of the EPS.

Further, the voltages at nodes 2 and 3 were determined using the obtained sensitivity models of the first and second orders for taken power line I capacity perturbations increments. In table 2 calculation results are demonstrated when power line I capacity changing on the levels: 1; 5; 9; 34; 48; 67; 86; 90; 95 % of the initial power line capacity. Calculation result when power line I is disabled (capacity perturbation is 100%) is shown in the last row of table 2.

DOI: 10.1051/e3sconf/20172503001

Table 2. Voltage deviation in test scheme nodes on power lines I-II capacity perturbation.

Power line capacity perturbation		U (init.mod.), kV U_2= U_3=	$\Delta_Y U$, kV $\Delta_Y U_2$= $\Delta_Y U_3$=	U *(lin.mod.)*, kV U_2= U_3=	$\Delta_{YY} U$, kV $\Delta_{YY} U_2$= $\Delta_{YY} U_3$=	U *(quad.mod.)*, kV U_2= U_3=
% of initial.	absolute value					
1%	$-5{,}87*10^{-5}$ + $i2{,}11*10^{-4}$	235,02 - i7,78 213,72 - i71,7	0,1 - i0,05 0,05 - i0,03	235,04- i7,78 213,72 - i71,72	0,13 - 0,06i 0,06 - 0,03i	235,17 -7.81i 213,75 -71,73i
5%	$-2{,}83*10^{-4}$ + $i0{,}10*10^{-2}$	235,32 - i8,08 213,89 - i71,8	0,47 - i0,25 0,24 - i0,12	235,41- i7,98 213,91 - i71,81	0,6 - 0,31i 0,3 - 0,16i	235,72 -8.13i 214,06 -71,89i
9%	$-5{,}4*10^{-4}$ + $i0{,}19*10^{-2}$	235,75- i8,22 214,1 - i71,81	0,91 - i0,47 0,45 - i0,23	235,85 - i8,2 214,12 - i71,92	1,15 - 0,59i 0,57 - 0,3i	236,42 -8.5i 214,41 -72,07i
34%	$-1{,}98*10^{-3}$ + $i0{,}71*10^{-2}$	238,41 - i9,9 215,5 - i72,21	3,32 - i1,72 1,66 - i0,86	238,26 - i9,45 215,33 - i72,55	4,22 - 2,18i 2,11 - 1,09i	240,37 -10,54i 216,38 -73,09i
48%	$-2{,}97*10^{-3}$ + i0,01	240,84- i11,51 216,79 - i72,62	4,98 - i2,58 2,49 - i1,29	239,92 - i10,31 216,16 - i72,98	6,32 - 3,27i 3,16 - 1,64i	243,08 -11,94i 217,74 -73,8i
67%	$-3{,}96*10^{-3}$ + i0,014	244,03 - i13,74 218,49 - i73,21	6,64 - i3,43 3,32 - i1,72	241,58 - i11,16 216,99 - i73,41	8,43 - 4,36i 4,22 - 2,18i	245,79 -13,35i 219,1 -74,5i
86%	$-4{,}95*10^{-3}$ + i0,018	248,4- i17,02 220,86 - i74,13	8,3 - i4,29 4,15 - i2,15	243,24 - i12,02 217,82 - i73,84	10,54 - 5,45i 5,26 - 2,73i	248,51 -14,75i 220,45 -75,2i
90%	$-5{,}4*10^{-3}$ + i0,019	250,98- i19,08 222,27- i74,74	9,05 - i4,68 4,53 - i2,34	243,99 - i12,41 218,22 - i74,03	11,5 - 5,95i 5,74 - 2,97i	249,74 -15,39i 221,07 -75,52i
95%	$-5{,}65*10^{-3}$ + i0,02	252,68 - i20,5 223,21 - i75,17	9,48 - i4,91 4,74 - i2,45	244,43 - i12,64 218,41 - i74,14	12,04 - 6,23i 6,02 - 3,12i	250,45 -15,75i 221,42 -75,7i
Disabled	$-5{,}94*10^{-3}$+ i0,021	254,79 - i22,31 224,38 - i75,73	9,96 - i5,15 4,98 - i2,58	244,9 - i12.88 218,65 - i74,27	12,65 - 6,54i 6,32 - 3,27i	251,22 -16,15i 221,81 -75,9i

Graphical interpretation of the results is shown on the fig.2, namely the values of the real part of the voltage in the second node. The values of the imaginary part of the voltage in the second node are shown on the fig.3. Fig. 4 and 5 show similarly values for third node.

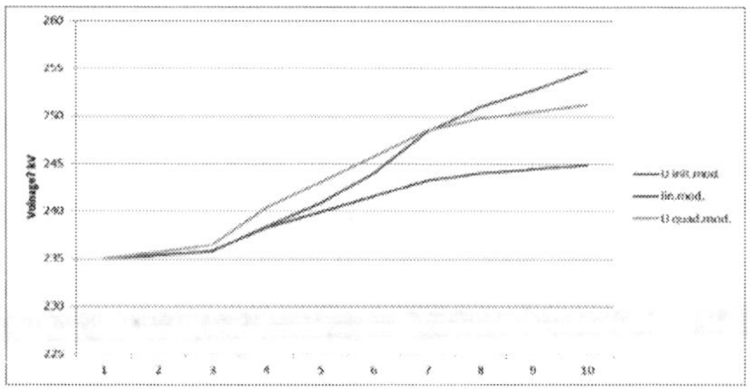

Fig. 2. The values of the real part of the voltage in the second node.

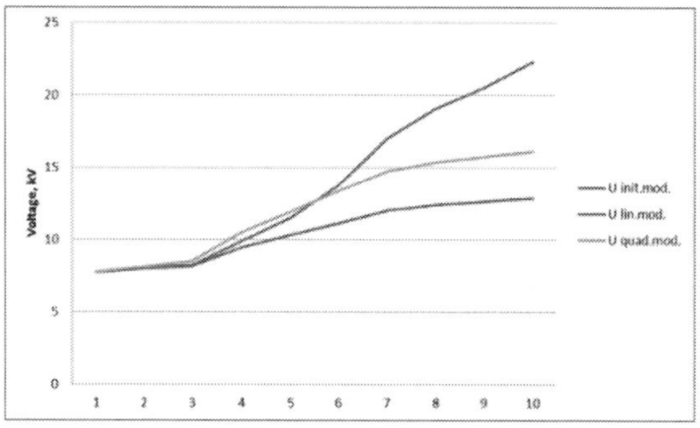

Fig. 3. The values of the imaginary part of the voltage in the second node.

DOI: 10.1051/e3sconf/20172503001

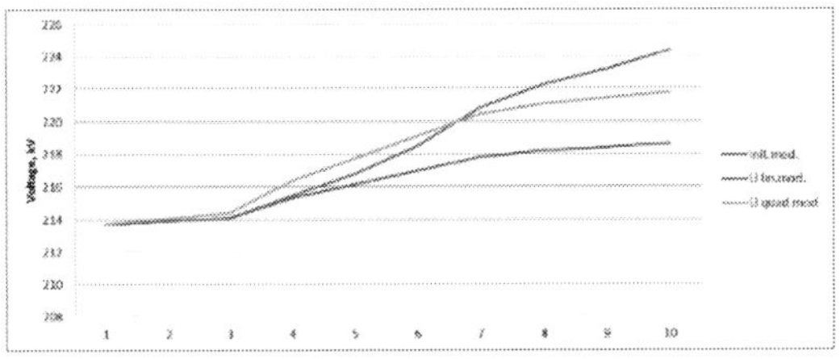

Fig. 4. The values of the real part of the voltage in the second node.

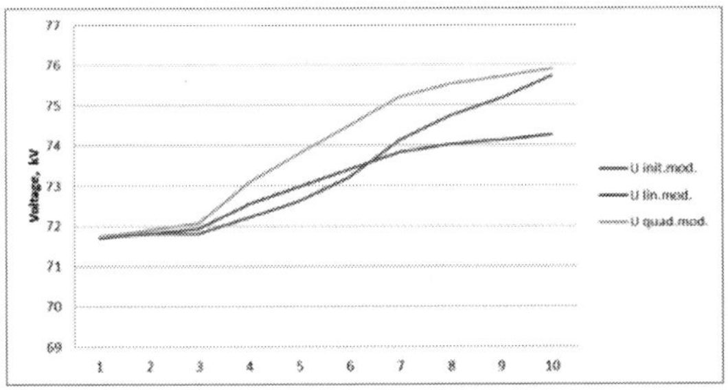

Fig. 5. The values of the imaginary part of the voltage in the third node.

As can be seen from the results, steady state linearized model using for investigated EPS with disabled power line 1 gives voltage deviation in comparison with calculation of steady state with disabled power line in 9.89kV.

When sensitivity model of second order was used deviation was 3,57 kV (1,4%) on the active part of power and 6,16 kV (27,6%) on the reactive part. The voltage deviation by modulus in the second node was 1.6%, in the third node it was 0.8%.

4 Conclusion

In this paper we consider the approximation task of the postfault EPS steady state calculation while assess its static security. Sensitivity models based on first-order and second-order differentials of EPS steady state are suggested for doing that. When security assess, it becomes necessary to calculate the set of EPS steady states because of equipment failures and deviations of customers load. Therefore, when using sensitivity models changings of powerlines capacity, customers load and generating capacity are taken as input parameters and voltages in EPS nodes and powerlines overflows as output. Some differentials of the first and second orders of the steady-state EPS are obtained.

Within the framework of a numerical experiment, investigations were carried out on a three-node EPS scheme where one of the powerlines was disabled. The values of the voltages in the EPS nodes were obtained as a result of steady state calculating on the initial model and on the sensitivity model, which characterizes the voltage deviation in the EPS nodes from the powerlines capacity changing.

The reported study was funded by RFBR according to the research projects No. RFBR 16-38-00312 мол_а

References

1. Y.N Kucherov, O.M. Kucherova, L. Kapoyi, Yu.N. Rudenko, *Reliability and efficiency of large transnational power plants. Methods of analysis: European dimension* (Novosibirsk: Science, Siberian Publishing Company, Russian Academy of Sciences, 1996).

2. Manov N.A., Chukreev Yu.Ya., *Methods and models of investigation of reliability of electric power systems* (Syktyvkar, 2010).

3. Pei-Qing Liu, Hua-Qiang Li, Yang Du, Ke Zeng *Risk assessment of power system security based on component importance and operation state, 2014 International Conference on Power System Technology*, (2014).

4. Panasetsky D., Tomin N., Voropai N., Kurbatsky V., Zhukov A., Sidorov D., *Proceedings of the 2015 IEEE Eindhoven PowerTech*, (2015).

5. Gamm A.Z., Golub I.I., *Sensors and weaknesses in electric power systems* (SEI SB RAS. 1996).

6. Domyshev A.V., Krupenev D.S. *Electricity*, **2**, (2015).

7. L.A.Ll. Zarate and C.A. Castro. *IEE Proc.-Gener. Transm. Distrib.*, Vol. **153**, (2006).

8. S. Chen, W. Huang, W. Lai, P. Shi. *Power System Fast Line Flow Calculation for Security Control by Sensitivity Factor Second International Conference on Innovative Computing, Informatio and Control* (2007).

9. Idelchik V.I. *Calculation of steady-state modes of electrical systems* (M. Energy, 1977).

10. Zorich V.A. *Mathematical analysis. Part 1* (M, 2002).

11. Trenogin V.A. *Functional analysis* (Moscow: Science. Main edition of physical and mathematical literature. 1980).

12. Epifanov S.P., Novitsky N.N. *Proceedings of the XIIIth Baikal International School-Seminar "Optimization Methods and their Applications"*, Irkutsk: ISEM SB RAS, Vol. **5**, (2005).

13. Epifanov S.P., Novitsky N.N., Borovin D.I. *Pipeline Energy Systems: Methodological and Applied Problems of Mathematical Modeling*, Novosibirsk: Science (2015).

14. Rainshke K. *Models of reliability and sensitivity of systems* (Moscow: "THE WORLD". 1979).

15. Magnus J.R., Neidekker H. *Matrix differential calculus with applications to statistics and econometrics* (Moscow, FIZMATLIT, 2002).

DOI: 10.1051/e3sconf/20172503002

Construction of equivalents of large equations systems of electrical network based on matrix annulators

Magomed Gadzhiev [1,*], *Misrikhan* Misrikhanov [1], *Vladimir* Ryabchenko [1], *Yuriy* Sharov [1]

[1]National research university "MPEI",Institute of Electrical Power Engineering, 14 Krasnokazarmennaya St, Moscow, Russia

Abstract. An original method of reducing the equations of node voltages is proposed with the aim of equivalence of the electric network. The method is based on the matrix transformations of the indicated equations with the help of matrix annulators. It is shown that this method, in comparison with the traditional one, makes it possible to improve the conditionality of the solved equations by an order of magnitude or more. This has a significant effect on the numerical stability of the resulting electrical network equivalents. The results of reduction of a small, large, and very large system of nodal stress equations are presented.

1 Introduction

It is well known that the feature of calculation modes in electric power systems (EPS) is their multiparametricity. The solution of complete systems of equations covering all the nodes and connections of large power systems, even with simplified models, because of the lack of reliable information on all elements of the network, poses a serious problem, which is complicated by the need to accumulate and store large amounts of information [1] and the corresponding increase in the requirements for the speed of computers [2].

The need to reduce the counting time appears most when carrying out multivariate and multimode calculations in operational control, as well as in planning problems for EPS regimes [3, 4]. Equivalence allows not only to reduce the time of solving the nodal electrical voltage equations (NEVE), but also to reconcile the amount of information and its error.

In a broad sense, the equivalence of EPS is in the transformation (reduction) of a complex mathematical model into a simpler one while preserving the most important (required) properties within a given accuracy. With this approach, the toolkit of Krylov subspaces is widely used [5]. In the narrower sense, with respect to the calculation of EPS modes, equivalence is reduced to the transformation of the replacement circuit and its parameters to a species having a smaller number of nodes and branches and suitable for modeling of the initial EPS modes.

The report proposes an original method of UUN reducing with the purpose to equate an electrical network. The method is based on the NEVE matrix transformations with the help of algebraic objects, called annihilators of matrices.

This method, in comparison with the traditional algorithm based on the Schur complement, makes it possible to improve significantly the conditionality of the solved equations, especially for large (up to 10 000 equations) and very large (up to 100 000 equations) systems of NEVE. It has a significant effect on the accuracy of the received electrical network equivalents and, consequently, on the correctness of numerical models.

2 The traditional approach for NEVE equivalenting

For equivalenting procedure excluded nodes (set *M*) and nodes stored in equivalent units (set *N*) are given [2].

The nodes were renumbered so that from the first node to the *n*-th node there were nodes *N*, then all *m* excluded nodes of the set *M*. In this case, the NEVE were structurally divided into blocks (block matrices and sub-vectors)

$$\begin{bmatrix} Y_{NN} & Y_{MN} \\ Y_{MN} & Y_{MM} \end{bmatrix} \begin{bmatrix} U_N \\ U_M \end{bmatrix} = \begin{bmatrix} I_N \\ I_M \end{bmatrix}. \qquad (1)$$

Here, U_N, U_M – the stress subvectors; I_N, I_M – current sub-vectors in the vector

$$\begin{bmatrix} I_N \\ I_M \end{bmatrix}. \qquad (2)$$

* Corresponding author: gadzhiev_mg@interrao.ru

Sub-matrix of conductivities Y_{NN} in the conductivity matrix

$$\begin{bmatrix} Y_{NN} & Y_{MN} \\ Y_{MN} & Y_{MM} \end{bmatrix} \tag{3}$$

is square and covers only the constraints on the set of nodes *N*, and the square submatrix of conductivities is only the connections on the set of nodes *M*. Note that Y_{NM} и Y_{MN} are rectangular matrices.

Expanding the equation (1) up to the selected blocks

$$\begin{aligned} Y_{NN}U_N + Y_{NM}U_M &= I_N, \\ Y_{MN}U_N + Y_{MM}U_M &= I_M, \end{aligned} \tag{4}$$

and expressing from the second equation (4) the stress vector of the excluded nodes U_M, it is not difficult to obtain the equation

$$\begin{aligned} \left(Y_{NN} - Y_{NM}Y_{MM}^{-1}Y_{MN}\right)U_N = I_N - \\ - Y_{NM}Y_{MM}^{-1}I_M \end{aligned} \tag{5}$$

which with the introduction of new designations

$$\begin{aligned} Y_{NN}^{\text{э}} &= Y_{NN} - Y_{NM}Y_{MM}^{-1}Y_{MN}, \\ I_N^{\text{э}} &= I_N - Y_{NM}Y_{MM}^{-1}I_M, \end{aligned}$$

is reduced to the equivalent form

$$Y_{NN}^{\text{э}}U_N = I_N^{\text{э}}. \tag{6}$$

Thus, the equivalence procedure considered above, for given excluded nodes (set *M*) preserves in the resulting equivalent solution (equivalent) only the nodes of the set *N*.

From a formal point of view, the described equivalence procedure is based on the well-known *Schur complement algorithm* [6, 7]. One of the drawbacks of this approach is the complexity and, often the impossibility of a preconditioning ("regulation" of the conditionality) of the solved equations.

In the next section of the paper, the original method is described, which is alternative to the *Shur complement algorithm*, and then the equivalence method is constructed on its basis.

This method, as it was said earlier, makes it possible to effectively impact on the condition of the NEVE and, as a consequence, reduce the computational errors and increase the correctness of the equivalent solutions obtained.

3 Mathematical justification of the alternative method

The following notation and definitions will be used: $\mathbf{0}_{n\times m}$ – zero size matrix $n\times m$; E_n – unit size matrix $n\times n$; $\cdot^{T}$ – the transposed matrix; $\cdot^{+}$ – a pseudo-inverse matrix according to Moore-Penrose; $(\cdot)^{\perp}$ – annihilator of the matrix of maximal rank of a given matrix; rank $\cdot$ – the rank of the matrix; size $\cdot$ – dimensions of the matrix (vector dimension); null $\cdot$ – basis of the null space of the matrix; cond $\cdot$ – number of conditionality of the matrix; $\|\cdot\|$ – a given vector norm [6, 7].

In this paper, the so-called left annihilator of matrices is used, which is called the annihilator. Recall [8, 9], that the left annihilator of the maximal rank of matrix M with the size $m\times n$ and rank r is called a matrix $M^{\perp}$, if the both conditions are correct in the same time $M^{\perp}M = \mathbf{0}_{(n-r)\times m}$, rank $M^{\perp} = n-r$.

For simplicity, we shall assume that the annihilators of zero satisfy the orthogonality condition $M^{\wedge}M^{\wedge\,T} = E_{n-r}$.

As a procedure for calculating of matrices annihilators, well-developed methods for computing the null space of a matrix can be used null(M) [6, 10]. In this case $M^{\perp} = \text{null}\ M^{T}\ ^{T}$.

We will consider the NEVE equation in the following block decomposition:

$$\begin{bmatrix} A_1 & A_2 \end{bmatrix} \begin{bmatrix} x_1 \\ x_2 \end{bmatrix} = \mathbf{b} \tag{7}$$

here A_1, A_2 – are rectangular submatrices of sizes

$$\text{size}\ A_1 = n\times n_1, \quad \text{size}\ A_2 = n\times n_2, \tag{8}$$

in this case $n_1 + n_2 = n$. The decomposition (7), (8) is clearly shown in fig. 1.

The statement: the solution of the linear equation (7) for an invertible block matrix $A_1 \mid A_2$ is determined by the equivalent formulas [9]

$$\begin{cases} x_1 = A_1^{+}\ \mathbf{b} - A_2x_2\ , \\ x_2 = \ A_1^{\perp}A_2\ ^{-1} A_1^{\perp}\mathbf{b}. \end{cases} \tag{9}$$

$$\begin{cases} x_1 = \ A_2^{\perp}A_1\ ^{-1} A_2^{\perp}\mathbf{b}, \\ x_2 = A_2^{+}\ \mathbf{b} - A_1x_1\ , \end{cases} \tag{10}$$

DOI: 10.1051/e3sconf/20172503002

here $A_1^{\perp}$, $A_2^{\perp}$ are the left annihilators of zero of the submatrices maximal rank A_1, A_2, respectively, A_1^{+}, A_2^{+} are the pseudoinverse matrices for the submatrices A_1, A_2.

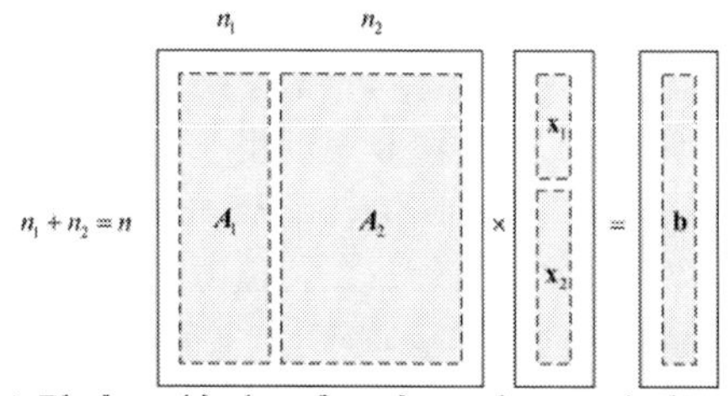

Fig. 1. Block partitioning of matrices and vectors in the matrix equation.

4 Reduction of NEVE based on matrix annihilators

Considering the NEVE (1) in the division into blocks, as it is done in the equation (7), we introduce the designations

$$Y_N = \begin{bmatrix} Y_{NN} \\ Y_{MN} \end{bmatrix}, \quad Y_M = \begin{bmatrix} Y_{MN} \\ Y_{MM} \end{bmatrix} \tag{11}$$

and write (1), taking into account (11). Obtained equation

$$\begin{bmatrix} Y_N & \vdots & Y_M \end{bmatrix} \begin{bmatrix} U_N \\ U_M \end{bmatrix} = \begin{bmatrix} I_N \\ I_M \end{bmatrix} \tag{12}$$

Annihilator $Y_M^{\perp}$ is introduced and it satisfies the following conditions:

$$Y_M^{\perp} Y_M = 0_{M\times M}, \quad Y_M^{\perp} Y_M^{\perp\mathrm{T}} = E_{M\times M} \tag{13}$$

Then, according to the first equation (10), from the theorem proved earlier we can write

$$Y_M^{\perp} Y_N U_N = Y_M^{\perp} \begin{bmatrix} I_N \\ I_M \end{bmatrix} \tag{14}$$

Introducing new designations

$$\widehat{Y}_{NN}^{\text{э}} = Y_M^{\perp} Y_N, \quad I_N^{\text{э}} = Y_M^{\perp} \begin{bmatrix} I_N \\ I_M \end{bmatrix} \tag{15}$$

The equation (14) can be rewritten in a generalized form

$$\widehat{Y}_{NN}^{\text{э}} U_N = \widehat{I}_N^{\text{э}} \tag{16}$$

The equations (6) and (16) are different, but they have the same solution U_N. The principal difference of the equation (16) is that it allows simultaneously with the equivalence to solve the problem of preconditioning in order to minimize computational errors.

It is well known [6, 11, 12] that in order to reduce the influence of errors in the initial data, to increase the accuracy of the solution, and to accelerate the convergence of the iterative methods, various algorithms are used that usually consist of elementary transformations of rows (columns) of matrices in equation (7): scaling, regularization, balancing, preconditioning (preconditioning, use of spectrally equivalent operators), etc.

With respect to the matrix equation (16), the problem of reducing errors will consist in minimizing the ratio [11]

$$\tau = \frac{\left\|\Delta U_N\right\| \cdot \left\|\widehat{I}_N^{\text{э}}\right\|}{\left\|U_N\right\| \cdot \left\|\Delta \widehat{I}_N^{\text{э}}\right\|} \tag{17}$$

However, the direct determination of the value t in terms of the coefficients of the matrices of the original equation is difficult due to the nonlinearity of the valuation operation. Therefore, it is preferable to use a qualitative characteristic called the matrix condition number [6, 10, 11]. In the case considered, this number is

$$\operatorname{cond} \widehat{Y}_{NN}^{\text{э}} = \left\|\widehat{Y}_{NN}^{\text{э}}\right\| \cdot \left\| \widehat{Y}_{NN}^{\text{э}\,-1} \right\| \tag{18}$$

and satisfies the inequality

$$\frac{\left\|\Delta U_N\right\|}{\left\|U_N\right\|} \le \operatorname{cond} \widehat{Y}_{NN}^{\text{э}} \le \left\|\Delta \widehat{I}_N^{\text{э}}\right\| / \left\|\widehat{I}_N^{\text{э}}\right\| \tag{19}$$

Taking into account (15), the ratio (19) is transformed to the form

$$\frac{\left\|\Delta U_N\right\|}{\left\|U_N\right\|} \le \operatorname{cond} Y_M^{\perp} Y_N \le \frac{\left\|\Delta \left(Y_M^{\perp} \begin{pmatrix} I_N \\ I_M \end{pmatrix} \right)\right\|}{\left\| Y_M^{\perp} \begin{pmatrix} I_N \\ I_M \end{pmatrix} \right\|}. \tag{20}$$

The larger the condition number (18), the greater the impact on the decision NEVE errors in the original data.

Reduction of the conditionality number (18) can be achieved by further transformation of the NEVE equivalent system (16) by introducing a new matrix D, which should be [12] as close as possible to $Y_M^{\perp} Y_N{}^{-1}$ easily computable and easily invertible. In this case, the NEVE will be replaced by the equation

$$D \widehat{Y}_{NN}^{\text{э}} U_N = D \widehat{I}_N^{\text{э}},$$

where

$$\operatorname{cond} D\widehat{Y}_{NN}^{\text{э}} < \operatorname{cond} \widehat{Y}_{NN}^{\text{э}} \cdot$$

5 Reduction of a small NEVE system

Let us consider the computational example [9].

Let the electric network shown in fig. 2 [2]. For convenience of counting, we take all the bond resistances to be the same r_{ij} = 10 Ohm (Y_{ij} = 0,1 S), except for the two bonds r_{13} = 20 Ohm (Y_{13} = 0,05 S), r_{24} = 5 Ohm (Y_{24} = 0,2 S).

The basic mode corresponds to the NEVE system

$$\begin{bmatrix} -0{,}25 & 0{,}1 & 0{,}05 & 0 \\ 0{,}1 & -0{,}4 & 0{,}1 & 0{,}2 \\ 0{,}05 & 0{,}1 & -0{,}25 & 0 \\ 0 & 0{,}2 & 0 & -0{,}2 \end{bmatrix} \begin{bmatrix} U_1^0 \\ U_2^0 \\ U_3^0 \\ U_4^0 \end{bmatrix} = \begin{bmatrix} -8{,}5 \\ 1 \\ -7 \\ 2 \end{bmatrix} \quad (21)$$

the results of the calculation are shown in Fig. 2.

Let us suppose that the set M of excluded nodes is 3 and 4. Then the matrix (3) has the following block decompositions:

$$Y_N \,\vdots\, Y_M = \left[\begin{array}{cc:cc} -0{,}25 & 0{,}1 & 0{,}05 & 0 \\ 0{,}1 & -0{,}4 & 0{,}1 & 0{,}2 \\ 0{,}05 & 0{,}1 & -0{,}25 & 0 \\ 0 & 0{,}2 & 0 & -0{,}2 \end{array}\right] \quad (22)$$

considering that

$$Y_M^{\perp} = \left(\text{null} \left(\begin{bmatrix} 0{,}05 & 0 \\ 0{,}1 & 0{,}2 \\ -0{,}25 & 0 \\ 0 & -0{,}2 \end{bmatrix}^T \right) \right)^T = \begin{bmatrix} 0{,}9646 & 0{,}0915 & 0{,}2295 & 0{,}0915 \\ -0{,}1837 & 0{,}6752 & 0{,}2333 & 0{,}6752 \end{bmatrix}$$

Indeed

$$Y_M^{\perp} Y_M = \begin{bmatrix} 0{,}9646 & 0{,}0915 & 0{,}2295 & 0{,}0915 \\ -0{,}1837 & 0{,}6752 & 0{,}2333 & 0{,}6752 \end{bmatrix} \begin{bmatrix} 0{,}05 & 0 \\ 0{,}1 & 0{,}2 \\ -0{,}25 & 0 \\ 0 & -0{,}2 \end{bmatrix} = \begin{bmatrix} 0 & 0 \\ 0 & 0 \end{bmatrix}$$

Performing further calculations, we obtain

$$\widehat{Y}_{NN}^{э} = Y_M^{\perp} Y_N = \begin{bmatrix} -0{,}2205 & 0{,}1011 \\ 0{,}1251 & -0{,}1301 \end{bmatrix},$$

$$\widehat{I}_N^{э} = Y_M^{\perp} \begin{bmatrix} I_N \\ I_M \end{bmatrix} = \begin{bmatrix} -9{,}8979 \\ -0{,}7473 \end{bmatrix},$$

$$U_N = \widehat{Y}_{NN}^{э\;-1} \widehat{I}_N^{э} = \begin{bmatrix} 85{,}0 \\ 87{,}5 \end{bmatrix},$$

which exactly corresponds to the values indicated in fig. 2.

It is important to note that the number of conditionality for the matrix (3) and the number of conductivity for original matrix (22) for the original matrix is $\text{cond}\; Y_N \,\vdots\, Y_M = 15{,}9373$, and for the equivalent

$$\text{cond}\; \widehat{Y}_{NN}^{э} = \text{cond} \begin{bmatrix} -0{,}2205 & 0{,}1011 \\ 0{,}1251 & -0{,}1301 \end{bmatrix} = 5{,}5204,$$

almost in 3 times less. This number can be further reduced if the $Y_M^{\perp}$ will be the matrix

$$Y_M^{\perp} = \begin{bmatrix} 6{,}6150 & 4{,}7836 & 3{,}2365 & 4{,}7836 \\ 3{,}5836 & 8{,}6518 & 4{,}1774 & 8{,}6518 \end{bmatrix} \quad (23)$$

The use of the annulant (23) in the calculations provides

$$\widehat{Y}_{NN}^{э} = \begin{bmatrix} -0{,}2205 & 0{,}1011 \\ 0{,}1251 & -0{,}1301 \end{bmatrix} \quad (24)$$

$\text{cond}\; \widehat{Y}_{NN}^{э} = 1{,}2444$, which is more than an order of magnitude less than $\text{cond}\; Y_N \,\vdots\, Y_M$. In this case, the matrix

$$Y_{NN}^{э} = \begin{bmatrix} -0{,}24 & 0{,}12 \\ 0{,}12 & -0{,}56 \end{bmatrix}$$

calculated by the traditional method, has a 3,5 times greater number of conditionality than the matrix (24).

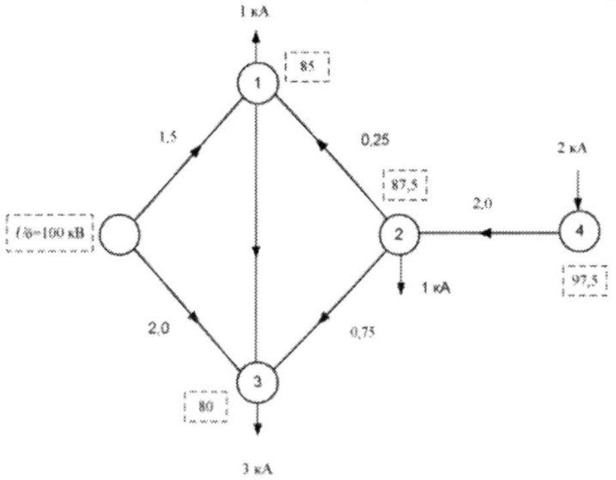

Fig. 2. The scheme of the electrical network in the basic mode.

6 Reduction of a large NEVE system

Let us consider the reduction of a large NEVE system, the basic mode of which corresponds to a size matrix

$$\text{size}\begin{bmatrix} Y_{NN} & Y_{MN} \\ Y_{MN} & Y_{MM} \end{bmatrix} = 1000 \times 1000$$

and vector of dimension

$$\text{size}\begin{bmatrix} I_N \\ I_M \end{bmatrix} = 1000.$$

The matrix (3) and vector (2) for the basic mode are densely populated arrays whose elements vary within the following limits: $-4{,}2;\ldots;5{,}7$. The total number of non-zero elements of the matrix from (3) is □ $9{,}91 \cdot 10^5$.

Let the set M of excluded nodes be equal 995, and, accordingly, the number of nodes left $N = 1000 - 995 = 5$.

Calculation by formulas (11) – (16) in the *Matlab* using orthogonal annihilators leads to the following results:

$$\hat{Y}^{э}_{NN} = \begin{bmatrix} 1,8571 & -0,7800 & 1,4434 & 0,4049 & -0,4329 \\ -0,7242 & 0,6592 & 1,5200 & 0,5827 & 1,6713 \\ 0,2350 & -0,3492 & -0,6982 & 0,9702 & -1,3692 \\ -0,6484 & -0,6965 & 0,0061 & -1,1241 & 0,5262 \\ -0,8512 & 0,2244 & 0,1312 & 1,1808 & 1,2175 \end{bmatrix}, \quad (25)$$

$$\hat{I}^{э}_{N} = \begin{bmatrix} 0,7807 \\ -0,1045 \\ -1,4519 \\ -0,2478 \\ -0,1782 \end{bmatrix}, \quad U_N = \hat{Y}^{э\,-1}_{NN}\,\hat{I}^{э}_{N} = \begin{bmatrix} 1,6907 \\ 0,4469 \\ -0,8643 \\ -0,3758 \\ 1,4110 \end{bmatrix} \quad (26)$$

In this case, the number of conditionality of the matrix (25) is 7,8019, and the Euclidean error rate of the solution (26) with respect to the exact value of the solution vector is $3,825310^{-13}$.

Calculations using conventional methods allow to obtain following matrix and vector

$$Y^{э}_{NN} = \begin{bmatrix} 41,7122 & 19,8072 & -2,6691 & 84,1475 & -29,5413 \\ -55,6173 & -48,4439 & 9,7584 & -51,1095 & 43,1407 \\ -84,5763 & -19,1018 & 14,5854 & -42,1711 & 75,9152 \\ 87,8123 & 63,1374 & -21,0482 & 19,9958 & -59,6218 \\ 49,3494 & -8,4865 & 36,5263 & 64,3591 & -27,9839 \end{bmatrix} \quad (27)$$

$$I^{э}_{N} = \begin{bmatrix} 8,3722 \\ -44,0346 \\ -41,1712 \\ 103,2302 \\ -15,6018 \end{bmatrix},$$

giving increased more than 3,5 times the Euclidean norm of the error with respect to the exact value of vector $1,465110^{-12}$. The number of the matrix conditionality (27) is 39,4146 and almost 5 times higher than the number of the matrix conditionality (25).

7 Reduction of a very large NEVE system

Let us suppose that a very large NEVE system is given, whose dimension is 10^4. In the matrix (3), there are no zero elements, thus the number of non-zero elements is 10^8, that is one hundred million ($100 \cdot 10^6$). In this case, the elements in the matrix and the vector (2) vary in the range from $-10,7$ to $11,5$. Supposing that the set of excluded nodes M is equal $9 \cdot 10^3$, that means that the number of nodes left is $N = 10^4 - 9 \cdot 10^3 = 100$.

As a result of the calculations, matrix with size 100×100 was obtained and it had a condition number $\text{cond}\ \hat{Y}^{э}_{NN} = 47,3464$, while the number of a matrix conditionality $Y^{э}_{NN}$ with the same size, was $\text{cond}\left(Y^{э}_{NN}\right) = 10390,4112$, that is more than 200 times higher.

We note that it is not possible to directly use the Schur complement algorithm for such a large matrix. In this case, the authors used the parallelization of the computation process.

8 Conclusion

The original algebraic method of the equations reduction of an electric network steady-state regimes on the basis of matrices annihilators with the purpose of obtaining equivalents is considered, allowing to carry out transformations of the substitution scheme and its parameters to a form having a significantly smaller number of nodes and branches. Numerical procedures for computing matrix annihilators are well developed for large ($100 \le n \le 1000$) and very large ($10^3 \le n \le 10^5$) matrices, while the variation of the annihilators properties makes it possible to significantly (by an order or more) improve the conditionality of the resulting equivalents of the nodal stress equations and thereby reduce computational errors and improve the correctness of the solution obtained.

References

1. Yu N., Shah S., Johnson R., Sherick R., Hong M., Loparo K. *ISGT, IEEE Power System (https://pdfs.semanticscholar.org)* (2015)
2. B.I. Ayuev, P.I. Bartolomey *Calculations of steady-state regimes in the tasks of operational and automatic EPS control*. (Ekaterinburg, publishing house UTI-UPI, 40, 1999)
3. M.Sh. Misrikhanov, V.N. Ryabchenko. *Automation and Telemechanics*. **5**, 133 – 141 (2006)
4. P.I. Bartolomey, A.V. Pasedin. *Automated dispatch control systems and optimization of EPS regimes*. (Ekaterinburg, publishing house UGTI-UPI, 1999)
5. N.E. Zubov, E.A. Mikrin, V.N. Ryabchenko. *Matrix methods of automatic control systems for aircraft in the theory and practice.* (M.: Publishing house of the Moscow State Technical University. Bauman, 666, 2016)
6. J. Golub, C. Van Lown. *Matrix calculations*. – (Moscow: Mir, 548, 1999)
7. F.R. Gantmakher *Theory of matrices*. (Moscow: Nauka, 576, 1967)
8. M.Sh. Misrikhanov, V.N. Ryabchenko. *Vestnik IEEU*, **5**, 196 – 240 pp. (2005)
9. M.Sh. Misrikhanov, V.N. Ryabchenko. Elektro, **1**, 15 – 19 pp. (2010)
10. J. Demmel. *Computational linear algebra. Theory and applications*. (Moscow: Mir, 430. 2001)
11. Kh.D. Ikramov *Numerical solution of matrix equations*. (Moscow: Nauka, 192. 1984)
12. M.Yu. Balandin, M.P. Shurina. *Methods for solving SLAE of large dimension*. (Novosibirsk: Publishing house of the NSTU, 65. 2000)

The transfer of energy from the solar-space power station to the Earth

William Kaganov[1*]

[1]Moscow Technological University - MIREA , Russia

Abstract. The resulting problem in the transmission of electrical energy from space to the Earth by means of microwave electromagnetic waves is discussed. The main parameters and principles for constructing the space and ground segments of such a solar power plant are given.

1 The essence of the problem

Solar power plants (SP) are the leading direction in alternative electric power industry, accounting for about 6% in the electric power balance of the leading countries in the world.

The specific power of solar radiation in outer space is 1.4 kW / m^2 or 12,000 kWh / m^2 of energy per year. Due to the reflection of sunlight in the atmosphere, the first indicator in clear weather in the equatorial zone of the Earth is reduced to 1 kW / m^2, and in other areas to an even smaller value. Approximately the average daily value of this indicator for different regions of the Earth does not exceed 0.3 - 0.4 kW / m^2.

Taking into account all the factors (atmospheric dispersion, weather effect, daily and annual indicators of the Earth's surface illumination) with the efficiency of solar cells 20% [1,2], the conversion of solar energy into electrical energy is 1%, i.e. the specific energy value per year is 120 kWh / m^2.

The view of such a small value of this parameter was the birth of a proposal for the removal of SPS into outer space where the influence of the atmosphere is excluded, and intensive irradiation of solar batteries occurs continuously throughout the day and the whole year [3]. Since then, there have been practically no publications on the problem submitted. However, at the present time, in connection with the appearance of powerful microwave transistors, with the help of which it is possible to convert the energy of solar batteries into microwave radiation, the idea of building a solar-space power plant becomes technically feasible. As calculations show, the efficiency of such solar-space power stations (SSP) can exceed the ground-based ones by 10-20 times and make up 1200 kW-h / m^2 of energy per year. All the data are summarized in Table 1.

One of the technical problems in the implementation of the idea for the creation of SSP is the transfer of the energy it generates to the Earth. The general scheme of SSPS, in which energy from space to the Earth is transmitted by microwave radiation, is shown in Fig. The scheme includes two segments: space and ground. Consider the principles of their construction and basic parameters

Table 1. Comparison of two types of solar power plants

Location: Solar Batteries	Specific Power Solar Radiation	Specific amount of radiated solar energy per year	Specific amount Electrical Energy per year with EFFICIENCY = 20%
Dimensionality	kW / m^2	kW-h / m^2	kW-h / m^2
In space	1,4	12 000 (100%)	1200 (10%)
On Earth	0,3 - 1	600 (5%)	120 (1%)

2 Space segment of the SSP

This segment includes:
- solar batteries, facing their flat surface to the Sun;
- a certain number of microwave semiconductor generators that convert the direct current energy generated by solar batteries into microwave electromagnetic oscillations;
- the same number of radiators forming a phased antenna array directed towards the Earth;
- two extreme systems of automatic control, the first of which provides maximum consumption of solar energy batteries, the second one - the continuous direction of the antenna array's directional pattern to the desired place on the Earth, where the receiving antenna is located.

The first of which provides maximum consumption of solar energy batteries, the second one - the continu-

* Corresponding author: Kaganovwil@yandex.ru

ous directivity pattern of the antenna array's directional pattern to the desired place on the Earth, where the receiving antenna is located.

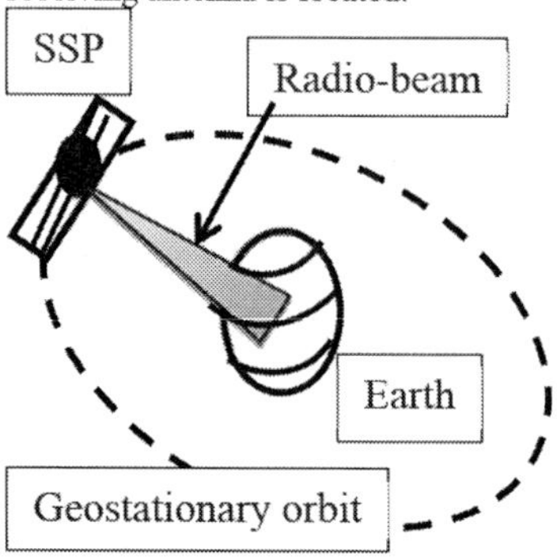

Fig.1. The scheme of transmission of microwave energy from space to Earth.

The general scheme for the transmission of microwave energy produced in a geostationary orbit to Earth is shown in Fig. 2, where the following designations were adopted: 1- solar array, 2- satellite in geostationary orbit, 3-phased array antenna, 4 - a set of receiving terrestrial antennas, DA - their total diameter. H- is the height of the antenna array above the Earth. The signal frequency emitted by the phased array, f = 6 GHz, i.e. the wavelength λ = 5 cm.

At H = 36 • 10^6 m (more accurately H = 35,786 • 10^6 m - height of the geostationary orbit), and D_A = 25 • 10^3 m, we obtain for the angle of the antenna pattern opening, located on the space object,

$$\alpha = 2 \bullet \left(\frac{180}{\pi} \right) \bullet \operatorname{atan}\left(D_A / 2H \right) = 0{,}044^0 . \qquad (1)$$

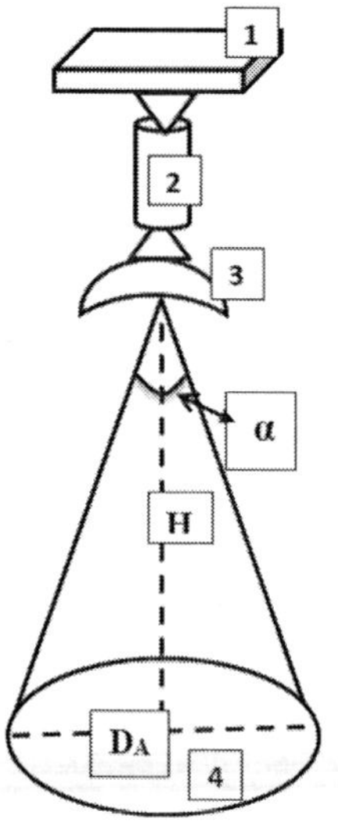

Fig.2. The scheme for determining the antenna's angle of directivity

The very diagram of the direction in one plane of the antenna array on two scales is shown in Fig. 3, where the angle is in degrees. Angle α = ΔΘ = $0{,}044^0$ is the width of the ray of the diagram of the direction of the 0.7 level with respect to the maximum value of the field strength.

To obtain such a directional diagram with a wavelength λ = 5 cm in each line of a flat phased array antenna, 1000 emitters should be located, and their total number is a one million. As the radiators of a phased array, it is possible to use micro strip antennas of the Vivaldi type (Fig. 4) [4]. The distance between two neighboring radiators is b = λ = 5 cm.

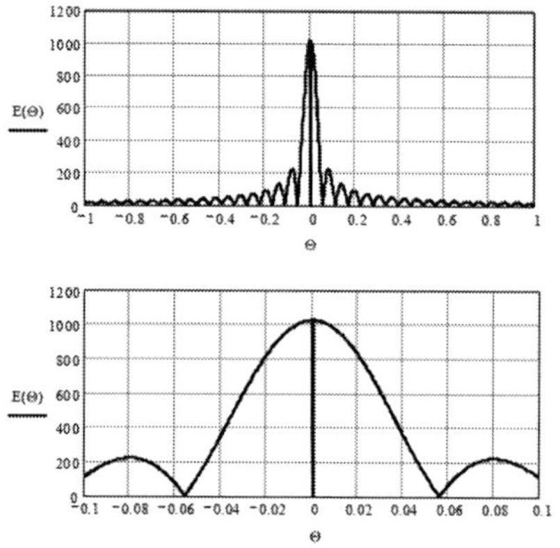

Fig. 3. Directivity diagram

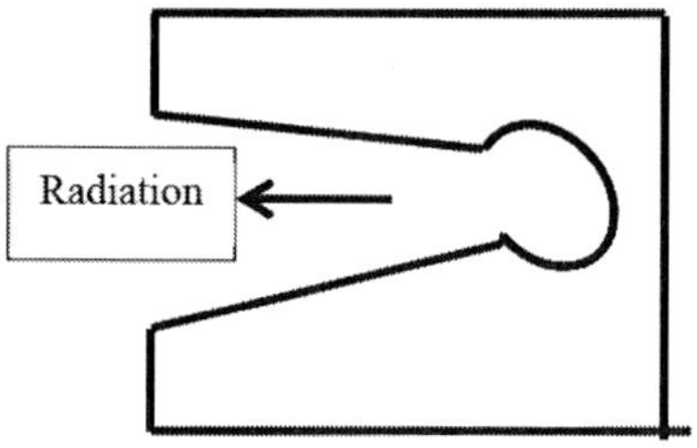

Fig.4. Outward visible Vivaldi

3 The terrestrial segment of the SSPP and the main parameters of the SSP

The ground segment includes:
- the antenna array receiving the microwave signal of the space segment;
- an ensemble of detectors that convert the received microwave signal into direct current;
- microprocessor system for automatic control and monitoring of operation of all elements of the facility.

When calculating the SPS-Earth radio link, it is necessary to take into account:
- the power and frequency of the microwave signal supplied to each of the radiators of the CFA,
- the total number of such emitters,
- the dimensions of the receiving terrestrial antenna,

 DOI: 10.1051/e3sconf/20172503003

- the following signal power losses in the radio link - due to the lobes of the transmitting and receiving antennas.

Attenuation of the microwave signal during passage through the atmosphere loss when detecting a signal received by a receiving antenna, the efficiency of the microwave generator connected to the radiator, the efficiency of solar cells.

The results of the calculation, carried out taking into account the listed factors with 20% efficiency of solar cells, a total emitter of 1 million and two values of the microwave power supplied to each radiator at 150 W, are shown in Table 2.

Table 2. The main parameters of the SSP

Parameter	Value
Total power of solar cells in the X-ray segment	430 MW
The area of the solar cell complex of the space segment	550*55 m^2
Microwave power of one radiator	150 W
The total power of the microwave signal emitted from the space segment	150 MW
Total DC power produced by the ground segment	100 MW
The density of the microwave energy flux near the Earth	200 MW/ m^2

4 Microwave generators feeding the antenna array, and their synchronization

The microwave transistor generators with a frequency of 6 GHz in the space segment can be powered directly from solar panels with a voltage of 20 to 30 V. Note that in the microwave range, two types of tracers with a power of up to 25 W are produced: field gallium arsenide (GaAs) and "gallium nitride Silicon (GaN-on-Si). [5].High reliability and durability, low value of the supply voltage (up to 30 V) predetermines their use as generators in the space segment of the SSP.

Schemes and the device of microwave transistor generators are considered, for example, in [4]. In the 1st version of the SSP (i.e with a microwave power supply feeding to one radiator, at 15 W), such a generator can include two stages: an auto generator and an amplifier. In the case of the 2nd variant, that is, at a power of 150 W, they should be supplemented with a scheme for summing the powers of the four transistors to obtain a microwave power of 150 W.

5 Synchronization of auto generators

As a result of the interaction between all auto generators, when certain conditions are met, a single rhythm of operation is established in the system, which means that the frequencies of all auto-oscillations are equal and the system is put into synchronization mode. In this case, the system can be multistage with integration at the 1st level, for example, at the first level 1000 auto generators, on the second - 10000 where 1 is an auto generator, 2 is a microwave power amplifier.

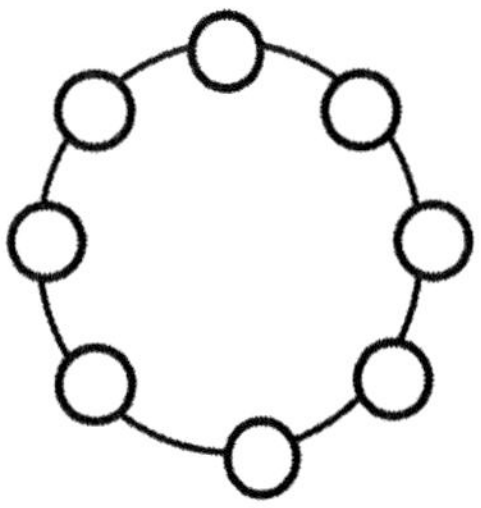

Fig.5. Structural scheme of synchronization of microwave oscillators

According to the scheme in Fig. 5, each of the auto generators interacts with two neighboring ones, which should be taken into account in the compilation of differential equations describing the operation of the system as a whole and consisting of N equations-according to the number of sources of oscillations.

$$\begin{aligned} &\frac{d^2u_1}{d\tau^2} - \mu_1\Phi_1(u_1)\frac{du_1}{d\tau} + \Omega_1 u_1 + \sum_{M_2}^{M_2} k_i u_i = 0 \quad , \\ &\frac{d^2u_2}{d\tau^2} - \mu_2\Phi_2(u_2)\frac{du_2}{d\tau} + \Omega_2 u_2 + \sum_{M_3}^{M_4} k_i u_i = 0 \quad , \\ &\ldots\ldots\ldots\ldots\ldots\ldots\ldots\ldots\ldots\ldots\ldots\ldots \quad , \\ &\frac{d^2u_n}{d\tau^2} - \mu_n\Phi_n(u_n)\frac{du_n}{d\tau} + \Omega_n u_n + \sum_{M_{n-1}}^{M_n} k_i u_i = 0 \quad , \end{aligned} \qquad (2)$$

where Φi (u_i) is a nonlinear function, k_{ii} - is the coupling coefficient of the auto generator with the other two, Ω_i - is the ratio of the deviation of the natural frequency of the i-th self-oscillator in the autonomous mode of operation to the frequency of self-oscillations in the system.

To coincide the initial phases of oscillations of synchronized oscillators, the distance between them should be equal to an integer number of wavelengths λ. One of the higher power generators in such a circuit can be the master. Computer simulation, conduct with 16 auto generators included in the system, showed that the initial discrepancy of their frequencies can reach 5%, one hundred does not prevent their entry into synchronism.

6. Remote charging of batteries unmanned aerial vehicle

In addition to the considered option of remote transmission of electrical energy in outer space, the solution

 DOI: 10.1051/e3sconf/20172503003

of this problem is important for some projects of today.

These include, in particular, remote charging of batteries unmanned aerial vehicles. The ground station such a system includes a microwave generator with a frequency of 2.45 GHz, a power of 1 kW from the mass-produced microwave oven; parabolic antenna with a diameter of 3 m and the device of its guidance on an unmanned aerial vehicle, hovering at a height of up to 100 m.

The equipment on board the unmanned vehicle includes: a phased array antenna with a number of 32x32 emitters, microwave detectors attached to these emitters, and a rectifier connected to the battery to charge it.

Thus, it is possible to carry out remote charging of batteries by an unmanned aerial vehicle without landing on the Earth, which is extremely necessary in many cases.

7 Conclusion

1. The conducted comparative analysis of two types of solar power plants ground and space - shows unquestionable advantages in terms of the energy performance of a power plant of the second type.
2. Within the framework of the first stage of the implementation of this project, it is necessary to solve the problem of today: remote charging of unmanned vehicle batteries.

References

1. K. Chopra, S. Das *Thin-film solar cells* (M. Mir, 1986).
2. T. Byers *20 designs with solar elements* (M: Mir, 1988).
3. P.E. Glezer *The Potential of Satellite Solar Power* (Proceedings IEEE, v.65).
4. S.E. Banks, et al., *Radio engineering and electronics*, **№12**, p.52-63, (2015).
5. *Electronics news*, **№7**, p.5-20, (2015).

Analysis of the security during power system expansion planning

Alexey Osak[1,*], *Daniil* Panasetsky [1], and *Elena* Buzina[1]

[1]Melentiev Energy Systems Institute, 130 Lermontov str., Irkutsk, Russia

Abstract. Increasing the intelligent level of the EPS control systems, caused by the implementation of Smart technologies, changes the structure and the properties of EPS and increases the importance of system reliability analysis. System reliability analysis includes two components – for the balance and for the regime. On the one hand, there is a large number of studies to assess the reliability of the power system, which examines various aspects and methods of solving this problem. On the other hand, in Russia there is no generally accepted methodology with clear criteria that could be used for feasibility studies of various technical solutions taking into consideration system reliability aspects. In practice, the security analysis is limited by the calculations of power flows, static and dynamic stability for a number of forecast periods for the normal and repair circuits considering the most severe disturbances. The existing approach allows defining the requirements and adjusting emergency control systems, but does not allow evaluating and comparing solutions for power grid constructions. The authors propose a new method for power system reliability evaluation, which is suitable for planning development and operation of power systems. The method includes a general description of the algorithm which allows to compare various development scenarios, as well as to assess the reliability level of their implementation. In particular, the method allows to determine where it is needed only the relay protection and emergency control system development, and where it is necessary grid, protection and control development and reconstruction.

1 Introduction

The rise of intelligent level of the power system (PS) control due to the implementation of Smart technologies and conception of intelligent power system (IPS) of active-adaptive grid (AAG) changes the structure and characteristics of PS [1-2], simultaneously increasing the importance of assessment of system reliability. System reliability (PS reliability) includes balance and regime components.

In Russia, in accordance with [3], power system security is understood as a property of the system to maintain the prescribed modes of operation when conditions change, element failures and sudden perturbations occur. In other countries, the term security has a similar meaning, which is understood as the ability of the system to withstand sudden disturbances [4-6]. There is distinguish between static and dynamic power system regime security. In assessing static regime security, dynamic transient processes in EPS are not taken into account, the model of the system is limited to the equations of steady-state. Dynamic regime security is related to the dynamics of EPS behavior, when a nonlinear model of transient processes in EPS or its linearized version can be considered.

There is a great number of studies on assessment of power system reliability (in particular [7-8]) dealing with various aspects and methods for solution of this problem. On the other hand, in Russia there is no generally accepted methodology with clear criteria which could be used for development of inter- and intra-system connections while choosing the schemes of power generation at power plants and schemes of power supply of customers taking into account system reliability.

In practice, the security analysis is limited by calculations of static and dynamic stability for a number of forecast periods for normal and repair circuits considering the most severe disturbances [9]. Such an approach allows us to define requirements and adjust emergency control (EC). However, it cannot assess and compare solutions for power grid constructions.

The authors propose a new method to determine application of various aspects of security at different stages of development planning and during operation power system. A technique and algorithm are proposed that allow to determine the magnitude of possible under-supply of electricity for various emergency situations, taking into account the emergency control and the expected actions of the dispatcher. This approach allows us to determine the mathematical expectation of annual damage from all accidents in EPS, resulting in a shortage of electricity.

* Corresponding author: osakalexey@mail.ru

2 Methodology

2.1 Reliability criteria for power system expansion planning

The most important criterion in planning of power system development is system reliability and PS viability. Moreover, both balance and regime aspects of system reliability should be provided [5-8]. Planning is carried out at various levels with different time interval. Therefore, various aspects of system reliability are considered for different types of planning (see Table 1). Measures for power grid construction are supported by an algorithm given in Fig. 1.

With regard to balance reliability, it is necessary to control it not only at the levels of power system (PS), unified power system (UPS) and integrated power system (IPS), but also at the level of large and medium power areas within PS and intersystem regions. At the level of UPS and PS, balance reliability is supported first of all by its own generation, whereas balance reliability of large and medium power areas is provided by transmission capacity of power grid connecting the power region with one or several PS. Total available reserve capacity (spinning and cold reserve capacity) of PS and adjoining PS and UPS is considered in assessment of balance reliability of PS or a power region. We distinguish two components in assessment of PS security:

Operational assessment of PS security during short-term post-emergency state of power system: evaluation of disturbance consequences (including static and dynamic stability analysis), operation of relay protection (RP), emergency control (EC), automatic voltage regulators (AVR), automatic frequency and power control (AFPC), sources of reactive power and on-load tap-changers (OLTC). Changes of generator power are recorded with primary regulation within the primary reserve.

Security during long-term post-emergency state of power system: evaluation of disturbance sequences implementing regime measures (20 and 40 minutes). Changes of generator power are recorded with primary, secondary and operational regulation within primary and secondary reserves.

Table 1: System reliability during different types of planning

Reliability aspects	Reliability effect
Long-term planning (over 10 years)	
Balance reliability	Measures for development of large generation and power grid construction of intersystem connections
Medium-term planning (1–10 years)	
Balance reliability	Planning of development of generation and main power systems
Security of long-term post-emergency state	Planning of power grid construction (transmission lines, static var compensator)
Operational security	Measures for EC development and, in some cases, for power grid construction (power generation at nuclear power plants (NPP), some thermal power plants (TPP) and power supply of essential customers)
Short-term planning (up to 1 year)	
Balance reliability	Planning of power grid and generation repair and energy supply for generation
Security of long-term post-emergency state	Planning of power grid and generation repair
Operational security	Adjustment of RP, EC and ACS devices
Operational planning	
Balance reliability	Choice of composition and load of operating power units
Security of long-term post-emergency state	Permit (ban) of joint repair and realization of regime measures
Operational security	Realization of regime measures

As a matter of priority, we assess the effect of these measures on the balance reliability of PS or power region during planning of power system development.

During planning of measures for power grid construction we assess the effect of these measures on the security in terms of long-term post-emergency PS state. Therefore, if security is not achieved because of some disturbances, we consider load limitation, energy undersupply and loss of lack of power overloading with corresponding probabilities of this disturbance, and feasibility of these or those measures for PS development is evaluated.

Finally, it is necessary to assess the effect of these measures on PS operational security during planning of power system development, including the comparison of some technical solutions, and during planning of measures for improvement of reliability of power grid within the framework of its operation.

To provide PS operational security, the following priority measures are applied: solutions of RP (including additional sets of the main protection and single-phase auto-reclosing with 220 kV), EC (including circuit-breaker failure protection caused by EC operation) and regime automation (including control of reactive power sources in the grid with 110 kV and lower). Installation of additional sources of reactive power in the grid (including those on bus with 110 kV and lower) is another measure for providing PS operational regime. Only in case of extremity, measures on additional power grid construction of new transmission lines (TL) are implemented for uninterrupted reliable power supply of customers or for supply of static or dynamic stability of NPP or TPP operation (if tripping of TPP energy units from EC causes long-term load limitation of customers).

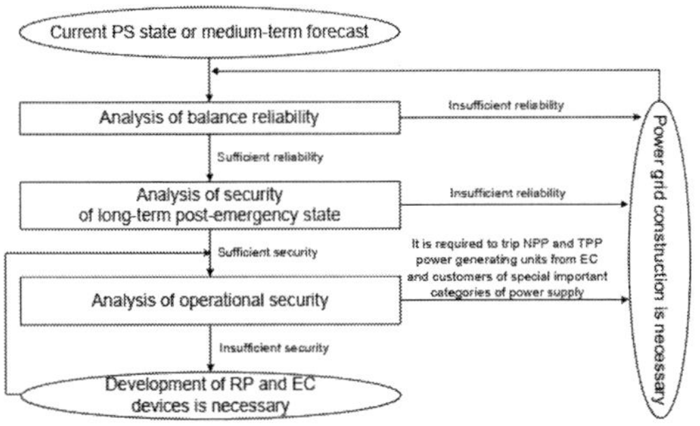

Fig. 1. Algorithm of feasibility studies for measures of power grid construction.

2.2 Methods, assessment criteria and characteristics of PS security

To assess PS security, we determine probability of reference disturbances [6] at different PS points. Reference disturbances should be subdivided into:

• Enumeration of reference disturbances with successful SPAR (single-phase auto-reclosing) / TPAR (three-phase auto-reclosing) or successful testing of voltage upon command of the operational-dispatcher personnel pointing to the probability for each elements of the grid (transmission line, transformer, power bus, power generating unit and 2 transmission lines). Data on reference disturbances are considered only in the analysis of PS security.

• Enumeration of reference disturbances with unsuccessful automatic and manual testing leading to long-term tripping of the element with probability for each element of the grid (grid element, power bus and power generating unit) as well as to the expected time of recovery. These reference disturbances are considered in the analysis of operational security and security during long-term post-emergency state of PS.

Different recovery (repair) time is needed for different emergency device damage. Therefore, besides medium duration of one repair and probability of emergency state of the grid element, it is recommended to predict probability of repairs of different duration. For example, repair time of one transmission line phase differs from time repair of transmission line pole with different probability of such damages. As for station equipment, time scale for reconstruction is larger, taking into account the fact that some damages need complete substitution of the damaged high-voltage equipment that could be either kept in the storehouse (emergency reserve when hours are needed for repair) or be absent (days or even weeks are needed for repair).

2.2.1 Methods, assessment criteria and characteristics of security during long-term post-emergency state of PS

We evaluate the ability of power system to resist disturbances associated with withdrawal of equipment for emergency repair taking into consideration the operation of RP, EC of automatic regulators and regime automatics and realization of regime measures by operational-dispatcher personnel during the first 20 (40) minutes after the first disturbance assessing the effect on power consumers.

At the moment of emergency disturbance by RP and EC operation, partial load shedding (power consumers) is possible as well as output of regime parameters (current, voltage and frequency) beyond permissible values. After RP and EC response, automatic regulators and regime automatics together with the operational-dispatcher personnel mobilize hot reserve at power plants. Regime parameters are inserted into permissible limits and power consumers disconnected during the emergency are connected. According to the rules [10, 11], it takes 20 minutes to carry out these measures. However, if this time is not enough, the forced regime of power system is introduced and other 20 minutes (totally 40 minutes from the beginning of the emergency) are spent for normalization of the post-emergency regime.

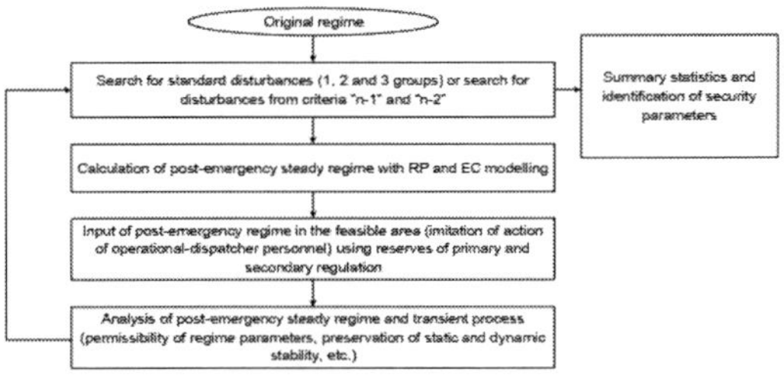

Fig. 2. Algorithm of security analysis during long-term post-emergency state of PS

During this type of security the possibility of normalization of post-emergency regime is evaluated with switching on all consumers tripped for 20 (40) minutes during the emergency. If it is provided for concrete pre-emergency scheme-regime conditions and for concrete emergency disturbance, the security does not decrease. If limitations for consumers remain after 20 (40) minutes, we determine the volume and undersupply of electrical energy taking into account mobilization of all kinds of reserves.

Analyzing all disturbances (Fig. 2) and integrating probable volumes of electrical energy undersupply, the following algorithm is defined:

1. Original calculating model is developed. Pre-emergency values of loads and generations in the nodes and state of network elements are determined.
2. Available ranges are specified for regulation of variable regime parameters of static var compensator (SVC), on-load tap-changers (OLTC) of transformers and autotransformers (if automatic or operational tap-change operation is allowed), booster transformers (BT),

fixed series compensation (FSC), thyristor-controlled series capacitor (TCSC), as well as generators with active and reactive power. These ranges can be as follows:

• For SVC:

- variable ranges of shunt of capacitor banks (*Bmin, Bmax*);
- variable ranges of reactive power (*Qmin, Qmax*);
- type of regulation – smooth or discrete (a number of steps or values of changing parameter for each step of control);

• For longitudinal FSC (TCSC) and controlled longitudinal reactors:

- variable ranges of longitudinal resistance (*Xmin, Xmax*);
- type of regulation – smooth or discrete (a number of steps or values of changing parameter for each step of control);

• For OLTC of power and phase-variable transformers (autotransformers):

- variable ranges of transformation coefficient (*Ktr.min, Ktr.max*);
- a number or a step of changing of transformation coefficient;

• For generators:

- permissible values of reactive power (*Qmin, Qmax* or in the form of permissible power);
- permissible values of active power (*Pmin, Pmax* – technological minimum and maximum);
- permissible values of changing of active power for 20- and 40-minute range (*dPmin.20, dPmax.20, dPmin.40, dPmax.40*).

3. Continuous permissible ranges are specified for regime limitations in nodes and branches of the calculating model:

• for voltage (Umin, Umax) and current (*Imin, Imax*).

4. Limitations are specified for capacity in the controlled sections:

• or in the form of values of maximum available load flow / emergency allowable load flow for concrete conditions of the grid elements (a list of switched on or tripped grid elements);

• or in the form of assignment of vector (trajectory) of weighting for calculation of maximum load flow (MLF) and emergency allowable load flow (EALF) of the concrete controlled section.

5. Controllers of EC device effect are specified:

• or with possible maximal volumes of decrease of load and generation;

• or with EC algorithms stating concrete controlling effects for each stage;

• load category is specified, which can be limited.

6. Possible controlling effects are specified for operational-dispatcher personnel (except specified in Item 2) in the following way:

• enumeration of possible maximal volumes of load shedding and generation of nodes with the help of

• operational-dispatcher personnel;

• load category which can be limited.

7. Repair schemes are specified (for a concrete calculation interval: winter/summer etc.):

• or tripping of one grid element (under repair);

• or enumeration of changes of the grid topology (tripping of the equipment under repair, switching on/tripping of the section breakers, etc.), and switching on/tripping of the branches of the calculating model);

• or enumeration of the network topology changes and changes of generation composition (switched on/tripped

• of generators and values of active generation power);

• in some cases, if the calculating model does not switch on distribution systems, besides changes of the grid topology and generation, the load redistribution between the nodes is specified (thus, changes of the distribution grid topology is considered including those of the grids of internal power supply of consumers with several supply centers).

8. Enumeration of emergency disturbances caused by long-term grid tripping is specified. Every disturbance is specified by:

• enumeration of tripped grid elements (branches of the calculating model, e.g. TL with tapping is modelled with several nodes and branches);

• probability of concrete emergency disturbance;

• expected time of recovery.

9. Search for emergency disturbances (Item 8) is specified for the original (normal) scheme and for all specified control schemes (Item 7) by turn:

• pre-emergency regime is assessed and changes are overlaid in the control schemes (Item 7, tripping/switching on of branches and changes of generation and load);

• emergency disturbances are searched by turn (Fig. 2) overlaying tripping specified for a concrete disturbance (Item 8);

• EC operation is modelled during estimation of the steady regime (Item 5, in case of its actuation for a concrete disturbance);

• if the iterative process is non-convergence during estimation of the steady regime, it is necessary to model EC (Item 5, using optimization method and method of directed or complete search) to reach post-emergency steady regime;

• if the obtained steady regime has load limitations (load shedding by EC, Item 5) and/or regime parameters go beyond permissible values (Item 3) or load flow along the controlled sections exceeds EALF (Item 4, specified or determined by the weighting for achieving post-emergency regime), it is necessary to introduce the regime in the allowable area (Items 3 and 4), including the realization of measures available for operational-dispatched personnel (Items 2 and 6) and if possible to recover the load tripped from EC (Item 5);

• if post-emergency regime together with realization of input measures meets all limitations (Items 3 and 4) and if the load tripped from EC but during the input it was possible to recover it, such disturbance does not reduce security;

• if post-emergency regime together with realization of input measures does not meet all limitations (Items 3 and 4), it is necessary to fulfil additional adjustment of control able to be applied by operational-dispatcher personnel (Item 6);

• if post-emergency regime together with realization of input measures meets all limitations (Items 3 and 4) but has non-regenerable load (dPlim.), it is necessary to calculate electrical energy undersupply (dWlim). Duration of electrical energy undersupply is calculated from the expected time of recovery of the tripped emergency element (Trep.), which can be decreased by time of generation mobilization (Tmob), if all permissible values of changes of active power for 40-minute range or together with mobilization of cold reserve (limitations on technological maxima and minima) allow completely or partially recovery of pre-emergency level of load.

10. Integral parameters of electrical power undersupply and damage together with probability are calculated from results of search for all disturbances for all calculating models. Moreover, category of consumers included in the limitation list is taken into consideration.

11. If several options of the development of power grid are compared, parameters specified in Item 10 are calculated for each option.

Item 9 points to the necessity of applying methods imitating EC operation as well as necessity of using methods of input into the allowable area [12-13]. Moreover, the result obtained should be balanced steady regime (the first and second Kirchoff's laws are followed), as well as limitations stated in Items 2, 5 and 6 and Items 3 and 4 for input into the allowable area are taken into consideration.

2.2.2 Methods, assessment criteria and parameters of operational security

Operational security is an aspect of system reliability reflecting ability of power system to resist sudden disturbances taking into account operation of RP, EC, ACS and automatic regulators without unforeseen effects on power consumers. Operational control includes static and dynamic security. Dynamic transitional processes in PS are not taken into account during assessment of static security. The system model is limited by equations of the steady regimes. Dynamic security is connected with dynamics of PS behavior. Moreover, nonlinear model of transitional processes or its non-linearized variant occurring in PS can be considered.

Priority measures for providing operative security of PS are solutions of RP (including those of additional sets of the main protection and SPAR with 220 kV), EC (including CBFP caused by EC), regimes automatic control device (including control of the available sources of reactive power with 110 kV and lower). All necessary devices of RP, EC and ACS have to be installed in full volume during planning of development and operation of power systems. All these devices do not have to permit disturbances of static or dynamic stability of PS during emergence of standard disturbances and also non-permissible in value and duration overload of devices, as well as deviations of voltage and frequency. Optimal options of realization of RP, EC and ACS should be chosen within the framework of fulfilment of the project work during construction and reconstruction of energy facilities.

In terms of long-term post-emergency state of PS, the analysis of security in accordance with the proposed algorithm and scheme of power generation at NPP and TPP allows us to reveal long-term limitation of power load for customers because of long period of putting into operation the power unit tripped from EC. In these cases, we need such a scope of power grid construction, which in combination with effective and sufficient RP and EC devices would not permit tripping of NPP or TPP power units cause by disturbance of static or dynamic stability during standard disturbances. Moreover, during standard disturbances EC (automatic stability control system or automatic overload protection) can use controlling effects that are not associated with tripping of power units, such as CBFP from EC, control of SVC from EC, impulse and long-term unload of turbines, etc.

If reliability of power supply of consumers with special needs for reliability and continuity of power supply is under consideration, possibility of their tripping from EC or under command of operational-dispatcher personnel is not foreseen. Uninterruptedness of power supply of such consumers is provided if necessary by additional power grid construction (including by SVC device), RP, EC and ACS devices.

2.3 Summary criteria (standards) of evaluation and parameters of security

To determine damage caused by electrical power undersupply, it is suggested to assess it from power capacity of gross domestic product (GDP) for the country as a whole and from power capacity of gross regional product (GRP) for concrete entities of the Russian Federation. GDP and GRP values as well as their quantitative parameters of the power consumed within the entire country and within the concrete region are quite objective criteria with high degree of their validity. Therefore, mean specific damage from the electrical power undersupply is determined from the equations:

$$Y_{0.RF} = GDP_{RF} / W_{RF\,year} \quad (1)$$

$$Y_{0.Reg} = GRP_{Reg} / W_{Reg\,year} \quad (2)$$

It is possible to introduce a multiplying coefficient in the equations (1) and (2) that take into consideration additional time and resource costs of power supply consumers for resumption of their economic activity after power restoration (lifting of limitations). However, due to complexity of the grounds for concrete values increasing the coefficient, it is recommended not to introduce them. For large or particularly critical industrial value of the damage is determined individually.

The duration of electrical power undersupply ($T_{lim.emerg..No.\ xxx}$) for a concrete emergency situation (see Item 2.2.1.) is determined from the minimum of mean repair time of the damaged device ($T_{rep.\ emerg.\ No.\ xxx}$) or mean time of mobilization of the generating power ($T_{mob.\ emerg.\ No.\ xxx}$):

$$T_{lim.\ emerg.\ No.\ xxxx} = min(T_{rep.\ emerg.\ No.\ xxx},\ T_{mob.\ emerg.\ No.\ xxx}) \quad (3)$$

 DOI: 10.1051/e3sconf/20172503004

Therefore, the value of electrical power undersupply ($dW_{lim.\ emerg.\ No.\ xxx}$) for a concrete emergency situation is determined from the value of power limitation ($dP_{lim.\ emerg.\ No.\ xxx}$) and duration of power undersupply ($T_{lim.\ emerg.\ No.\ xxx}$):

$$dW_{lim.\ emerg.\ No.\ xxx} = dP_{lim.\ emerg.\ No.\ xxx} \cdot T_{lim.\ emerg.\ No.\ xxx}) \qquad (4)$$

The expectation value of the annual monetary loss (ML) from the concrete situation ($ML_{ann.\ emerg.\ No.\ xxx}$) depends on failure flow for a year for concrete damaged equipment during this emergency ($w_{emerg.\ No.\ xxx}$):

$$ML_{ann.\ emerg.\ No.\ xxx} = w_{emerg.\ No.\ xxx} \cdot dW_{lim.\ emerg.\ No.\ xxx}\ Y_{0.RF} \qquad (5a)$$

$$ML_{ann.\ emerg.\ No.\ xxx} = w_{emerg.\ No.\ xxx}\ dW_{lim.\ emerg.\ No.\ xxx}\ Y_{0.\ Reg.} \qquad (5б)$$

The expectation value of the annual loss from all emergencies in the power system associated with the electrical power undersupply will be determined from the arithmetical sum of expectation values of annual losses from concrete emergencies ($ML_{ann.\ emerg.\ No.\ xxx}$):

$$ML_{ann.emerg} = \sum_{i=1}^{n} ML_{ann.emerg\ №i} \qquad (6)$$

Besides the loss of the expected electrical power undersupply, it is necessary to take into account a probable loss from unforeseen tripping of consumers or generation. It is suggested to consider this loss only in those cases and volumes which are stipulated in the agreements (agreements of power supply or technological connection). If these aspects are not stipulated in the agreement, the loss is not considered. The maximum allowed time of electrical power undersupply for concrete consumers should correspond to agreement conditions and category of consumers (according to the decrees of the Russian Federation Government).

3 Conclusion

This paper presents methods for assessment of system reliability during planning of power system development. The method proposed contains general description of the step-by-step algorithm with the help of which it is possible to compare options and scenarios of power system development and to assess the level of reliability control for validation of decision making.

The most significant assessment of security is the evaluation of losses during long power limitation of consumers (over 20 and 40 minutes). The algorithm proposed allows the calculation of probable losses for comparative analysis of different options of PS development. Operational security should be provided, first of all, with adequate and effective RP and EC devices. Only in those cases when even short-term tripping of generation or consumers by EC operation are unacceptable (due to specific conditions), security should be provided with the power grid construction.

References

1. N.I. Voropai, D.N. Efimov et al., *Smart Grid, IEEE Transactions on*. Vol. 4, **no. 3**. P. 1732-1740. (2013)
2. N.I. Voropai, A.B. Osak, *Energy Policy*. **No. 5**. P. 60-63. (2014).
3. Eds. N.I. Voropai, *Reliability of energy systems (Collection of recommended terms)* (Moscow: "Energia", 2007. 192 p).
4. *Reliability concepts in bulk electric power systems* (New York: North American Electric Reliability Council, 1985. 96 p).
5. N. Balu, T. Bertram, A, Bose et al., *Proceedings of the IEEE*. Vol. 80, **No. 2**. P. 262-280. (1992).
6. R.J. Marceau, J. Endrenyi, R. Allan et al., *Electra*. **No. 175**. P. 49-77. (1997).
7. Ed. N.A. Manov, *Methods and models for studying power system reliability* (Syktyvkar: Komi SC UrB RAS. 2010. 292 p).
8. Eds. N.I. Voropai, G.F. Kovalev, *Concept of reliability control in electric power engineering* (M.: LLC PH "Energy". 2013. 304 p).
9. *Guidelines for stability of power systems*, approved by Order No. 277 of 30.06.2016 of Russian Ministry of Energy.
10. *STO 59012820.29.240.007-2008. Rules for prevention of development and disturbance liquidation of the normal state of electrical annex of power systems*. Standard JSC "SO UPS".
11. *STO 59012820.27.010.002-2016. Rules of transition to operation in forced regime in controlled sections of the dispatcher center JSC SO UPS*, Standard JSC "SO UPS".
12. A.B. Osak, A.I. Shalaginov, D.A. Panasetsky, E.Ya. Buzina, *Book of reports of international conference "Modern tendencies of System Development of Relay Protection and Automation of Power System"*, Sochi. 1-5 June 2015.
13. A.B. Osak, A.I. Shalaginov, D.A. Panasetsky, E.Ya. Buzina, *Methodic problems of studies of reliability of large power systems*, Syktyvkar: LLC "Komi Republican Printing House", p. 337-346. (2016).

Consideration of uncertainty of information on load and generation values under daily reconfiguration of primary distribution network

Irina Golub[1,*], *Oleg*Voitov[1], *Evgeny* Boloev[1] and *Lyudmila* Semenova [1]

[1]Melentiev Energy Systems Institute, 664033 Lermontov str., 130, Irkutsk, Russia

Abstract. Intelligent electric networks make it possible to remotely control switchings, thus preventing overloads and enhancing reliability of power supply to consumers. In the paper, the daily reconfiguration of a primary distribution network is considered as a means to reduce energy losses in the network. An algorithm applied for the reconfiguration is based on the methods of the theory of graphs and includes a high-speed program for load flow calculation. The research is focused on the impact of renewable generation and active demand with loads changing depending on daily variation in electricity price on the reduction in losses at reconfiguration. An algorithm is proposed to optimize load curves of load-controlled consumers. The calculation of probabilistic load flows is applied to assess the impact of the uncertainty of nodal power forecast on energy loss reduction at reconfiguration. The results of the research demonstrate the effectiveness of the proposed approaches, and are illustrated by a 33-node test distribution network.

1 Introduction

Most of the energy losses in the electric power systems are known to occur in distribution networks. These losses make up 10-13 % of all the generated electric energy. The problem of energy loss reduction in the distribution network that consists of a medium voltage distribution network (primary) and low voltage distribution network (secondary) remains important for engineers and researchers. This paper is concerned with an issue of reducing energy losses in the primary distribution network which feeds secondary distribution network transmitting energy to end consumers.

The primary distribution networks are weakly closed, however owing to normally open tie switches between feeders, they operate as open. Apart from the normally open tie switches, there are normally closed sectionalizing switches that can disconnect one of the feeder sections. Closing of a tie switch and opening of a respective closed switch allow us to obtain a new radial configuration of the distribution network. Such a procedure called reconfiguration algorithm enables us not only to enhance power supply reliability but also to reduce energy losses and hence energy consumed from supply network, currents in tie lines, voltage losses in distribution network, and, moreover, more fully involve renewable generation in the case it is used in the network.

In a traditional passive distribution network power flows go from primary distribution substation being the only supply source along the branches of a tree scheme to a leaf node. Voltage deviations increase as the distance between source and the dangling node rises. Thus, voltage behavior in the traditional distribution network is predictable and its monitoring is unnecessary.

Transition to active distribution network is related to the use of distributed generation, energy storage systems, and active demand. Directions of flows in the branches of the active distribution network vary during a day. The load nodes can become generator nodes and voltage deviations can exceed admissible values. Therefore, such networks must be monitored.

In passive distribution network the flows and voltage are measured only at the primary distribution substation. The conditions of the secondary distribution substation are normally unknown [1]. In the future intelligent distribution networks active distribution networks will be part of advance metering infrastructure [2] and operator of distribution network will receive data on loads and generation in real time. In this paper, however, we assume that the distribution network operator has a short-term forecast of active and reactive loads and active power generation of renewable energy sources represented by wind turbines and photovoltaics. Daily load curves and generation schedules of renewable energy sources are taken from the research [3] which is also devoted to the problem of energy loss reduction under hourly reconfiguration of distribution network. Energy losses per hour are numerically equal to average power losses during an hour. Therefore, in an analysis of losses for a concrete hour we will consider power or energy losses.

Since the forecasts of loads and generations contain errors, the most pressing problems are the assessment of an impact of forecast errors on the errors of energy losses in the distribution network under its daily

*Corresponding author: golub@isem.irk.ru

 DOI: 10.1051/e3sconf/20172503005

reconfiguration, and the assessment of reconfiguration justifiability under uncertain initial information.

In our research reconfiguration is considered as the main tool of reducing power losses in the distribution network. A great many algorithms for solving this problem are presented in publications. According to [4] they include the algorithms of mixed integer and nonlinear programming, and heuristic methods such as Genetic Algorithms, Artificial Neural Networks, Ant Colony, Harmony Search, and Tabu Search. The other algorithms involve linear load flow to calculate losses at network reconfiguration, because nonlinear load flow is considered to be time-consuming. In [4], the authors solving the problem of distribution network reconfiguration apply both the algorithms for the construction of a maximum spanning tree and the simplified approaches to specification of currents and power losses. It is probably this simplification that did not allow the researchers to find better solutions which were obtained with the help of other algorithms, for example, in [5].

The proposed reconfiguration algorithm is based on the methods known in the theory of graphs and intended for the construction of a maximum spanning tree [6] and determination of branches of independent loops by their chords [7]. These methods are included in the high-speed program of steady state calculation, which is the main advantage of the proposed algorithm compared to the reconfiguration algorithm in [4]. The main stages of the algorithm that does not take into consideration the presence of several generation sources in the distribution network are presented in Section 1.

Section 2 is focused on the study of the impact of renewable generation and/or reconfiguration on losses. The impact of renewable generation on the number of switchings at reconfiguration is studied, and the possibility of determining a scheme of switchings providing the minimum daily losses within the entire range of variations in nodal power is demonstrated. The assessment of losses was illustrated along with the expected characteristic of voltage, whose deviations decline with reduction in losses. Moreover, the impact of renewable generation and reconfiguration on the currents in distribution network is analyzed.

Section 3 is concerned with another possibility of reducing losses of energy and currents in tie lines. This is regulation of hourly energy consumption by load controlled consumers, which leads to a change in the daily load curve. An overview of the methods for active demand control is presented in [8]. The authors of [8] solve this problem by the linear programming method.

In Section 4 the linear analytical method of probabilistic load flow is applied to assess the impact of uncertainty of the day-ahead forecast of loads and generation on energy losses caused by the distribution network reconfiguration. In [9], the impact of uncertainty of the initial information on losses due to reconfiguration is studied by using the probabilistic load flow based on the point method.

2 A reconfiguration algorithm

Power losses in a closed network are normally lower than in an open network [10]. This condition is not met in the event of circulating and interchange currents in the loop. The circulating currents are caused by phase-shifting devices in the loop and the interchange currents are caused by several supply sources. Nevertheless, in the case of reconfiguration, it is necessary to make where possible power losses in the open network close to the losses in the closed network.

This condition can be met if we construct a spanning tree with the minimum sum of power losses in its chords in the closed network. Such a criterion however may prove unacceptable, if resistance in the branch with high current virtually equals zero. Therefore, the criterion of the minimum sum of absolute values of currents in chords is more reliable. The maximum spanning tree with the minimum currents in chords can be constructed using the known method of the theory of graphs [6]. In this method, the network graph branches are arranged in a descending order of absolute values of currents in them. In a cycle, depending on the number of branches, the branches, in which either one or both nodes are not yet included in the tree, are successively connected to the spanning tree. In the event that both nodes of the branch enter the tree, such a branch is called a chord. The algorithm divides all the branches of the network graph into the branches of the spanning tree and the chords.

To determine branches of each independent loop by its chords, we construct a submatrix of a block of trees of the second incidence matrix [7]

$$N_t = -M_{ch}^T \left(M_t^T\right)^{-1}. \qquad (1)$$

This matrix contains the number of rows equal to the number of chords and the number of columns equal to the number of tree branches, where M_t and M_{ch} – are submatrices of the first incidence matrix, that correspond to the branches of the tree and chords. Expression (1) serves as a basis for the topological algorithms [10] used to construct the first incidence matrix $\left(M_t^T\right)^{-1}$ inverse to the block of trees and the second incidence matrix N_t

The algorithm consists of two steps. In the first step, we determine the composition of disconnected chords. In the beginning, all the switches are closed. Then, the load flow is calculated in the number of iterations equal to the number of independent loops in the graph of the distribution network and a chord with the minimum current to be open is determined.

In the second step, we check the possibility of reducing power losses by replacing the chords opened in the first step by the branches of the spanning tree in a loop connected to the chord. For each of the chords considered in the first step we do the following: close the next chord; identify the branches of the loop connected to it; calculate the load flow; mark the nodes to the left and to the right of the chord nodes that have the degree above two; simulate a successive disconnection of branches situated between such nodes, with calculation of load flow and determination of total losses; open the previous or new chord that corresponds to the minimum total losses. If the loop does not have nodes with the degree above 2, then the disconnection of all branches of the loop is simulated. It is recommended to repeat the second step of the algorithm once again.

3 Hourly reconfiguration of distribution network

Power losses under optimal configuration, which were taken for some conditions, will not be optimal in the entire range of variation in the nodal powers. This means that reconfiguration should follow changes in loads and generation. We will illustrate the efficiency of applying the proposed algorithm of distribution network reconfiguration to reduce losses at each hour of a daily load curve and generation.

The algorithm performance and subsequent analysis of factors that have an effect on the losses are illustrated by a 33-node scheme [11] (Fig. 1), including 37 sectionalizing switches and 5 tie switches. Daily curve of an hourly variation in active, Fig. 2a, and reactive power loads at nodes 1-33 and variations in active power generated by renewable energy sources at nodes 6,9,13, 32, Fig. 2b, are taken from [3]

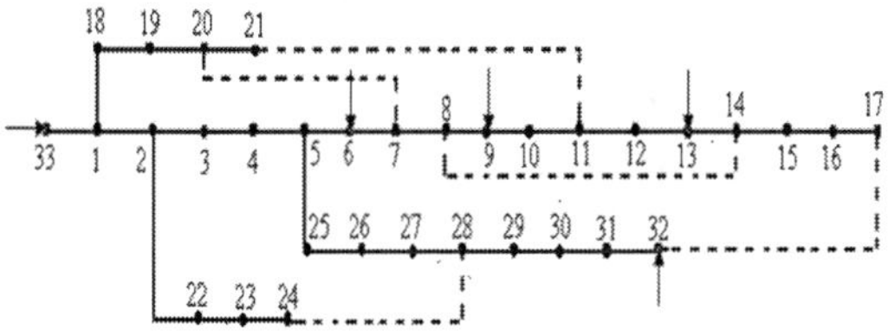

Fig. 1. Scheme of a 33-node distribution network. The lines with the tie switches are shown by dotted lines.

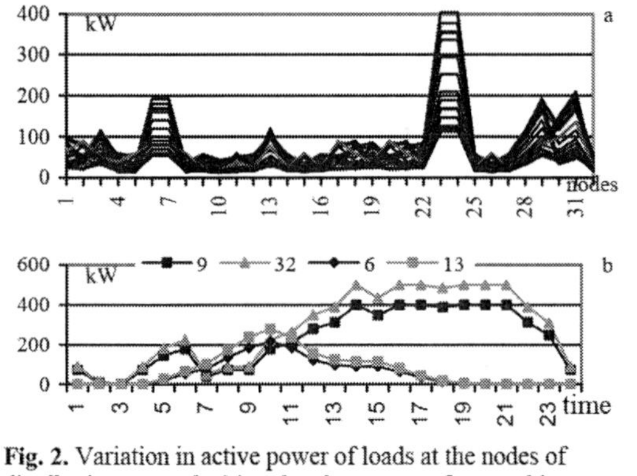

Fig. 2. Variation in active power of loads at the nodes of distribution network, (a) and active power of renewable energy sources (b) during a day.

Figures 3 a, b demonstrate the performance of the reconfiguration algorithm for a test network for loads and generation at hour 11. A reduction in the losses for the option with renewable generation after opening three tie lines on the first step is caused by the interchange currents.

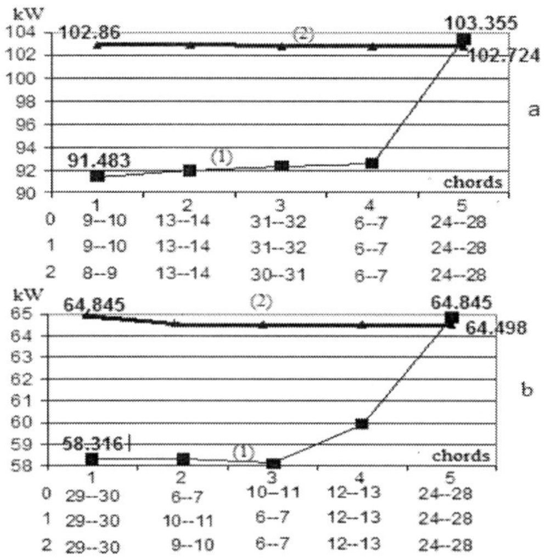

Fig. 3. Change in power losses at reconfiguration of a test network, without (a) and with (b) renewable generation, 0 – an initial composition of chords, arranged in a descending order of currents in them; 1 – chords obtained after the first step (1); 2 – chords obtained after the second step (2).

Figure 4 presents the values of daily energy losses and curves of variation in hourly energy losses in a distribution network with and without reconfiguration and renewable energy sources.

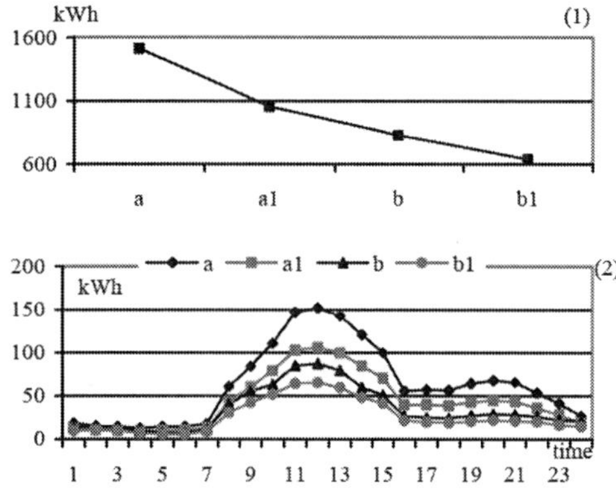

Fig. 4. A change in the daily (1) and hourly (2) energy losses in distribution network with and without reconfiguration and presence of renewable energy sources, a – without renewable generation and without reconfiguration, a1 – without renewable generation and with reconfiguration, b –with renewable generation and without reconfiguration, b1 – with renewable generation and with reconfiguration.

Comparison of the losses made it possible to draw the following conclusions:

- Renewable generation has a greater effect on losses than network reconfiguration. The daily energy losses make up 829.37 kWh and 1055.95 kWh, respectively;
- Simultaneous use of renewable generation and reconfiguration provides more than a two-fold reduction in losses equal to 636.98 kWh compared to the losses in the distribution network without renewable generation and reconfiguration, 1513.97 kWh;

 DOI: 10.1051/e3sconf/20172503005

- The involvement of renewable energy generation leads not only to a change in power flows in lines, but also to a change in their directions, and as a result, to an increase in the total number of switchings at daily reconfiguration. Their number without renewable energy sources is 16 and with renewable energy sources – 76.

A great number of switchings can lead to a switching cost commensurate with or exceeding the cost of reducing energy losses at reconfiguration. Figure 5 illustrates the possibility of selecting a constant distribution network configuration ensuring the minimum daily energy losses.

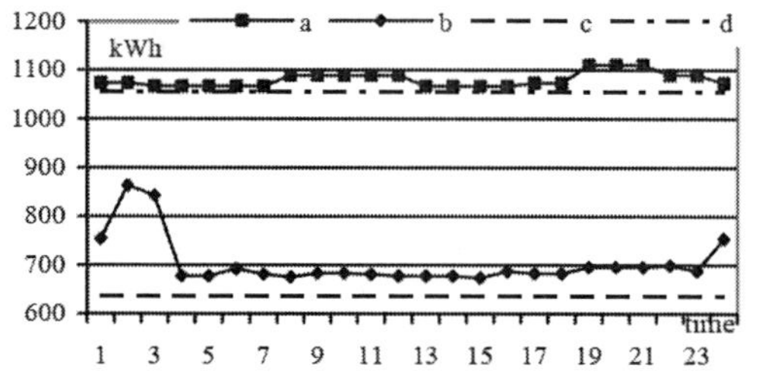

Fig. 5. Comparison of daily energy losses at constant configuration of distribution network without renewable generation – a, and with renewable generation – b, with the daily losses obtained for the optimal reconfiguration without renewable generation – c and with renewable generation – d.

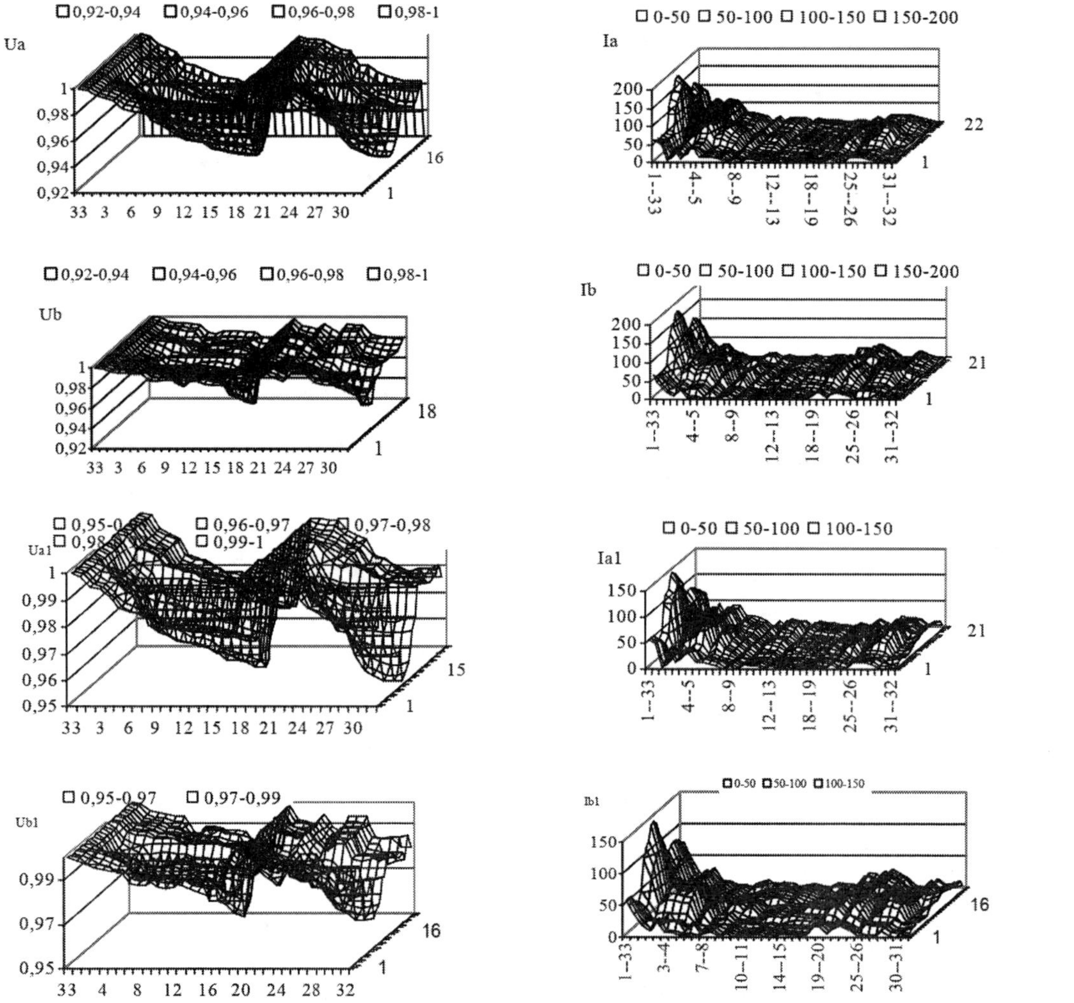

Fig. 6. Hourly variations in voltage at nodes and current in branches of the test network with and without reconfiguration and renewable generation, Ua, Ia – (voltage and current) without renewable generation and reconfiguration, Ub, Ib – without renewable generation and with reconfiguration, Ua1, Ia1 – with renewable generation and without reconfiguration, Ub1, Ib1– with renewable generation and with reconfiguration.

Figure 5 shows the values of daily losses under constant distribution network configuration coinciding with each of the hourly configurations obtained for it with and without renewable energy generation in the network. In the former case such losses are compared with daily energy losses in the network without renewable generation but with reconfiguration, that are equal to 1055.95 kWh, and in the latter case, they are compared with the losses in the network with renewable generation and with reconfiguration, that are equal to 636.9 kWh.

Comparison of daily losses shows that under constant configuration of distribution network with renewable generation that corresponds to hour 15, losses (673.6 kWh) will be higher than the optimal losses (636.9 kWh) by 36.7 kWh. Thus, a loss reduction by 36.7 kWh required 76 switchings. In the case study without renewable generation the minimum losses under constant daily configuration coinciding for hours 3-7 and 13-16 differ from the losses under optimal configuration only by 10.5 kWh and require 16 switchings. The examples are indicative of the need to assess the cost-effectiveness of reconfiguration, which can be obtained by comparing the costs of switching and reduction in power purchase costs at loss reduction.

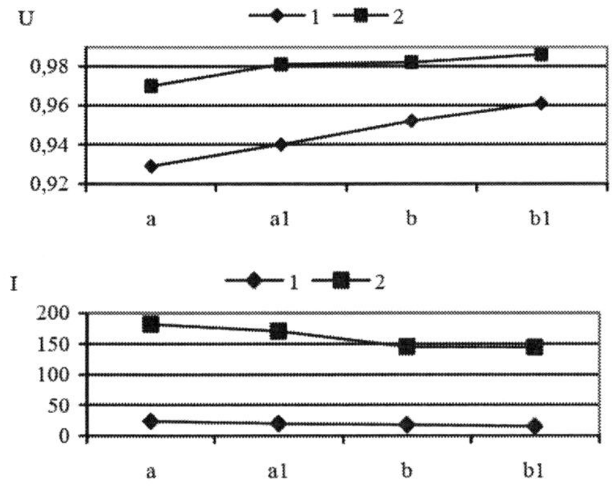

Fig. 7. Minimum 1 and average 2 voltage (U), average 1 and maximum 2 currents (I) in tie lines, a – without renewable generation and without reconfiguration, b - without renewable generation and with reconfiguration, a1– with renewable generation and without reconfiguration, b1- with renewable generation and with reconfiguration.

Hourly reconfiguration and involvement of renewable generation in the distribution network have a positive effect on both the reduction in losses and the decrease in the deviations in voltage and currents in the distribution network branches. Figure 6 demonstrates the plots characterizing hourly variations in voltage at nodes and currents in branches of the test network with and without reconfiguration and renewable generation. Their analysis made it possible to make the following conclusions, with their numerical confirmation demonstrated in Fig.7.

- Reconfiguration leads to a reduction in medium and maximum currents of feeders and fosters an increase in the number of nodes with high voltage.
- Renewable generation has a stronger effect on the reduction in average current and increase in voltage than reconfiguration. This is caused first of all by the fact that the distance of power transmission from source to load is reduced. Figure 8 shows a scheme with disconnected tie switches with power flows going from the network and renewable generation sources.
- The maximum reduction in current and increase in the minimum voltage are achieved when both renewable generation and changes in the distribution network topology by reconfiguration together affect the operation.

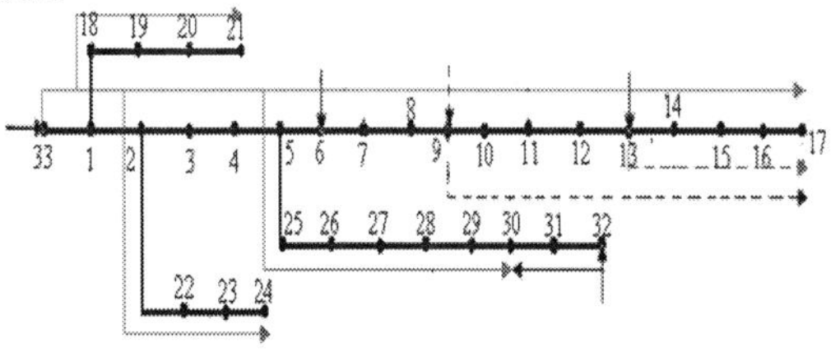

Fig. 8. Active power flows from network and renewable generation sources to the load nodes.

4 An algorithm for optimization of daily load curves

Optimization of daily load curves of load-controlled consumers is an additional effective way to minimize losses and reduce currents in distribution network. The problem of linear programming with constraints is applied to minimize the cost of electricity purchased at prices varying during the day

$$min \sum_{t=1}^{n_t} \sum_{i=1}^{n_p} c_t P_i^t \tag{2}$$

where n_t – the number of hourly intervals during the day; n_P – number of load-controlled consumers; c_t -- power price at hour t; P_i^t – load of the i-th load-controlled consumer at hour t.

The constraints for each time interval include: equations of total balances of active and reactive power under constant power factor; equations of invariability of daily power consumption by each load-controlled consumer; and inequalities, determining feasible ranges of variation in power of load-controlled consumers and a slack node.

All the necessary information for solving problem (2) that lies in the determination of optimal hourly values of powers P_{i*}^t of load-controlled consumer, is determined by calculating feasible load flow [12].

The repeated calculation of a feasible load flow is made after solving problem (2) either. This is necessary for both the assessment of feasibility of all variables obtained after optimization of load curves and the assessment of an impact the optimization has on the reduction in daily energy losses. In the event that the constraints on the state variables in interval t_k are not met the upper $\overline{P_i^{t_k}}$ or lower $\underline{P_i^{t_k}}$ constraints on load $P_i^{t_k}$ are corrected:

$$\overline{P_i^{t_k}} = P_i^{t_k} + s_d^{t_k}\,(P_{i*}^{t_k} - P_i^{t_k}),\ if\ P_{i*}^{t_k} > P_i^{t_k};$$

$$\underline{P_i^{t_k}} = P_i^{t_k} + s_d^{t_k}\,(P_{i*}^{t_k} - P_i^{t_k}),\ if\ P_{i*}^{t_k} < P_i^{t_k},$$

where $s_d^{t_k} > 0$ – maximum feasible step of a variation in loads of load-controlled consumer in the direction of vector whose i-th component equals $(P_{i*}^{t_k} - P_i^{t_k})$. After the limiting values in the intervals with unmet constraints are adjusted, problem (2) is solved again.

Figure 9 shows the saving costs of energy purchase at the prices varying during the day and variations in the daily load curves of load-controlled consumers at nodes 23 and 24 of a test network with the maximum total daily energy consumption equal to 10844.4 kWh.

The range of an feasible reduction in power of a load-controlled consumer from hour 8 to hour 17 is set equal to from 6 % to 26 %.

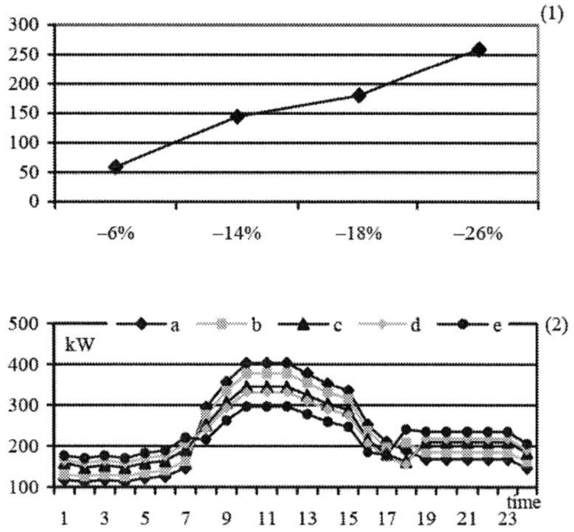

Fig. 9. A saving costs on energy (1) in conventional units and change in the daily load curves of load-controlled consumers (2) at nodes 23 , 24 under a feasible reduction in the maximum load at hours 8 -17 of the day by 6% - b, 14% - c), 18% - d) and 26% - e) compared to the initial load curve a).

Figure 10 presents daily and hourly energy losses that make it possible to assess the contribution of the 26% reduction in maximum loads of the load-controlled consumer to the losses: in the distribution network without renewable generation (without reconfiguration and with reconfiguration), with renewable generation (with and without reconfiguration). Comparison of the results demonstrates that load-controlled consumers have a lesser impact on the losses than renewable generation and reconfiguration. The daily losses equal to 613.91 kWh obtained by simultaneously using reconfiguration, renewable generation and load-controlled consumers are only by 23.07 kWh lower than the losses under the joint use of renewable generation and reconfiguration. The influence of load-controlled consumers on voltage losses in distribution network was also little. The introduction of load-controlled consumers in the cases with renewable generation without reconfiguration and renewable generation with reconfiguration reduced maximum currents in the distribution network from 145.34 A to 134.78 A and from 144.275 A to 133.716 A, respectively.

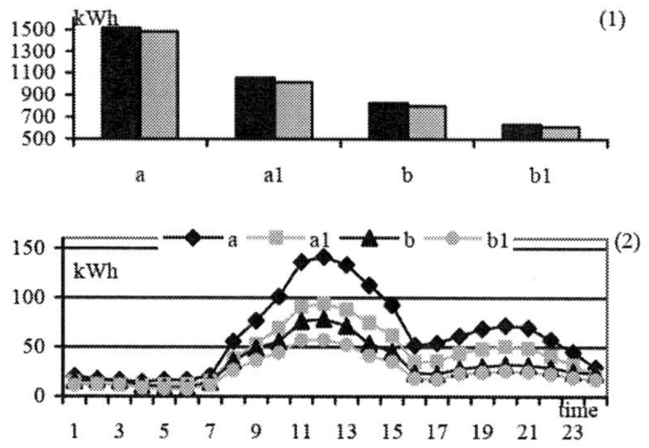

Fig. 10. Values of daily energy losses without load-controlled consumers (black) and with load-controlled consumer (grey) with a feasible reduction in their maximum power by 26% (1) and variant in hourly energy losses in the distribution network with load-controlled consumer (2) a – without renewable generation and without reconfiguration, a1 – without renewable generation and with reconfiguration, b – with renewable generation and without reconfiguration, b1 – with renewable generation and with reconfiguration.

5 Assessment of the influence of the uncertainty of load and generation data on reduction in power losses at reconfiguration of distribution network

The mathematical mean of power losses is known to increase with a rise in the uncertainty of nodal power data [13]. The question arises if the errors of determining hourly energy losses caused by the errors in the load and generation forecasts are commensurate with the payoff from the expected decline in losses under reconfiguration of the distribution network.

To answer this question we first determined optimal switchings and hourly and daily energy losses corresponding to them for the forecast values of nodal powers for each hour of a daily load curve and renewable generation. Then, the data on hourly loads, generation and hourly configuration of the distribution network were used to calculate the probabilistic load flow, which involved the determination of the standard deviations of hourly energy losses.

The standard deviations of loads and generation of renewable generation necessary for probabilistic load flow were determined by using an inverse function of errors for a specified error of load and generation data, equal to 5%, 10%, 15%, 20% of their forecast values representing mathematical means.

 DOI: 10.1051/e3sconf/20172503005

The inverse function of errors enables the determination of standard deviations σ_x of normally distributed random variable x, by a specified value of interval $\Delta\varepsilon_x$, in which this variable will be with a set probability p, which in our case is equal to 0.95,

$$\sigma_{x_{p=0.95}} = \Delta\varepsilon_x / \left(\sqrt{2} erfinv(p)\right) = \Delta\varepsilon_x / 1.96$$

The probabilistic load flow was calculated by linear analytical method of moments. According to this method the mathematical means and covariances of nodal power at the point of solution to the nonlinear systems of steady state equations are used to calculate mathematical means and covariances of absolute values and phases of voltage, power flows and losses [14].

Interval $\Delta\varepsilon_x$ of a potential change in any of the indicated variables x, including that of energy losses, was determined by their standard deviations $\Delta\varepsilon_x = \sqrt{2} erfinv(p)\sigma_x$. An interval of a potential variation in the random value as a percentage of mathematical mean μ_x was determined as $\Delta\varepsilon_x\% = \sqrt{2} erfinv(p)\sigma_x \cdot 100\% / \mu_x$.

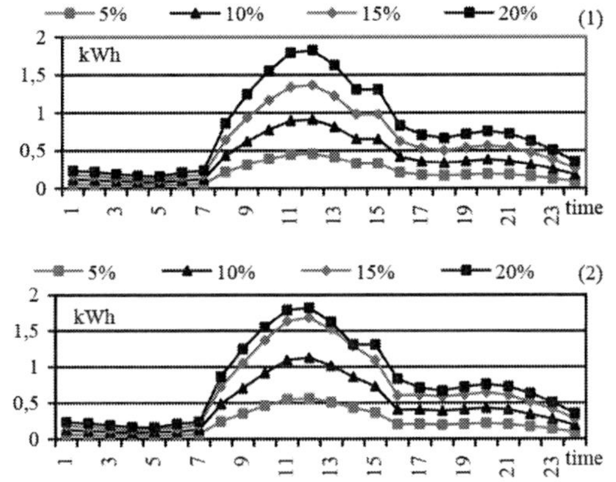

Fig. 11. Standard deviation of hourly energy losses for the scheme with renewable generation and with reconfiguration (1), and without renewable generation and with reconfiguration (2).

Figure 11 demonstrates standard deviations of hourly energy losses for distribution network with reconfiguration (with and without renewable generation). The deviations show that the hourly deviations of energy losses in both cases do not exceed 2 kWh. At the same time the hourly errors of determining energy losses, i.e. intervals of deviation of hourly energy losses from their mathematical means (Fig.12), for a 20% error of the nodal power forecast do not exceed 8%.

The standard deviations and errors of daily energy losses presented in Fig.13 show that at the maximum error in the forecast they do not exceed 6 kWh and 1.5% of the mathematical mean of losses.

The presented results confirm that with an increase in uncertainty of data on nodal power, hourly and daily energy losses rise, which cannot serve as grounds for the rejection of the reconfiguration aimed at reducing losses in the network.

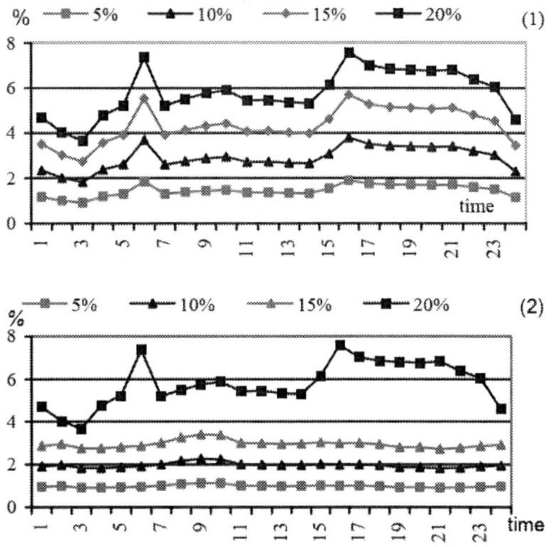

Fig. 12. Errors in the determination of hourly energy losses for the scheme with renewable generation and with reconfiguration (1), and without renewable generation and with reconfiguration (2).

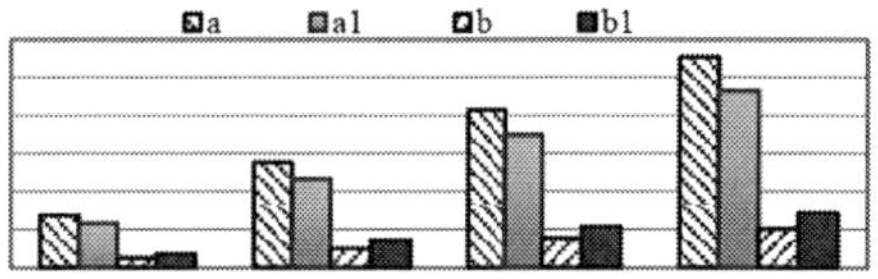

Fig. 13. Standard deviations (a, a1) (kWh) and errors (b, b1) (%) of daily energy losses for the network with renewable generation and with reconfiguration (a, b), without renewable generation and with reconfiguration (a1,b1).

6 Conclusions

1. A topological algorithm for distribution network reconfiguration is generated to reduce power losses, and the possibility of its application to the hourly reconfiguration of the network, in particular with available renewable energy sources in the distribution network is shown.
2. An algorithm is suggested to optimize daily load curve of a load-controlled consumer by the criterion of minimization of energy purchase costs, which also makes it possible to reduce the losses in distribution network.
3. The paper is focused on the study of the influence of renewable energy sources, hourly reconfiguration, optimization of load curve of load-controlled consumers, and their joint use on the daily energy losses.
4. The possibility of choosing an invariable distribution network configuration which provides the daily energy losses comparable with the losses determined at the hourly reconfiguration of distribution network is demonstrated.

DOI: 10.1051/e3sconf/20172503005

5. The numerical results confirmed the effectiveness of the suggested algorithms.
6. The study shows that the errors in the calculation of losses rise with an increase in the errors of the nodal power forecast, but they do not have a decisive influence on the reduction in losses at reconfiguration.

References

1. O.N. Voitov, I.I. Golub, L.V. Semenova, Elektrichestvo **9**, 38 (2010)
2. M. Emmanuel, R. Rayudu, JNCA **74**, 133 (2016)
3. M.R. Dorostkar-Ghamsari, M. Fotuhi-Firuzabad, M. Lehtonen, A. Safdarian, IEEE Trans. on Power Systems **31,** 1879 (2016)
4. H. Ahmadi, J.R. Marti , SEGAN **1**, 1 (2015)
5. D. Zhang, Z. Zhang, I. Fu, Electr. Power Sys. Res. **77**, 10 (2007)
6. E. Mainika, *Algorithms for optimization on networks and graphs* (Mir, Moskow, 1981)
7. N.A. Melnikov *The matrix method for analysis of electric circuits, M: Energiya.* 288 (1972) (in Russian)
8. M. Behrangrad, Renew. Sustain. Energy Rev. **47**, 270 (2015).
9. C.L. Su. IEEE Trans. Power Syst. **25,** 786 (2010)
10. A.A. Glazunov, A.A Glazunov. *Electric networks and systems* (Gosenergoizdat, Moskow, 1960)
11. M.E Baran, F.F. Wu, IEEE Trans. Power Del. **4**, 1401 (1989)
12. N.A. Murashko, Y.A. Okhorzin, L.A. Krumm *Analysis and control of steady states of electric power systems* (Nauka, Novosibirsk, 1987)
13. V.Z. Manusov, A.V. Mogilenko, Elektrichestvo, **3**, 2 (2003)
14. E.V Boloev, O.N. Voitov, I.I. Golub, Akta Energetica **2(23),** 91 (2015)

DOI: 10.1051/e3sconf/20172503006

Improving the principles of short-term electric load forecasting of the Irkutsk region

Vladimir Kornilov[1], *Victor* Kurbatsky[2], and *Nikita* Tomin[2]

[1] JSC "SO UPS" Irkutsk Regional Dispatch Office (RDO), Irkutsk, Russia

[2]Melentiev Energy Systems Institute, 130 Lermontov str., Irkutsk, Russia

Abstract. Forecasting of electric load (EL) is an important task for both electric power entities and large consumers of electricity [1]. Large consumers are faced with the need to compose applications for the planned volume of EL, and the deviation of subsequent real consumption from previously announced leads to the appearance of penalties from the wholesale market. In turn, electricity producers are interested in forecasting the demand for electricity for prompt response to its fluctuations and for the purpose of optimal infrastructure development. The most difficult and urgent task is the hourly forecasting of EL, which is extremely important for the successful solution of problems of optimization of generating capacities, minimization of power losses, dispatching control, security assessment of power supply, etc. Ultimately, such forecasts allow optimizing the cash costs for electricity and fuel or water consumption during generation. This paper analyzes the experience of the branch of JSC "SO UPS" Irkutsk Regional Dispatch Office of the procedure for short-term forecasting of the EL of the Irkutsk region.

1 The existing functional scheme for forecasting

At present, this task is carried out with the use of the corporate software "Software of the hierarchical forecasting system for planning and scheduling of regimes" (HSP), developed by Energostat. [2]

The first step is the launch of the statistical forecast module. As a result of the module work, EL predictions are generated for the period of the corresponding planning cycle. The following models of statistical forecasting are used in HSP software:

• HSP forecasting model taking into account seasonality and weather factors.

• model of the previous average day, taking into account meteorological factors;

At the same time, the forecast is carried out taking into account the forecast of meteorological factors obtained from the Hydrometeorological Center of Russia, and if they are not available, then a statistical forecast model of meteorological factors based on the seasonal temperature curve and the latest actual data is used for their formation. In each branch of RDO of JSC "SO UES" forecasting is carried out for each territory of the operational area of the given branch and all areas of forecasting of subordinate branches (up to power areas of power systems).

The forecasting models are set up by the developer who, according to certain methods, makes changes in the tuning coefficients of the forecast models, depending on the predicted temperature of the ambient air. Earlier, classical techniques for daily average temperatures were used, when the average daily temperature was averaged over the last 3 days: for the autumn-winter period (AWP) with weights 0.5, 0.3, 0.2; for summer - with weights 0.7, 0.2, 0.1. However, this weighted classical averaging of the average daily and hourly temperatures does not accurately simulate the dynamics under sudden temperature fluctuations. Proceeding from this, there are difficulties in modeling processes in hot weather, when it is necessary to take into account the fact of long-term preservation of high temperatures of outside air. This requires the use of large values of the lag parameter. At the same time, there are situations when it is necessary to instantly increase the conditioning loads (at temperatures above 30 degrees), which already requires a low lag parameter value. In addition, in hot weather, cloudiness can lower EL, because conditions for cooling the radiators of air conditioners improve.

According to the developer's technique, the tuning factors are selected from the condition of providing the best forecasting accuracy. The temperature delay parameters are defined for the AWP - 23 hours, for the summer period - 5 hours. At the same time, a decrease in the delay in the effect of temperature in the currently used models can lead to a decrease in the overall accuracy of the prediction. On certain days the forecast will be fulfilled perfectly, but on other days it can be much worse.

2 Forecast of hourly EL of Irkutsk region

Historically, the forecasted hourly EL of the Irkutsk region per day X-1 (i.e., one day before the start of the operational day) is defined as the sum of the forecasts of three major energy areas: Irkutsk-Cheremkhovo, Bratsk-Ust-Ilimsk and Bodaybo (Figure 1). For hourly forecasting of EL of these regions, there is a fairly large array of archival information, on the basis of which automated prediction models having sufficient accuracy have been created.

As can be seen in Fig. 2., the predictive models used based on the "seasonal curves" method [17] give a good forecast accuracy, but there are certain shortcomings revealed during the daily use of the method used. This is particularly pronounced in the unsatisfactory operation of models with sufficiently sharp and significant temperature changes, with a deviation from the average statistical values in the days in question.

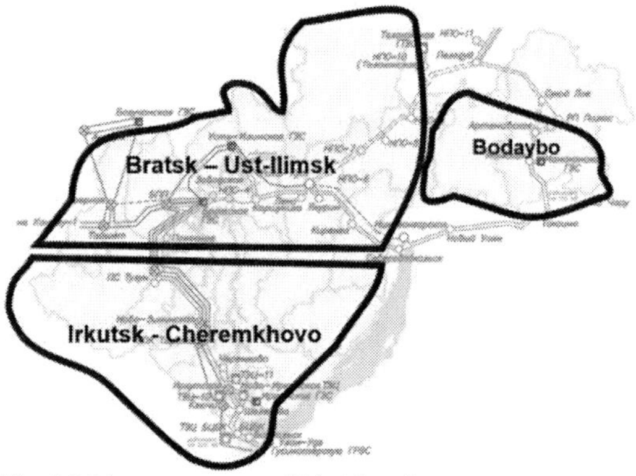

Fig. 1. Major energy areas of Irkutsk region

The annual change in the EL values of these regions for 2016 is shown in Fig. 2.

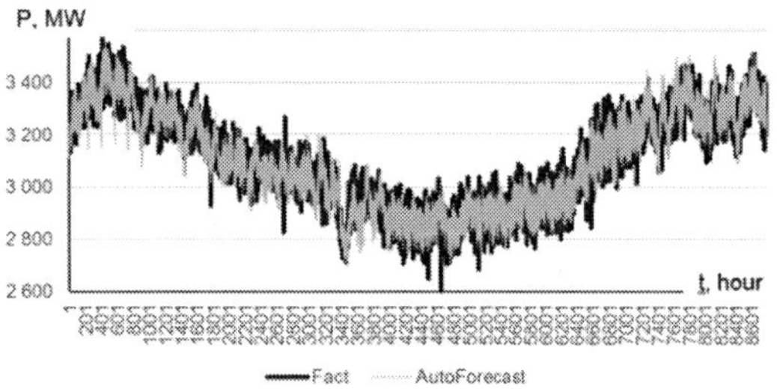

a. Hourly load of Bratsk-Ust-Ilimsk area

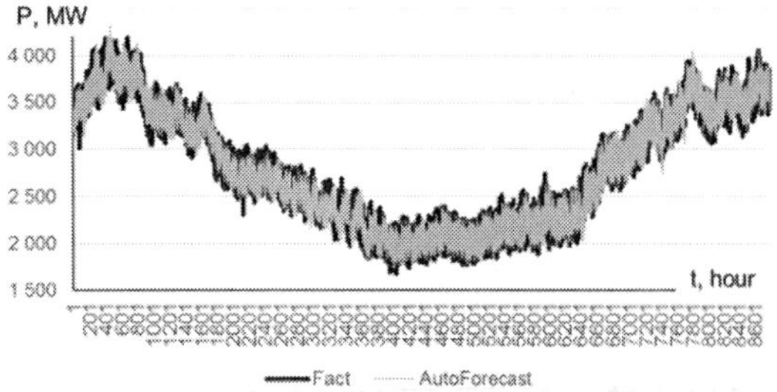

b. Hourly load of Irkutsk-Cheremkhovo area

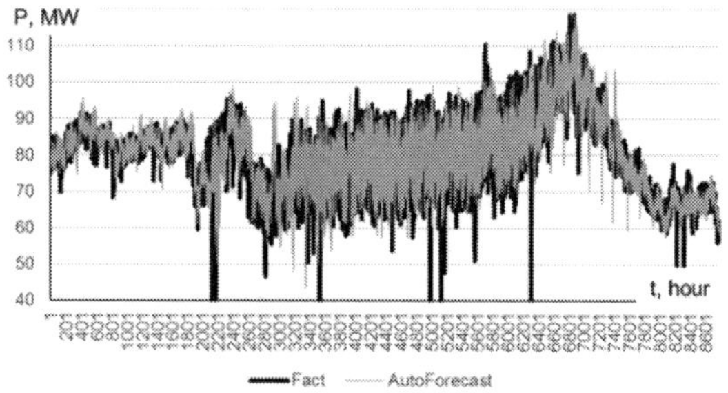

c. Hourly load of Bodaybo area

Fig. 2. Annual EL graphs of Irkutsk region

The method of "seasonal curves", based on the correlation and regression analysis of data, allows to analytically describe the fluctuations in the load and weather factors in the during of the year. The method assumes additivity of the load:

$$P(i) = P_0(i)+P_{season}(i)+\delta P(i)+\gamma P(i), \quad (1)$$

where i - hour of the day (1-24); P(i) is the actual load of the power system; $P_0(i)$ - the basic component, determined by stable production cycles, daily and weekly uneven load schedule; $P_{season}(i)$ - seasonal component (seasonal curve), determined by seasonal fluctuations in the load during of the year. This component is due, first of all, to the deep seasonal fluctuations of meteorological factors - temperature and illumination; δP (i) - the component determined by irregular fluctuations of meteofactors (irregular deviations of meteofactors from stable seasonal cycles are considered irregular); $\gamma P(i)$ - residual component, determined by the influence of unaccounted factors.

To exclude the effect of random emissions and to obtain a mathematical model of the seasonal component, P_{season}, are approximated by Fourier series:

$$P_{season}(n) = \frac{A_0}{2} + \sum_{k=1}^{N}(A_k \cos k\omega n + A'_k \sin k\omega n), \quad (2)$$

where A_o, A_k, A'_k are the Fourier coefficients; $\omega = \frac{2\pi}{T}$;
T = 365; N is the number of terms in the Fourier series.

The coefficients are estimated using the method of least squares. The order N is determined experimentally by a gradual increase in the order of the polynomial. As a criterion for stopping the increase in degree is checking the variances at each step. The result of the calculations is the matrix of Fourier coefficients of dimension (25•(2N-1)•7). In practice, the optimal degree of polynomial for describing the seasonal curve for different hours of the day ranges from 4 to 15.

3 Case Study

Let's consider the model of the EL forecast for August 10, 2016. The peculiarity of these days was a significant and sharp decrease in the temperature of the surrounding air. At the same time, the decrease in temperature was not predicted with sufficient accuracy by the supplier of

DOI: 10.1051/e3sconf/20172503006

temperature forecasts (Fig. 3). In Fig. 4 shows the temperature change in Irkutsk by the hours of August 2016, also confirming an abnormal temperature change.

In Fig. 5 shows the results of the work of forecast models. It can be seen that the increase in EL was not predicted with sufficient accuracy. The forecast of the EL, performed on the forecasted temperature for one district, gave deviations of up to 3.75% (in the Forecast1 chart). At the same time, the forecast made on the next day, using the actual temperature, increased the deviation to 4.95% (in the Forecast1+1 chart). The forecast of the EL performed on the predicted temperature in three regions gave deviations to 4.22% (in the Forecast3 chart). In turn, the forecast made the next day using the actual temperature increased the deviation to 4.42% (in the Forecast 3+1 chart).

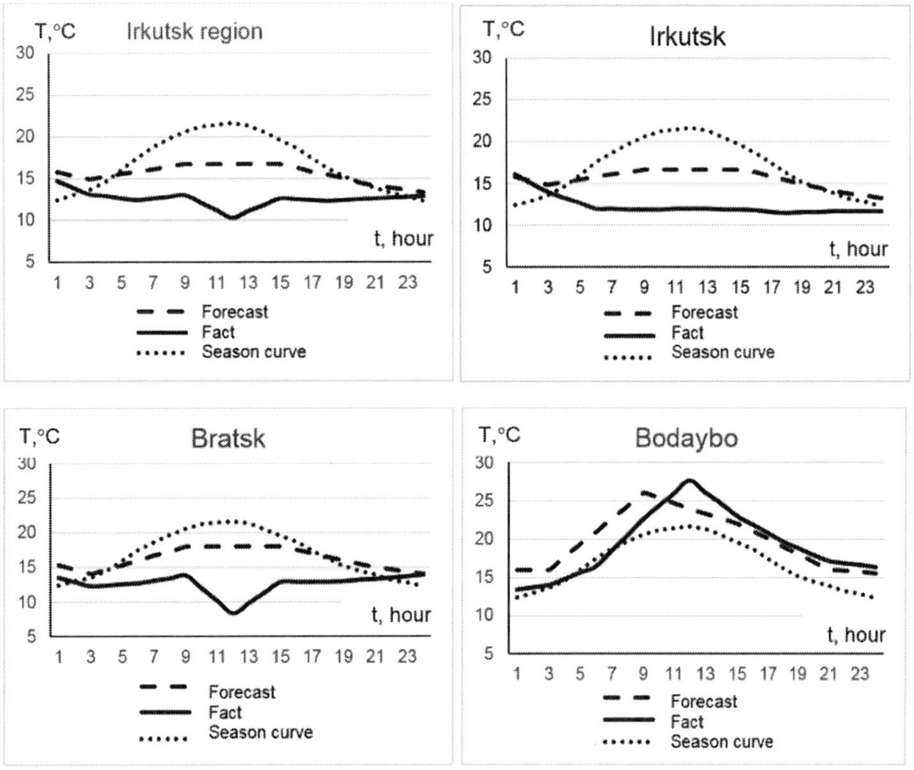

Fig. 3. Forecasts, facts and seasonal temperatures of 10 august 2016

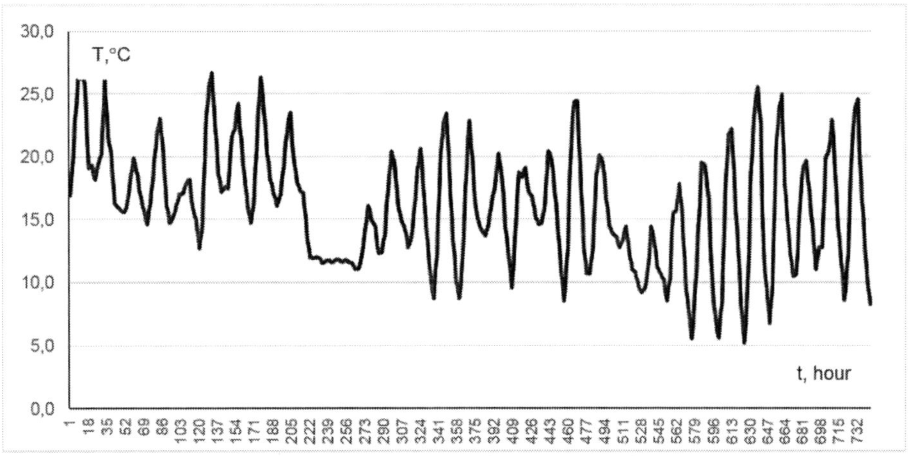

Fig. 4. Temperature in Irkutsk in august 2016, hourly

DOI: 10.1051/e3sconf/20172503006

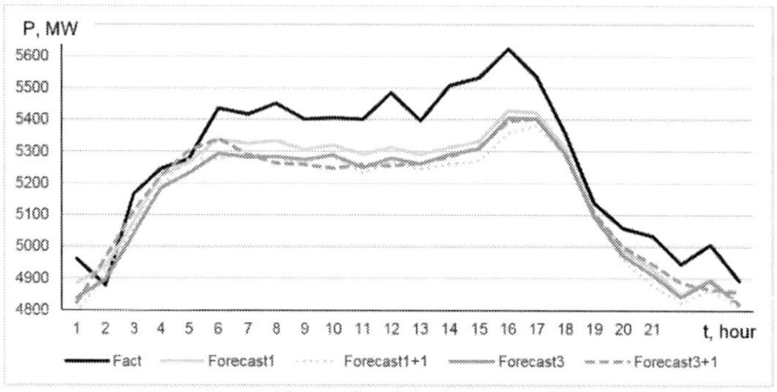

Fig. 5. The results of the forecast models

From the example, it is evident that the currently accepted "boundary" value of 4% accuracy is not always achievable with existing models and the forecasting system. The problem is worse by the fact that the settings that affect the accuracy of forecasting the model parameters are currently "closed" for the direct specialist who performs the EL predictions. At the same time, an adequate correction of the temperature forecast from the supplier presents a certain problem and is not realized at the moment by the standard procedures of the software used.

Then, in the existing mathematical model, a number of problems have not been solved, which are partially described above. Therefore, at the present moment, the task arises to improve the performance of forecasting functions for hourly EL.

4 Consideration of alternative methods for predicting EP

Existing market management principles, first of all the current rules of operation in the wholesale electric power and capacity market [3], determine the special significance of mathematical and software models for short-term forecasting of EP. As a rule, a deviation in EL of more than 4% (depending on specific regulatory requirements [4]) from the planned leads to additional costs. At the same time, the situation is complicated by the need to link the forecasting method used to specific real conditions, including "irregular" effects (meteorological factors, holidays, bad data, etc.), which are often adjusted in "manual mode". As a result, specialists of various RDUs face the difficult task of compiling a reliable EL forecast.

Despite the commonality of the task of short-term EL forecasting facing each energy system, there are many different methods in the scientific literature, most of which are a combination of different statistical approaches. These approaches can be divided into classical (deterministic, stochastic) and intellectual (algorithms of machine learning, expert systems, fuzzy logic). Deterministic methods are classical causal models of load and weather parameters (primarily ambient air temperatures). Such methods include curve fitting, data extrapolation, and smoothing methods [5-7]. Stochastic methods model the load behavior as a kind of stochastic process. This category includes: Kalman filtration, autoregressive integrated moving average, etc. [7, 8].

However, qualitative simulation of EL in modern conditions is a complex process due to nonlinear relationships between dependent factors and the load burdened by possible distortions and incompleteness of incoming data. Often these circumstances are an insurmountable obstacle to classical approaches. In connection with this, in recent years, methods of machine learning, such as artificial neural networks (ANN) [9-11], the support vector method (SVM) [12], decision trees [13] have been widely used in EL prediction tasks. Machine methods succeeded in overcoming these difficulties due to the best approximating abilities, primarily non-linear functions [9]. Their main difference from classical algorithmic approaches is that such models are taught by examples.

The prognosis model learns to "draw conclusions" without human help on the basis of the experience gained in the learning process and the original abilities to generalize (working with new data). The effectiveness of this approach for the task of short-term forecasting of EL is confirmed by a large number of studies [14, 15]. In particular, on the example of the data of the branch of JSC "SO UPS" of the Rostov RDO, the ANN models gave errors of 3.04-4.78%, the model of the SVM - 1.22-3.22% [16].

It is assumed that the newly developed forecasting system will not be a built-in HSP module. The new system can be used in parallel with HSP software. At the same time, the results of the new system will be for the performer to prognosticate an additional source of an alternative forecast of EL (including using alternative sources of ambient air temperature forecast). As a result, the contractor can manually change the final values of the predicted EL in the HSP software at the final stage.

5 Further directions of research

Proceeding from the mentioned problems of short-term forecasting of EL based on the HSP software, it is possible to single out several research directions for improving forecasts in the practice of the Irkutsk Regional Dispatch Office:

1. Decomposition of EL areas on the basis of the analysis of the presence of telemetry and the possibility of fixing EL in more "shallow" areas of the Irkutsk region (use of telemetry data, use of Counters data, etc.), as well as additional verification of the temperature forecasts at the performer level.

2. Development of alternative forecasting models that allow automating the prediction process and obtaining more accurate predictions. An effective technology for this is machine learning, in particular, the algorithms of ANN and decision trees.

3. Analysis of the importance of using additional input data for forecasting (cloudiness, precipitation, wind, etc.)

6 Conclusion

1. The modern hierarchical forecasting system HSP in most cases has an acceptable accuracy of the EL forecast for the Irkutsk power system.

2. During the off-season periods, with significant deviations in ambient air temperature from the average annual, the forecast errors of existing models exceed 4%.

3. It is advisable to conduct further research on the possibility of improving the accuracy of forecasting using machine learning techniques, in particular, ANN algorithms and decision trees, and to explore the possibility of applying additional input parameters for the performance of EL predictions.

References

1. M.M. Migranov, A.V. Melnikov *ELECTRICITY. Transmission and distribution*, **№3**, 60-64, (2017).

2. *Software for a hierarchical system for predicting power consumption for planning the EEC regimes (HSP) Version 3.14 Volume 1.* User's manual. (Edition of 01.10.2016) [Electronic resource], Moscow, (2016), CD-ROM.

3. *Rules of the wholesale electricity and capacity market*, http://www.consultant.ru/document/cons_ doc_ LAW_112537/

4. D. Singh, *Jour. Electric Power Components and Systems*. **No. 30**, 443-456, (2002).

5. W. R. Christiaanse, *IEEE Trans. On Power Appar. Syst*, **PAS-3**, 900-911, (1988).

6. A. D. Papalexopoulos, T. Hasterberg, *IEEE Trans. On Power Systems*, **Vol.5, No.4**, 1535-1544, (Nov. 1990).

7. S. R. Haung, *IEE Proc. Gener. Trans. Distrib.*,**Vol. 144, No.5**, 477-481, (Sept.1997).

8. H.S. Hippert, C. Eduardo Pedreira, R. Castro Souza, *IEEE Trans. On Power Systems*, **Vol. 16. - No. 1,** (2001).

9. G.P. Shumilova, N.E. Gotman, T.B. Startseva, *Forecasting of electric loads in the operational management of electric power systems based on neural network structures* (Ekaterinburg: UrB RAS, 2008).

10. V.G. Kurbatsky N.V Tomin, *Electrical Engineering*, **No. 7.,** 26-32, (2006).

11. Nadtoka II, Al-Ziheri Balasim M. *Modern problems of science and education*. **No. 6.**, 1-7. (2013).

12. Tomin N., Zhukov A., Sidorov D., Kurbatsky V., Panasetsky D., Spiryaev V. *International Journal of Artificial Intelligence*), **Vol. 13, No. 1,** 211-228, (2015 Spring (March).

13. *How Effective are Neural Networks at Forecasting and Prediction? , Journal of Forecasting , J. Forecast.* **17,** 481-495, (1998).

14. Kurbatsky V.G. *Automation and telemechanics.*, **No. 5.,** 143-158. (2014).

15. Al Zichri Balasim Mohammed. *Increase in the accuracy of short-term forecasting of electric load of consumers in the region, taking into account metofactors on the basis of the support vector method: dis. Cand. tech. Sciences*: 05.14.02, Al Zichieri Balasim Mohammed; Novocherkassk. YURPU (NPI), (2015)

16. A.S. Gordeev, *Issues of modern science and practice*. **No. 2 (12), Volume 2**, 33-36, (2008).

17. B. Makoklyuev *Analysis and planning of power consumption*, (M.: Energoatomizdat, 2008).

DOI: 10.1051/e3sconf/20172503007

Analysis and forecast of ensuring the reliability of the Kyrgyz energy system in the draft of "Concept for the development of energy in the long-term perspective

Valentina Kasymova[1]*, *Anna* Arkhangelskaya[2], and *Roza* Kurzhumbaeva [1]

[1]Kyrgyz National technical university named after I. Razzakov, 66 Mir av., Bishkek, Kyrgyz Republic
[2]Governmetal committee of industry, energy and subsoil use of the Kyrgyz Republic, 720000, 2 Erkindik str., Bishkek, Kyrgyz Republic

Abstract. In this article, based on the work done to prepare the project of the Concept of development of the fuel and energy complex of the Kyrgyz Republic until 2030 identifies the main directions for the development of the country's energy sector in the medium-term perspective, including ensuring sustainable and uninterrupted electricity and fuel supply for the development of the economy and population needs, the introduction of new technologies using renewable energy sources, institutional reforms, financial recovery of energy companies, etc. In addition, the main priorities of the state energy policy, priority tasks and policy measures on scenarios are outlined.

1 Introduction

The Kyrgyz Republic (KR) has sufficient reserves of fuel and energy resources (FER). However, the potential of the fuel and energy complex (FEC) is not being realized in sufficient measure. The fuel and energy complex of the Kyrgyz Republic, which is the main sector in ensuring sustainable social and economic development of the country, is currently in crisis [4]. An assessment of the state's energy policy shows the presence of threats to the country's energy security, including regions, caused by both external and internal factors. In addition to rising global prices for hydrocarbon fuels, disagreements in the Central Asian region are one of the main external factors, especially disagreements with neighboring countries on the introduction of economic mechanisms in the management of water and energy resources in the basin of the Naryn and Syrdarya rivers. Internal factors of threats are the irrational structure of the country's fuel and energy balance (FEB); low economic indicators of the development of enterprises in the fuel and energy sector, the manifestation of corruption schemes, physical and moral deterioration of the main equipment at the facilities, high energy losses, chronic financial shortages, increased accidents in power networks and, as a result, shutdown and shortage of energy carriers to consumers , which leads to lost profits in the country's economy and the growth of social tensions in society.

2 Main statements of the 'Concept for the Development of the Fuel and Energy Complex of the Kyrgyz Republic until 2030'

In order to increase the efficiency of the development of the fuel and energy complex in pursuance of the Detailed Plan for Implementing Measures to Counter Systemic Corruption in the Energy Sector of the Kyrgyz Republic, the draft Concept for the Development of the Fuel and Energy Complex of the Kyrgyz Republic until 2030 (hereinafter referred to as the Concept) was drafted on March 5, 2014.

The concept was developed in accordance with the Laws of the Kyrgyz Republic "On Energy", "On Electric Power", "On Oil and Gas", "On Coal", "On Energy Conservation", "On Renewable Energy Sources", regulating the country's fuel and energy complex and the National Energy Program for 2008-2010 and the strategy for the development of the fuel and energy sector until 2025; and also on the basis of the methodology for strategic planning of sustainable development approved by the order of the Minister of Economy of the Kyrgyz Republic No. 45 of February 27, 2015.

Over the past 10 years, the state policy on the development of FEC is carried out in accordance with the National Energy Program of the Kyrgyz Republic for 2008-2010 and the strategy for the development of the fuel and energy complex until 2025, the Medium-Term Strategy for the Development of the Electric Power Industry of the KR for 2012-2017, the National Strategy for Sustainable Development of the KR for the period 2013 -2017 years.

* Corresponding author: valentinakasymova@gmail.com

Due to the expiration of their term and methodology of system research in power engineering, the concept and development strategy should be updated (every 5 years) based on the real situation in the social and economic development of the state and challenges in the external energy policy, as well as global environmental problems.

The high capital intensity and long terms of the construction of power facilities makes the outstrip development of energy in order to ensure the industry, agriculture, construction, transport and social needs of the population with uninterrupted energy and fuel supply.

In addition, the annual complex situation with reliability associated with the unstable provision of energy resources to the consumers of the republic requires the adoption of strategic decisions on the development of fuel and energy industries for the short, medium and long term (10-15 years). [3]

In the draft Concept, the main sectoral and cross-sectoral programs and plans of social and economic development of the country and regions were used.

In accordance with the fundamentals of systemic research in the energy sector [1] and the Strategic Planning Methodology for Sustainable Development (hereinafter the Methodology) [2], the draft Concept contains the following sections: analysis and assessment of the current situation in providing fuel and energy energy policy; priority tasks and policy measures; expected results; assessment of risks and challenges; assessment of financial resources and sources of funding; mechanisms and responsible bodies for the implementation of the concept.

At the same time, the requirements of the Methodology on accounting for regional, social, environmental factors, ensuring energy security and energy efficiency of the real sector of the economy, and advancing along the path of sustainable development of the country and regions were taken into consideration. Proceeding from this, we determined the purpose of the Concept - to ensure the energy security of the country and regions, the availability of energy for each consumer in terms of quality and prices, and improving the living standards of the population, achieving energy efficiency of the real economy and sustainable development of the country and regions in the future.

Based on the goal, priorities are formulated and tasks are set, with successful implementation of which it is possible to achieve the goal and obtain results.

The following priorities are recognized as the main priorities of the state energy policy:

- management of demand for energy resources and achievement of energy efficiency of the economy, formation of a rational structure of the country's and regions' FEBs,
- Sustainable energy development: ensuring reliability and energy security, as well as reducing the risks of climate change;
- international cooperation and strengthening of the external energy policy
- financial recovery with the introduction of an economically sound tariff policy, ensuring financial stability and efficiency of energy companies and energy companies;
- energy saving and minimizing the technogenic impact of energy on the environment, public health and progress towards sustainable development of the country and regions;
- institutional reforms: strengthening of strategic management, regulatory legal framework and innovative development of energy,
- improving the quality of education and research.

Priority objectives and policy measures

1. Priority tasks "Managing energy demand and achieving energy efficiency of the economy, formation of a rational structure of the country's and regions' FEBs include, first of all, ensuring the energy efficiency of the real sector of the economy, which will be achieved with the projected growth rates of consumption in the FEC - 150% in 2020 and 210% in 2030 to 2005, which is below the GDP growth rate of 197% in 2020 and 320% in 2030 to 2005 (in comparable prices in 2005), with an annual decline in the energy intensity of GDP by 1.5%, and for the period 2015-2030 on 20%. At the same time, it is possible to achieve energy savings in the amount 11.1 million tons of tones by 2030.

The growth rate of electricity consumption should also be lower than GDP growth, with an annual decline in the GDP electric capacity by 1-1.6%, and for the period 2015-2030 on 15-19% under the different scenarios. At the same time, energy savings in volume 3.65-4.1 billion kWh by 2030 is expected.

Ensuring a decrease in the growth rates of fuel and energy consumption, in comparison with the growth rates of GDP and, accordingly, the regularity of the annual rate of reduction in energy intensity and GDP electrical capacity, meets the principles of energy-saving policies and the development of the "green economy". If the GDP growth rate increases, the forecasted demand for fuel and energy resources will also increase, in this case the degree of increase will depend on the prices for energy carriers - the more their growth, the less the degree of increase in demand. [4]

2. Priority tasks "Sustainable energy: ensuring energy security and preventing global warming" should be directed to:

2.1. In the electric power industry:

- the development of hydropower, which is an integral part of the "green economy" and the most important prerequisite for sustainable development, with access to modern, environmentally friendly and inexpensive energy services;
- creating a system for assessing the state of the country's EE, which requires the development and approval of indicators and their thresholds - crisis and pre-crisis - as an important tool for timely warning of threats;

- outstripping the pace of commissioning new capacities of power plants and guaranteeing the level of the reserve of installed HPP capacities in the power system;
- a phased implementation of a number of projects on the construction of hydroelectric power stations on its territory that are environmentally friendly energy sources, which is a sovereign right to development to meet the pressing needs of the people of Kyrgyzstan, including future generations;
- further integrated use of water resources in the Naryn river basin with gradual transition to the project mode of operation of the Toktogul HPP in the interests of energy and irrigation and not allowing the Toktogul reservoir to run to a critical level by implementing projects for the construction of the Kambarata HPP-1 installed capacity of 1860 MW in the power mode to cover the demand of the country's consumers in the autumn-winter period. As a result, the winter energy releases of water from the Kambarata HPPs will accumulate and will be re-regulated by the Toktogul reservoir under the irrigation schedule for water consumption, that is, the operation of Kambarata HPP-1 will correct seasonal inflow into the Toktogul reservoir and, therefore, allow sufficient quantities in the most droughty seasons water adjacent countries-lower reaches;
- acceleration in 2017-2019 years installation of the second aggregate of Kambarata HPP-2 with a capacity of 120 MW;
- Considering the possibilities of construction in the middle reaches of the Naryn River basin of the Kazarman cascade of HPPs in 2025-2030 years with a total capacity of 1050 MW, as well as the Susamyr-Kokemerens cascade of HPPs with a total installed capacity of 1,305 MW;
- development of the hydropotential of the Sary-Jaz river basin;
- Construction in 2019-2025 years Higher-Naryn cascade of HPPs with an installed capacity of 237.7 MW;
- Creation of guaranteed base capacity in the power system with the construction in 2018-2022 years Kara-Keche TPP with an installed capacity of 600 MW, followed by expansion to 1200 MW;
- Completion of the reconstruction in 2017 of TPP-1 in Bishkek with the installation of 2 boilers of 150 MW each and an increase in power to 812 MW;
- Conducting and completing the reconstruction of the Uchkurgan HPP – on 36 MW, Atbashi HPP - by 1.68 MW, Toktogul HPP to increase capacity by 240 MW;
- Development of a procedure for the allocation of land and the implementation of tariff policy in accordance with the Law of the Kyrgyz Republic "On Renewable Energy Sources";
- Increasing the efficiency and reliability of the district heating system in Bishkek, carrying out the rehabilitation work of the existing main heating network and constructing new heating networks and pumping stations in the operation area of TPP-2 in Bishkek.

Based on the forecasts of the growth rates of consumption of the fuel and energy resources for their coverage, the forecasts of the development of the fuel and energy sectors were developed according to scenarios that showed:

In the electric power industry: according to the baseline scenario, the growth in electricity production can be achieved through the active involvement of renewable energy sources and the construction of small HPP and solar power plants (SES), taking into account the construction and commissioning of the Higher-Naryn cascade of HPPs and Kambarata HPP-1, the growth rate of electricity production is 119% by 2020, 140% by 2025 and 246% by 2030? or up to 16.7, 23.8 and 27 billion kWh, respectively, over the growth rate of its consumption since 2020 or 119% to 140% by 2025, and 166% by 2030.

At the same time, the ratio of the growth rates of production over the growth rates of consumption will amount to 0.9 by 2020, 1.11 by 2025 and 1.48 in 2030, in this case it will be possible to provide a basic reserve of capacity in the power system within up to 48%, that is, the reserve in the power system is expected to exceed the threshold threshold value of 1.15.

According to the optimistic scenario, new approaches to forecast the development of the regions of the country have been taken into account by accelerating the development of enterprises for processing plant and livestock products, developing mineral and raw materials resources and forming mining, sports, tourist and agricultural clusters, as well as the rates of growth in prices and tariffs for energy, using econometric forecasting models under the STATA program, and additional expert assessments. The demand for electricity will increase annually and will reach the level of 14.87 billion kWh by 2020, 19.5 billion kWh by 2025 and up to 23.0 billion kWh by 2030, or, respectively, over the years, growth rates are expected - 124%, 163% and 193% compared to 2017. The growth rate and the volume of electricity production during the construction of the Kazarman HPPs will increase and will reach 172% by 2025 and 253% by 2030 or up to 27 and 33.3 billion kWh, respectively. The ratio of the growth rates of production over the growth rates of consumption will increase from 2020 and are expected to be 1.1 by 2025 and 1.31 to 2030, and in this case it will be possible to provide a basic capacity reserve in the power system within 30%, which will completely eliminate the energy security risks and cover the growing energy consumption rates in the regions of the country.

In the structure of electricity production, the share of RES will increase from 1.1% in 2015 to 5.0% by 2030.

To ensure the reliability of power supply in remote mountain areas, it is possible to build small thermal power plants when introducing new technologies for coal combustion with minimal GHG emissions. When constructing small TPPs in cogeneration mode in the region of Sulukta coal fields with a capacity of 75 MW in the Batken region, Tashkumyr with a capacity of 75 MW in Jalal-Abad region, Uzgen, Kyzyl-Kiya with a capacity of 50 MW in the Osh region, it is possible to generate additional electricity within 1,5-2 billion kWh by 2030 with the growth of their share also up to 5.4% in the electricity balance.

2.2. Accelerated development of renewable energy sources for diversification of electric power sources that can largely provide energy resources to both large and individual consumers, especially in rural areas, is provided by: construction of solar and wind power plants in the regions of the country with a gradual increase in electricity production by them to 1.2 billion kWh by 2030; introduction of biogas plants to produce biogas and with the associated use of waste as fertilizer for farms.

To solve the problems of hot water supply and heat supply in residential houses, the installation of solar collectors, heat pumps, and the use of heat from geothermal sources for the heat supply of social and cultural facilities and the population in the areas where they are concentrated is necessary.

In parallel with the search for investors in the construction of solar and wind power plants and the construction of small hydro power plants, it is necessary to determine the sources of coverage for the difference between electricity tariffs and tariffs, taking into account the coefficients in accordance with the Law of the Kyrgyz Republic "On Renewable Energy Sources" in order to pay for RES projects.

2.3. In order to adapt and mitigate to climate change, the development of small and large HPPs with efficient use of solar energy, biomass, geothermal sources and heat pumps with replacement of hydrocarbon fuels can reduce GHG emissions by 20% by 2020 and the attainment of the set goals for achieving by 2050. Specific emissions in CO_2 equivalent, not exceeding 1.56 tons per person according to the Climate Center under the SAEPF with the RCC.

For these purpose it is needed:

- the introduction on an ongoing basis of climate indicators for the energy sector and their monitoring;
- regular preparation of plans to fulfill commitments to the UNFCCC on reducing greenhouse gas emissions;
- intensification of activities to attract climate finance to ensure energy security.

3. Priority tasks - international cooperation and strengthening of the external energy policy will take place by entering the international electricity markets, while ensuring a surplus balance of electricity. [3]

Thus, according to the baseline scenario, the forecast shows a tight balance until 2020, as well as the possibility of export only by 2025 and 2030 in the volume of over 4 billion kWh, including 1,7 billion kWh for the CASA-1000 project and over 2 billion kWh into the energy systems of the EEA countries to the wholesale electricity market and (or) the PRC; according to the optimistic, as under the baseline scenario, reduction of imports up to 2018, further it is possible to increase exports to 0.23 billion kWh by 2020, 4.4 billion kWh by 2025 and up to 6.3 billion kWh by 2030, which will also allow entering the unified electricity market of the EAPS with an export volume of over 3 billion kWh and the Central and South Asia market for the CASA-1000 project - 1.7 billion kWh and 1.5 billion kWh in the PRC.

There is also an opportunity to increase exports during the construction of the Susamyr-Kokomerene cascade of HPPs with a design capacity of 1,305 MW, generating 3.32 billion kWh per year and a hydroelectric power station on the Sary-Jaz river with total design capacity up to 1200 MW with generation up to 3.3 billion kWh per year. However, given the lack of pre-project documentation, these HPPs were not included in the prospective electricity balance until 2030 and can be considered when developing an energy strategy until 2040.

The necessary investments for the construction and commissioning of the capacities of prospective HPPs and TPPs and 500 kV transmission lines are:

- Under the base scenario of development - $ 6597 million;
- According to the optimistic scenario of development - $ 8878.85 million.

Necessary investments in the development of transmission electric networks for 2017-2030 amount of $ 420.6 million.

The need for investment in small hydropower plants is estimated in $ 350 million, at the rate of $ 1000 per 1 MW of capacity, small-scale thermal power plants - $ 250 million.

4. Tasks for financial recovery with the introduction of an economically sound tariff policy, ensuring financial stability and efficiency of energy companies should be aimed at:

- ensuring economic efficiency and financial sustainability of energy companies and enterprises in the coal and oil and gas industries;
- introduction of a system of key indicators of energy security, efficiency and financial stability, increasing the efficiency of energy companies for their regular monitoring and media coverage for the public and the public;
- conducting a portfolio analysis of energy companies using matrices from international practice, identifying external and internal threats, weak and strong points, finding reserves and opportunities to withdraw them from the financial crisis;
- Introduction of an automated system of management accounting in energy companies, taking into account international experience for ensuring transparency and counteracting corruption;
- development and approval of the Methodology for the formation of tariffs for regulated energy

carriers with a transparent mechanism for establishing purchase tariffs for electricity and heat between the energy companies for the production, transmission and distribution of electricity and heat, ensuring their break-even and full reimbursement of objectively necessary costs, with the elimination of the existing practice of income distribution and cross- subsidizing;

- development and approval of the method of technological connection of consumers to electric and heat networks and fees for connection to power supply and heat supply systems for new facilities - consumers;

- maintenance of a share of the per capita income of the population spent for payment of power resources, is not higher than threshold values of energy security (20-30%);

- Conducting a revaluation of fixed assets with an increase in depreciation charges for the technical modernization and reconstruction of transmission and distribution lines and reducing the level of depreciation of fixed assets at facilities for the production, transmission and distribution of electricity.

5. Priority tasks "Institutional reforms: strengthening of strategic management, regulatory framework and innovative development of energy" should focus on Realization of the program of stage-by-stage reforming of organization and management of the electric power industry of the KR:

- development of the draft Concept of energy security and implementation of the Strategy for the Development of the Electricity Industry of the KR for 2018-2023 and long-term perspective up to 2030-2040;

- development and implementation of the Strategy for the development of the fuel industry of the KR for 2018-2023 and long-term perspective up to 2030-2040;

- development and implementation of the Strategy for the Development of RES of the KR for 2018-2023 and long-term perspective up to 2030-2040;

- Conducting systematic monitoring of the state of energy security by indicators and compliance with threshold values approved in legislation;

- Improvement of strategic management and management in NEFIC with optimization of costs in the activities of energy companies.

Concentration on priority and innovative directions of development of the energy science, strengthening of the scientific and personnel potential by:

- creation of a system for introducing the results of scientific and technical activities into the practice of the development and operation of the electricity and energy industry using various forms of public-private partnerships, supporting small and medium-sized businesses;

- introduction of innovative technologies (ASIS, SCADA, CIP wires, relay protection and automation, smart meters, etc.) in order to increase the efficiency of transmission and distribution of electricity to the consumer, reduce losses and improve the quality of electricity, including for urban and the rural population in the regions;

- reduction of the technological consumption of electric power to their standard values for transmission (5%) and distribution (10%).

6. The tasks on the priority of improving the quality of education and scientific research should be achieved by:

- Increasing on a systematic basis the qualification of engineering, technical, economic and financial personnel of energy companies of the KSTU. Named after I. Razzakov together with leading universities of Russia, to ensure crisis management and management of energy companies;

- Increase the level of training in higher education institutions of the country by way of exchange for teaching students and improving curricula and programs in the disciplines of specializations; and the restoration of the training of engineers in universities and the opening of new areas of magistracy;

- Strengthening of scientific system research in the energy sector and their constant updating, using economic and mathematical methods for forecasting the development of energy systems and optimizing EBs and software in the Science research institute on energy and economics and KSTU with the assistance of the Institute of Power Systems named after. L.A. Melentiev.

3 Conclusion

Given the limited nature of the publication, we focused only on the main directions of the draft Concept. It should be noted that, in order to realize it, expected results are expected with an assessment of the necessary financial resources, an Action Plan and a Matrix of indicators have been developed, and the forecasted values of indicators for the future have been modeled.

The Draft Concept presented for consideration and approval is a strategic document that provides a set of mechanisms for public administration and regulation for the efficient and rational use of energy resources, ensuring the advanced development of energy capacities, including renewable energy sources, the implementation of energy-saving policies and sustainable development goals.

The approval of the draft Concept by the Government of the Kyrgyz Republic will help confidence of potential investors that the strategically important objects of interest are included in the state program and are a guarantee of return of the capital invested with profit.

DOI: 10.1051/e3sconf/20172503007

References

1. L.A. Melentiev, *Optimization of development and management of large energy systems*. - M: High School, (1976).
2. *Methodology for strategic planning of sustainable development of the Kyrgyz Republic. Approved by the order of the Minister of Economy of the Kyrgyz Republic* No. 45 of February 27, 2015.
3. *Russia's energy policy at the turn of the century; Priorities of energy policy: from energy security to energy diplomacy.* (T.2. - M .: "Papyrus PRO", 2001).
4. V.M. Kasymova, *Energy policy. energy security and energy efficiency of the Kyrgyz Republic*. - Bishkek, KSTU named after I.Razzakov, publishing house "BarakElde", (2015).
5. V.M. Kasymova, A.V. Arkhangelskaya R.B. Kurzhumbaeva, *Scientific foundations of the state concept of energy policy and development strategy of the fuel and energy complex of the Kyrgyz Republic until 2030*. KSTU named after I.Razzakov, publishing house "Barak Elde", Bishkek, (2017).

E3S Web of Conferences **25**, 03008 (2017) DOI: 10.1051/e3sconf/20172503008
RSES 2017

Increase the reliability of operation of differential protection due to the use of the method of double entry

Aleksandr Kulikov[1,*], *Boris* Papkov[2], *Anton* Kolesnikov[3], and *Vladimir* Vukolov[1]

[1]Nizhny Novgorod State Technical University n.a. R.E. Alekseev, 603950, Nizhny Novgorod, Russian Federation
[2]Nizhny Novgorod State University of Engineering and Economics, 606340, Knyaginino, Russian Federation
[3]Branch office ESP-NN-SESH, Elektroshchit - TM Samara, GK, CJSC, 603000, Nizhny Novgorod, Russian Federation

Abstract. Double entry it is method of registration of economic operations accounting, applied to detect errors. The application of the method in differential protection leads to increased reliability of technical solutions. By providing redundant processing of information about currents measured at the ends of the protected object, and is formed in a special matrix that implements the reliable operation of the differential protection, eliminating its excessive action when damaged the current transformers or communication lines. To assess the reliability of protection used the method of Markov chains. The obtained results allow to make a conclusion about the prospects of implementing a differential protection using the method of double entry in the electrical networks.

1 Introduction

One of the promising directions of development of system of relay protection (SRP) is the establishment of the centralized differential protection (CDP). Typical examples of the organization of the CDP are method of differential rings [1], which used in the ship power systems 13,8/4,16/0,48 kV and method of multi-zone differential protection [2]. Also has practical experience multiterminal line differential protection, which is a special case of CDP [3].

Consider the principles of operation protection on the example of the part of the 110 kV network.

In accordance with the method of differential rings, the network is divided into several regions (fig.1)

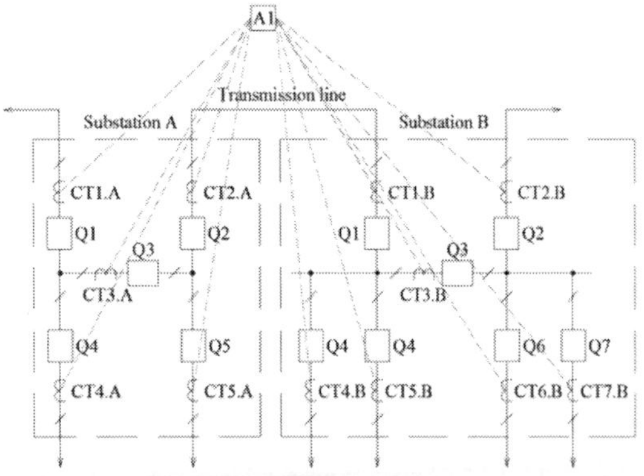

Fig. 1. Method of differential rings.

Region is formed by current transformers (CT) on the boundary of the protected area, is checked for fault in the network (region bounded by CT1.A, CT2.B, CT4.A, CT5.A, CT4.B, CT5.B). In case of detection fault, starts checking of differential protection of individual installations, and if fault is detected, isolates damaged item. Also in the process of functioning of protection there is a permanently diagnosis CT fault by checking on the presence of fault regions formed by the expanding of the zones of differential protection of installations to the next CT on both ends of the protected object.

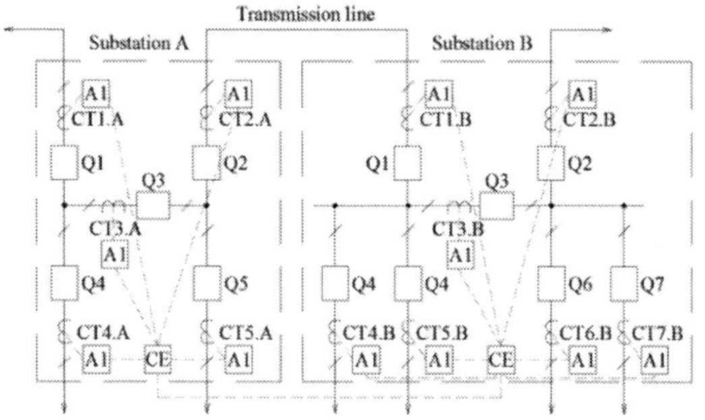

Fig. 2. Method of multi-zone current differential protection
A1- terminal; CE - central equipment; Q1-Q7- circuit breakers; CT1- CT7 - current transformers.

When using the method of multi-zone differential protection (fig. 2), terminals collect the measured values of currents and status of circuit breakers (CB). Established local central equipments (CE) collect data from the terminals on their own and adjacent substations. Once the local CE obtains all data required for the CDP, it will make the protection decision (trip or block) and transfer commands to the local CE of the adjacent substation.

* Corresponding author: inventor61@mail.ru

 DOI: 10.1051/e3sconf/20172503008

When CT fault is detected, protection zone will be extended to bigger due to the use of the data received from next CT.

A common disadvantage of the proposed approaches, from the point of view of the operation of modern electrical networks with changing configurations is the difficulty of the technical implementation.

The purpose of research is the development of highly reliable CDP with adaptation to the changes in the configuration of the electrical network. This problem is important in power systems with distributed generation, where the protection must react to dynamically changing network operation condition. Developed CDP may be applied both as main and as a backup.

2 Realization of proposed protection

CDP, proposed by the authors, based on the differential principle, in combination with the method of double entry [4]. Double entry - registration method of economic operations accounting, under which every change of the state funds recorded in two accounts, ensuring the overall balance.

When implementing the new CDP apply graph theory and the theory of matrices.

To illustrate the operation of protection, consider the part of the electrical distribution network shown in fig. 3a

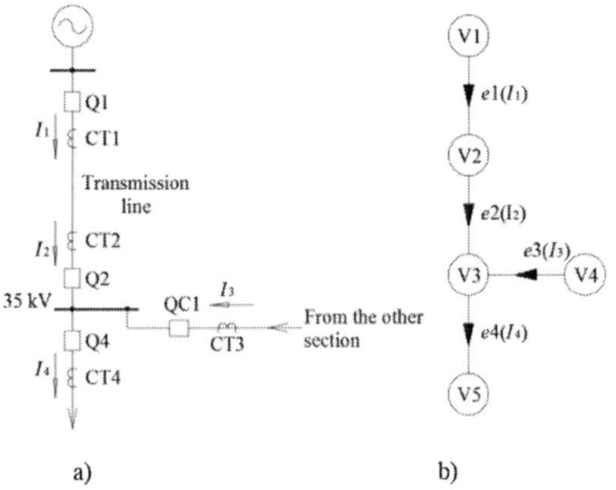

Fig. 3. Scheme of fragment of distribution network and an unidirectional graph.

The structure of the scheme displayed as a unidirectional graph (fig. 3b) with arcs representing the high-voltage units and the vertices representing the protected items – busbar and transmission line. Then, each vertex of the graph is represented by the special matrix of currents whose rows represent the vertices of the graph, the columns - the separation of arcs connecting the vertex on the directed to it and directed away from it.

For the present example, the matrix of currents will have the view:

$$MV1=\begin{bmatrix}0&0\\0&I_1\\0&0\\0&0\\0&0\end{bmatrix}\quad MV2=\begin{bmatrix}I_1&0\\0&0\\0&I_2\\0&0\\0&0\end{bmatrix}\quad MV3=\begin{bmatrix}0&0\\I_2&0\\0&0\\I_3&0\\0&I_4\end{bmatrix}$$

$$MV4=\begin{bmatrix}0&0\\0&0\\0&I_3\\0&0\\0&0\end{bmatrix}\quad MV5=\begin{bmatrix}0&0\\0&0\\I_4&0\\0&0\\0&0\end{bmatrix}$$

For determining the location of fault, protection compares the currents on differential principle for particular regions and detects potentially damaged item. For all graph vertices with degree >1 compile equations in matrix form for the sum of the currents at the nodes.

Then it is determined whether to trigger starting on fault in the region, or there is a CT or communication lines (CL) fault. The presence of a fault is determined using the method of double entry. Since each arc of the graph is reflected with the same weight (current value) in the matrices twice: as an arc, associated with the vertex directed to one matrix, and as the arc associated with the vertex directed from another matrix, in the case getting wrong values is the sum of incoming and outgoing currents in the two matrices becomes incorrect, but the total differential current of the whole network remains equal to zero. In case of observance of rule of detection of the fault and maintaining a sum of currents for the entire network is equal to zero, a CT or CL fault is detected.

Signs of the functioning of the differential protection for different ratios of currents and the results of performing matrix operations for this scheme is shown in Table 1.

Table 1. Results matrix operations and signs of the functioning of the protection.

The resulting value matrix operations			Signs of the functioning of the differential protection
SMV2	*SMV3*	*SUM*	
0	0	0	Fault is missing, CTs is functioning
≠0	0	≠0	Fault on the element, corresponding to vertex V2
0	≠0	≠0	Fault on the element, corresponding to vertex V3
≠0	≠0	0	CT Fault on CT, common to V2 and V3 (arc $e2(I_2)$)

In table 1 variables *SMV2, SMV3*, have been received by the resulting matrix operations, define the conditions of the existence of a fault in the protection zone. Variable *SUM* defines total differential current of the network.

Detection of CT fault of separate fider can be implemented by additional technical diagnostics or automatically by forming the resultant currents that characterize them operability. For example, by adding additional conditions to those shown in table 1 presents the additional matrices currents.

For example, for considered scheme, assuming that the feeder of circuit breaker Q4 represents the connection of power transformer, for inclusion in the protection zone of the CDP, need to connect the windings of CT of the transformer to the protection. In accordance with the principle of the protection, transformer representing an additional matrix *MV6*. Accordingly, changing in the structure of scheme will changes the dimensions of all matrices with 5x2 to 6x2 and a description of the matrix MV5, the degree of which is equal to 2.

The advantage of the proposed CDP is the flexibility of the system of the protection when changing network configuration due to the fact that the resulting matrices are generated automatically, in contrast to the protections discussed above, where for each network configuration, you must set special algorithms. Depending on the ratio of currents on the part of electrical network and results of operations on matrices of currents can be realized the reliable operation of the centralized differential protection network. This provides not only the action of protection in case of fault on each region, but excluded its excessive action when CT or CL faults.

3 Quantitative assessment of the reliability of the proposed protection

To quantify the advantages of the proposed technical solutions from the point of view of reliability, using the method of Markov chains. This method is often used to describe the processes of failure and repair with the elementary streams, and is most suitable for calculating reliability of SRP [5-8]. In this paper applies the mathematical model of the functioning of the SRP, presented in [6]. The distribution laws of failure and repair will accept exponential.

Taking into account specificity of the analyzed differential protection, define two types of protection failures:

- undesired-tripping protection failures (in the absence of fault on the protected object);
- fail-to-operate protection failures (in case of fault on the protected object).

This approach is used in several papers, e.g. [9].

In fig. 4 presents three variants of organization SRP: differential line protection (DLP) and differential busbar protection (DBP); CDP as a busbar protection and as a line protection; total system, consisting of distributed DLP, DBP and CDP, giving a signal, blocking the work of the DLP and DBP upon the occurrence CT or CL fault. Calculate reliability indices for all three cases on the basis [6, 10], and information from the manufacturers of relay protection.

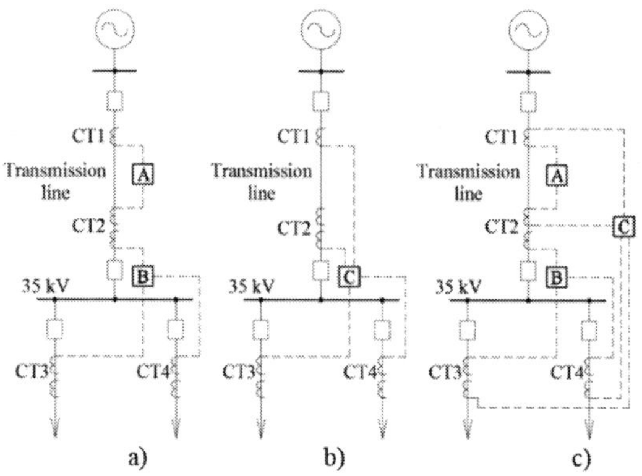

Fig. 4. Part of a network with connection of protections to CT a) the system includes a DLP (A) and DBP (B), b) SRP based on the CDP (C), c) SRP with the use of CDP (C) and distributed protection DLP (A) and DBP (B).

3.1 Mode 1: absence of fault on the protected object

In fig. 5 shows graphs of states and transitions in an absence of fault on the protected object mode. Here: Ew – state SRP without failures; A_1, B_1, C_1, $CT1_1$... $CT4_1$ – the state of the SRP in the presence of defects that could lead to undesired-tripping protection failures of devices of relay protection A, B or C, and, as a result, failure of the SRP in general; μ_1, μ_2 – repair rates of the relay protection and CT.

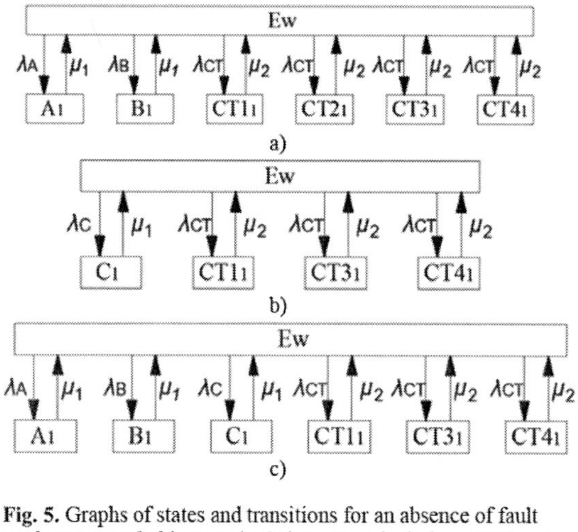

Fig. 5. Graphs of states and transitions for an absence of fault on the protected object mode a) the system includes a DLP and DBP, b) SRP based on the CDP, c) SRP with the use of CDP and distributed protection DLP and DBP.

Convert the graphs (fig. 5) to the form shown in fig. 6, where E_1 - failure state of SRP, λ_1 – the resulting failure rate of protection.

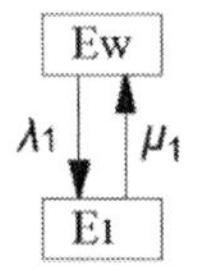

Fig. 6. Converted graph of states and transitions for an absence of fault on the protected.

Define the probability $P_1(t)$ that the system is in state E_1, which is an emergency, for random time t. Form a system of differential equations describing a graph, where $P_w(t)$ - the probability of location the system in state without failures E_W.

$$\left.\begin{aligned} \frac{dP_w(t)}{dt} &= -\lambda_1 \cdot P_w(t) + P_1(t) \cdot \mu_1 \\ \frac{dP_1(t)}{dt} &= \lambda_1 \cdot P_w(t) - P_1(t) \cdot \mu_1 \end{aligned}\right\} \quad (1)$$

Normalizing expression, the meaning of which is that the researched system located in state E_1 or E_W as constituting a complete group of events, has the form

$$P_w(t) + P_1(t) = 1$$

As at the initial moment of operation of the system at $t = 0$ the system is in state without failures:

$$P_w(0) = 1, \; P_1(0) = 0$$

As a result of solve the system of differential equations a function of the unreadiness probability

$$q(t) = P_1(t) = \frac{\lambda_1}{\lambda_1 + \mu_1} \cdot (1 - \exp[-(\lambda_1 + \mu_1)t])$$

Make an assessment of the probability of failure-free operation of the SRP. Probability of failure-free operation - is the probability that within a given work or a given time interval, the failure of the object does not occur [6].The graph for the calculation will look similar as the graph in fig. 6 with the difference that it will not be possible to transition from state E_1 to E_w, that is excludes repair rate μ_1. This circumstance is due to the fact that when system fails (transition to absorbing state E_1), the experiment is finished – the system cannot leave this state.

Due to absence of repair instead of a system of differential equations (1), the result will be an expression of the probability of failure-free operation (PFFO), with exponential distribution law:

$$R(t) = P_w(t) = \exp[-(\lambda_1 \cdot t)] \quad (2)$$

In fig. 7 shown the dependences of the function of the unreadiness probability of SRP, and functions of the PFFO on the average time between checks.

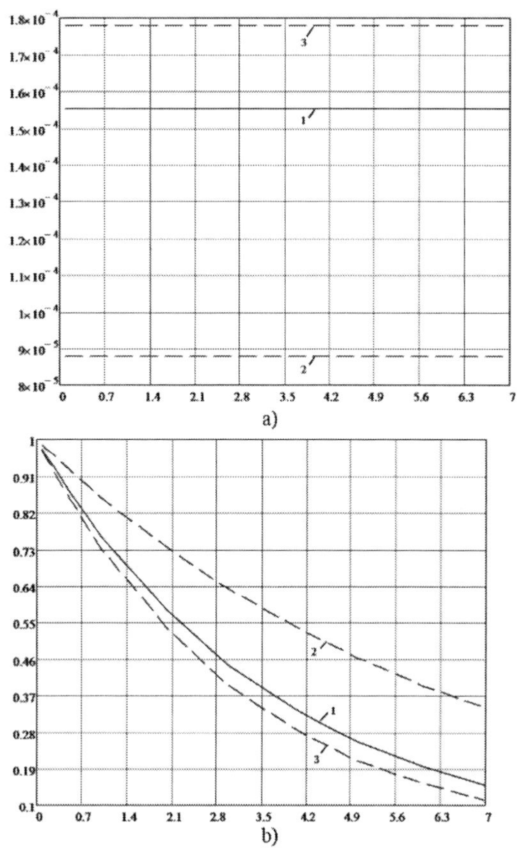

Fig. 7. Function of the unreadiness of SRP (a) and function PFFO of SRP (b). 1 - the system includes a DLP and DBP; 2 - SRP based on the CDP; 3 - SRP with the use of CDP and distributed protection DLP and DBP.

The analysis of graphs in fig. 7a shows that the functions of unreadiness probability are straight lines, the values of which are close to zero. This is due to the fact that for the considered case, the repair process can be considered instantaneous.

3.2 Mode 2: in case of fault on the protected object

In fig. 8 shows graphs of states and transitions in case of fault on the protected object mode. Here: E_w – state SRP without failures; A_1, B_1, C_1 - the states of the SRP in the presence of defects that could lead to fail-to-operate protection failures of devices of relay protection; A_1C_1, B_1C_1- states with defects, that happens at the same time, in two terminals A and C or B and C; μ_1, μ_2 – repair rates.

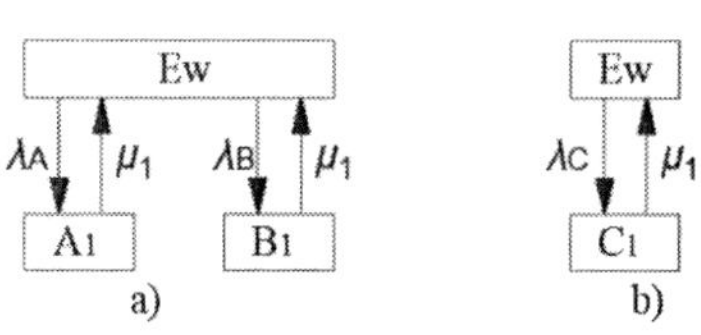

 DOI: 10.1051/e3sconf/20172503008

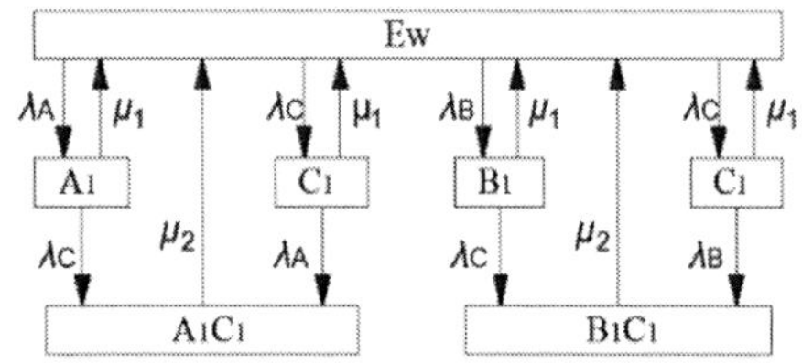

c)

Fig. 8. Graphs of states and transitions in case of fault on the protected object mode a) the system includes a DLP and DBP, b) SRP based on the CDP, c) SRP with the use of CDP and distributed protection DLP and DBP.

Convert the graphs (fig. 8) to the form shown in fig. 9. Since the damage and CT cables cannot lead to fail-to-operate protection failures, they are not included in the scheme.

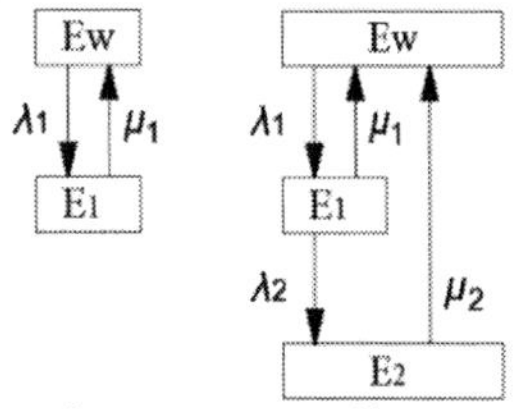

a) b)

Fig. 9. Converted graphs of states and transitions in case of fault on the protected object mode a) the system includes a DLP and DBP; SRP based on the CDP, b) SRP with the use of CDP and distributed protection DLP and DBP.

The system of differential equations for the system includes a DLP and DBP and SRP based on the CDP describing the graph (fig.9, a), are similar as (1) and, accordingly, the expressions for functions of unreadiness probability and probability of failure-free operation will be similar.

The calculation for the system includes a DLP and DBP and SRP based on the CDP.

Determine the probability that the system is situated in the state E_2. The system of differential equations describing the graph:

$$\left.\begin{aligned} &\frac{dP_w(t)}{dt} = -\lambda_1 \cdot P_w(t) + \mu_1 \cdot P_1(t) + \mu_2 \cdot P_2(t) \\ &\frac{dP_1(t)}{dt} = \lambda_1 \cdot P_w(t) - (\lambda_2 + \mu_1) \cdot P_1(t) \\ &\frac{dP_2(t)}{dt} = \lambda_2 \cdot P_1(t) - \mu_2 \cdot P_2(t) \end{aligned}\right\} \quad (3)$$

Solving the system of equations (3), the unreadiness probability

$$q(t) = P_2(t) = 0{,}057 \cdot e^{-0{,}491 \cdot t} + 0{,}07 \cdot e^{-0{,}409 \cdot t} + 0{,}013$$

Graph for the calculation of assessment of the probability of failure-free operation will look similar as the graph shown in fig. 9, b, with the difference that it will not be possible transition from state E_2 в E_w.

The system of differential equations describing the graph:

$$\left.\begin{aligned} &\frac{dP_w(t)}{dt} = -\lambda_1 \cdot P_w(t) + \mu_1 \cdot P_1(t) \\ &\frac{dP_1(t)}{dt} = \lambda_1 \cdot P_w(t) - (\lambda_2 + \mu_1) \cdot P_1(t) \\ &\frac{dP_2(t)}{dt} = \lambda_2 \cdot P_1(t) \end{aligned}\right\} \quad (4)$$

Solving the system of equations (4), the probability of failure-free operation

$$R(t) = 1 - P_2(t) = 1{,}009 \cdot e^{-0{,}0045 \cdot t} - 0{,}009 \cdot e^{-0{,}509 \cdot t}$$

In fig. 10 shown the dependences of the function of the unreadiness probability of SRP, and functions of the probability of failure-free operation on the average time between checks.

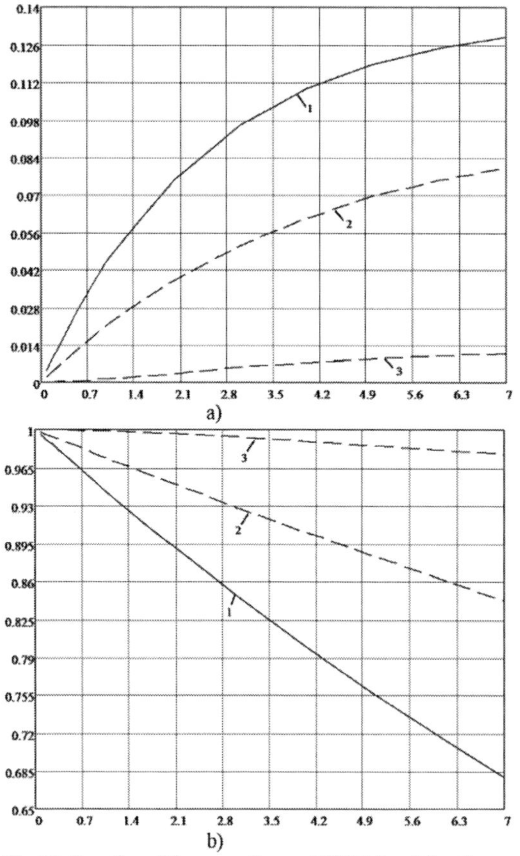

Fig. 10. Function of the unreadiness of SRP (a) and function PFFO of SRP (b).1 - the system includes a DLP and DBP; 2 - SRP based on the CDP; 3 - SRP with the use of CDP and distributed protection DLP and DBP.

Analysis of numerical calculations of indicators of reliability of the differential relay protections (fig. 7, 10) shows:

- developed protection part of electrical network using the method of double entry, in both modes in general, has higher indicators of reliability of than technical solution with DLP and DBP;
- to increase reliability in an absence of fault on the

protected object mode in variant of the total system of with the use of CDP and distributed protection DLP and DBP, it is recommended to apply the scheme with the voting;
- the obtained calculation ratio can be used in the methods of practical reliability analysis of centralized relay protection of power substations.

4 Conclusions

Proposes a new principle of organization of the centralized differential protection of part of electrical network using the method of double entry.

This protection can adapted to changes in the configuration of the electrical network and suitable for use in the power supply circuit with sources of distributed generation.

Developed technical solutions allow to detect the current transformers and communication channels faults, and having high reliability.

References

1. J. Tang, P.G. McLaren, IEEE Trans. Power Del., **21 (3)** 1183 (2006)
2. H. Guo, K. Kangvansaichol, P. Crossley, Proc. of int. conf. *Actual trends in development of power system protection and automation* (2013)
3. Z. Gajić, I. Brnčić, T. Einarsson, B. Ludqvist, Proc. of int. conf. *Relay protection and substation automation of modern power systems* (2007)
4. A.L. Kulikov, V.Ju. Vukolov, A.A. Kolesnikov, *Vestnik NGIEI*. **2 (69)**, 71 (2017)
5. B.H. Smeets, M.H.J. Bollen, *Stochastic modelling of protection systems: comparison of four mathematical techniques* (Eindhoven, TU/e, 1995)
6. A.I. Shalin, *Nadezhnost' i diagnostika relejnoj zashhity jenergosistem* (Novosibirsk, NGTU, 2002)
7. B.V. Papkov, A.L. Kulikov, *Teorija sistemy i sistemnyj analiz dlja jelektrojenergetikov* (Jurajt, 2016)
8. Je.P. Smirnov, Elektrichestvo, **9**, 44 (1965)
9. A. Khurram, H. Ali, A. Tariq, O. Hasan, Proc. of int. workshop *Formal Methods for Industrial Critical Systems*, 169–183 (2013)
10. Ju.B. Guk, *Teorija nadezhnosti v jelektrojenergetike*, (Leningrad, Energoatomizdat, 1990)

 DOI: 10.1051/e3sconf/20172503009

Disintegration of power grid as part of the task of increasing functionality of electric system

Bekzhan Mukatov[1*], *Alexander* Fishov[2]

[1]JSC "KEGOC", National Dispatch Center of the System Operator, Z00T2D0, Astana, Kazakhstan
[2]Novosibirsk State Technical University, 630073, Novosibirsk, Russia

Abstract. When operating electric power systems there are often situations in which the power systems operation is inevitable with reduced reliability or, otherwise, with incomplete functionality where functionality is the set of functions provided by the power system and the quality of their performance. With the mass input of distributed small generation in grids of almost all voltage classes it is necessary to solve the problem of ensuring stability in previously passive distribution networks. The traditional approach based on the "struggle" to maintain synchronism between power plants in the distribution networks is associated with a number of difficulties, which causes to apply another approach to control modes in distribution networks. Complication of the power grid, automatic devices, increase in possible variations of modes, and tendency to maximize the use of production assets lead to an increase in the complexity of tasks solved by dispatch centers. In this regard, it is important to note that availability of cascade failures in power systems speaks of the urgency of the task of ensuring the survivability of energy supply systems both globally and locally. The paper shows how disintegration of the power grid can solve the task of ensuring the functionality of traditional power systems and help to create favorable conditions for distributed small generation integration into the integrated electric power system.

1 The study of potential of power grid disintegration

In traditional power grids functionality is provided by the integrity of structure and system mode, i.e. stability of parallel operation of all power plants. To do this restrictions are imposed on the power transmitted over the network in the form of capacity reserves, and a complicated emergency control system is maintained. In such electric power systems (EPS) disintegration (division of the power grid into parts) refers to exceptional measures of impact on the mode in order to ensure survivability of EPS.

Currently, development and operation of power systems are based on the invariance of the network switching scheme to their modes. At the same time, the network disintegration potential is underused to ensure reliability of the power supply system and ensure its survivability. Disintegration of network is used to separate the district or power plant for isolated work with an approximately balanced load in a predetermined section, as well as to terminate the asynchronous mode in section that connects non-synchronous parts.

In networks with an adaptive structure the reliability and quality of power supply is more determined by the system's ability to balance the separation and post reduction of the system integrity. Functionality may to a considerable extent be provided by changing the configuration of power grid with allocation of balanced subsystems to isolated work depending on the current mode. Unlike traditional EPS where the integral structure is a carrier of reliability, in adaptive EPS due to the development of the power system survivability feature preventive disintegration becomes the basis of reliability [1,2].

Disintegration of power grid with reduced functionality is capable of raising its level or restoring it completely due to the effective use of EPS internal reserves of structural nature. These reserves include:

- the use of the regulating effect of load in frequency and voltage;
- increase in permissible power flows corresponding to the regulatory stability reserves which can be achieved by reducing irregular fluctuations, by reducing (up to complete elimination) the total consumption in one of the intersectional subsystem;
- forced power flow to relieve overload on the electric power line due to the load of other unloaded elements.

Disintegration of EPS can be used to re-commute the circuit in such a way that the system survivability increases under severe system conditions (for example, deep frequency reduction in the region), as well as in modes where n-1 criterion is not met. This is especially important in the power system operating conditions with reduced stability reserves.

Currently, with a decrease in functionality due to non-compliance with the requirements for stability reserves, i.e. in increased risk modes (IRM), the

* Corresponding author: mukatov@kegoc.kz

possibility of implementing such mode, as a rule, is provided by Emergency Control Automatics (ECA) with an impact on load shedding or generation. Thus, in case of system integrity and line shutdown in section 1 or generator in subsystem 2 (Fig. 1a), the Stability Control Schemes (SCS) will unload the section 1 due to the load shedding in subsystem 2.

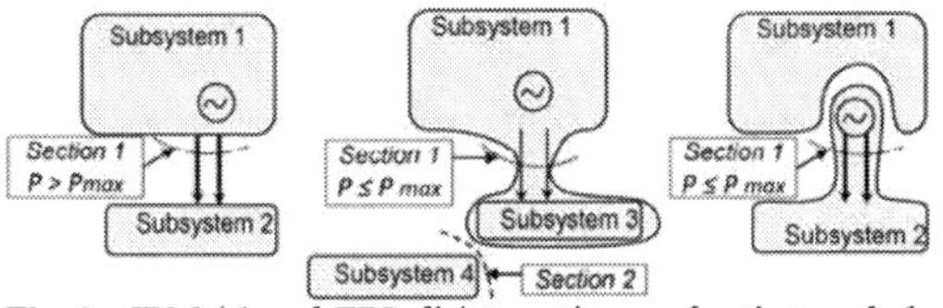

Fig. 1. IRM (a) and EPS disintegration modes that exclude IRM (b, c), where P – section loading in IRM, P max – the maximum permissible flow in section.

IRM can be excluded or passed with more functionality due to the power grid disintegration. In Fig. 1 shows the options of power grid disintegration to demonstrate the effectiveness of its application to improve functionality in IRM.

According to [3] permissible overflow on static stability with normative reserves on active power (k_p) in section is calculated according to the formula (1):

$$P = P_{\lim} \cdot (1 - k_p) - \Delta P_{if} \qquad (1)$$

where $P_{\lim}$ – active-power flow that is limit according to non-periodic static stability in the reviews section; P – the flow in section in the reviewed mode, ΔP_{if} – the amplitude of irregular fluctuations of active power in this section.

If the actual data is missing ΔR_{if} shall be calculated according to the formula:

$$\Delta P_{if} = k \cdot \sqrt{\frac{P_{l1} \cdot P_{н2}}{P_{l1} + P_{l2}}}, \qquad (2)$$

where P_{l1}, P_{l2} – total load capacities on each side of the section, k - coefficient that is equal to 1.5 for manual control and 0.75 for automatic control (restriction) of power flow in section.

After reconfiguration with division of subsystem 2 into two subsystems (3 and 4), ΔP_{if} in (1) decreases as the total capacity of the subsystem 3 is less than the power of the original subsystem 2. In case of power unbalance in subsystem 4, the necessity for disconnecting consumers can be completely or partially eliminated by using the regulating effect of the load in frequency and voltage.

Table 1 shows the volumes of the ECA control actions with reliability provided by traditional SCS, as well as disintegration according to the schemes shown in Fig. 1b), c) for the original characteristic mode of the Kazakhstan power grid (Figure 1a), where subsystem 1 is the northern part of Kazakhstan power grid, 2 is the southern part of Kazakhstan power grid and the EPS of Uzbekistan and Kyrgyzstan.

Calculations were carried out using "RastrWin" and "Mustang" software complexes for modes calculation.

Table 1. The volume of control actions in case of designed disturbances

Designed disturbances		**Power transmission line shutdown in section 1**			**Shutdown of 200 MW generation in subsystem 2**		
Method of reliability assurance		SCS	**Reconfiguration**		SCS	**Reconfiguration**	
			According to the scheme of Fig.1 b)	According to the scheme of Fig.1 c)		According to the scheme of Fig.1 b)	According to the scheme of Fig.1 c)
Control and compensating actions, MW	Shutdown of generators	0	0	200	0	0	0
	Load shedding	400	250	0	200	50	0
	Compensation of unbalance due to regulating effect of load	0	0	200	0	0	200

2 Decentralized reconfiguration of active electrical distribution network

Active increase in the number of distributed small generation (DSG) facilities with their inclusion in parallel operation of the power grid poses the problem of development of technologies for modes management to ensure their effective integration into EPS [4-8].

Input of DSG in the previously passive power distribution networks (DN) radically alters their original properties, turning the passive DN into active power grids - the power systems [9-13].

In view of the existing difficulties in implementation of centralized management of modes of the active distribution network, the existing traditional management technology is too costly for systems with DSG and unable to fully solve the tasks. Taking into account the large number of participants of the electricity and related services market acting for their own sake, it is appropriateness of organizing distributed management of power grid modes with DSG based on multi-agent technologies [14-19].

The source [20] proposes decentralized algorithms for searching sections of the network and re-commutating the circuit while preserving integrity of the power system to improve functionality of the power supply systems. Then, in the enlarged form, the results of testing of distributed algorithms for reconfiguring electrical network on specially designed software modeling behavior of agents of the multi-agent system (MAS) are presented.

2.1 Multi-agent reconfiguration when network element is overloaded

According to example of the circuit representing 220-110 kV electric networks of the Western part of the EPS of the Republic of Kazakhstan (Fig. 2), a multi-agent search of the network structure excluding overload of the power lines was performed.

DOI: 10.1051/e3sconf/20172503009

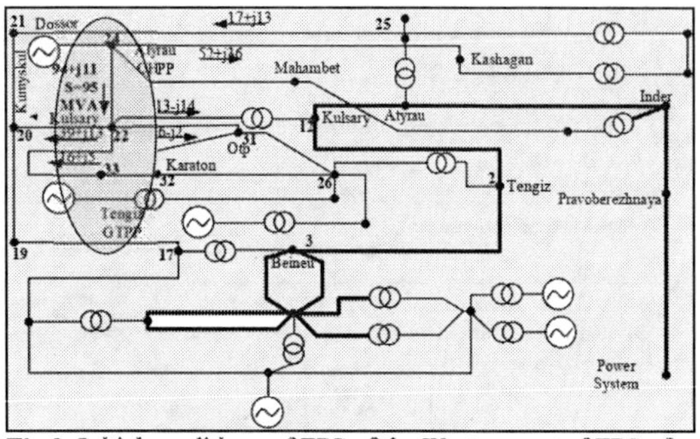

Fig.2. Initial conditions of EPS of the Western part of EPS of the Republic of Kazakhstan.

In the original scheme, the overhead line between nodes 22 and 24 is overloaded (Atyrau TPP – Kulsary SS).

The MAS performed reconfiguration, as a result of which bars were electrically separated at the Kulsary substation and part of the overhead line was switched to 1 busbar, the other part to 2 busbar. Taking into account the possible combinations in distribution of loads on the bars, the minimum possible load was transferred to the 2 busbar with preservation of communication with transformer of Kulsary substation. The result of reconfiguration is shown in Fig.3.

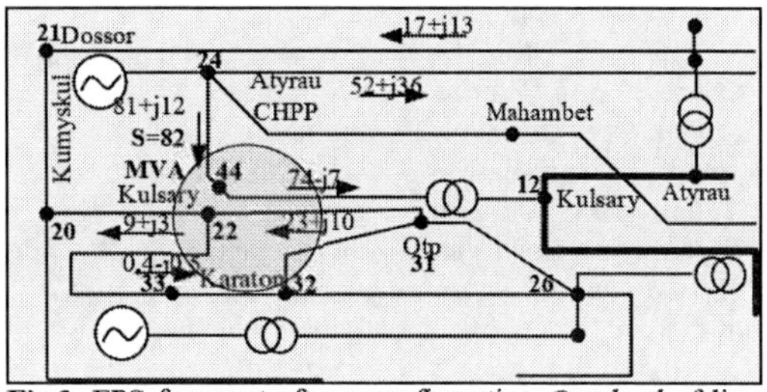

Fig.3. EPS fragment after reconfiguration. Overload of line of Atyrau TPP – Kulsary SS was eliminated.

2.2 Multi-agent disintegration of the power grid

To test adequacy of the MAS operation in case of adaptive network division, a study was conducted using the example of electrical network of the Kazakhstan's power grid southern regions.

Fig. 4 shows the result of simulation of actions of agents with perturbation in external network in a characteristic winter maximum of consumption.

Dotted lines mark out the nodes that form areas passing into the insular mode of operation. To ensure the balanced state of one of the subsystems bars of Mirgalimsay substation were separated.

The simulation results show that fission cross sections change within a day. Division by bars was performed in three nodes (Chulak, Zhambyl, Shelek substations) at the hour of night minimum of consumption.

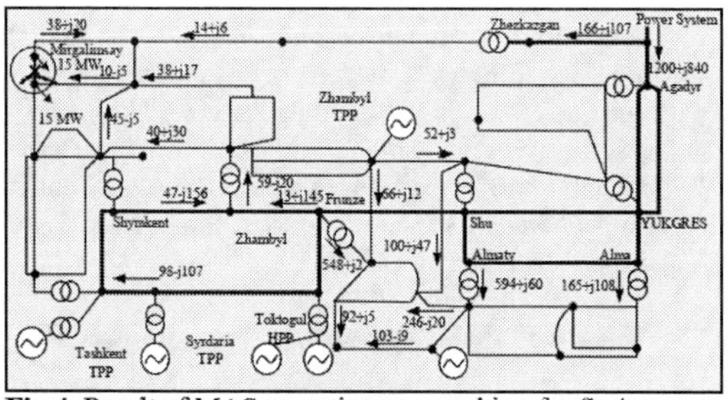

Fig.4. Result of MAS operation on searching for fission cross sections per hour of maximum.

To compare the results of MAS operation with traditional automatics, the fission cross sections (double continuous line) performed by the centralized division automatics are shown in the same figures.

In the scheme of a certain MAS division, the amount of disconnected load in the southern part of the Kazakhstan's power grid is about 350 and 200 MW at the time of maximum and minimum, respectively. The volume of restrictions during operation of the traditional division automatics in similar hours is more than 2 times higher.

3 Preventive disintegration in active electrical networks

An analysis of transient modes when connecting small generation to electric network and short circuit in the power system [2] showed that due to strong connections between subsystems there is a high probability of occurrence of unacceptable shock moments on synchronous machine shafts, disturbances of dynamic stability that are capable of causing their damage. Moreover, the switched-off short-circuit currents increase significantly, relay protection becomes more complicated [20].

These negative phenomena can be minimized due to the new way of emergency control – the accelerated termination of synchronous operation of DSG from integrated electric power system (preventive division) to disconnection of the short-circuit places.

Figure 5 shows the values of cutoff currents of short-circuit and dynamic moments on the shafts of generators G1, G2, obtained as a result of simulation for various control methods in case of short-circuit.

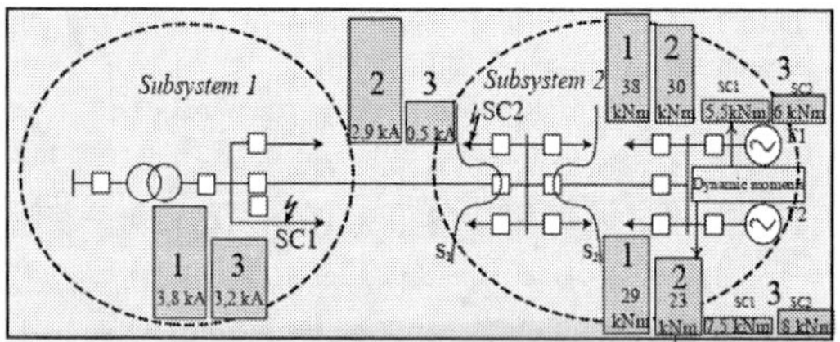

Fig.5. Cutoff currents of short circuit and dynamic moments: 1- in case of short circuit 1 without network division, 2 - in case of short circuit 2 with network division by automation of the termination of an asynchronous mode action, 3- in case of short circuit 1, 2 with preventive network division in sections S1, S2

Results of the research show that the preventive balanced disintegration of the network of power systems connection according to fixed sections ensures reduction of disconnected short-circuit currents, does not change the conditions determining the operation of the relay protection of a high-power electric system in the absence of parallel operation, eliminates the threat of unacceptable moments on synchronous generator shafts, prevents the emergence of asynchronous modes, reduces the volumes of load shedding to prevent the development of accidents.

4 Conclusion

The potential of the disintegration of the electric grid was studied using the example of Kazakhstan's power system. The results of the research demonstrate the possibility of reducing the amount of disconnected power consumers in case of emergency disturbances by 400 MW, by increasing acceptable load flow in the overloaded sections and using the regulating effect of the loads obtained during network reconfiguration.

The conducted researches of multi-agent algorithms of the electric networks reconfiguration show the prospects and efficiency of their application for providing the functionality of power systems. For the examples considered, the separation of subsystems in the sections received by MAS in emergency load flows, makes it possible to reduce the amount of disconnected power for consumers up to 350 MW, comparing to traditional automatics.

The results of simulation of the automatic preventive islanding demonstrate its effectiveness in systems with DSG when emergency disturbances can cause shock dynamic moments and a threat to dynamic stability. In the low-power subsystem, a six-fold decrease in the disconnected short-circuit current is observed, as well as at least a three-fold decrease in the dynamic moments on the synchronous machine shafts.

References

1. A.G. Fishov, B.B. Mukatov *Bulletin of Tomsk Polytechnic University. Engineering of geological resources*, **№9**. - P. 143-152. (2015).
2. B.B. Mukatov, A.G. Fishov, N.A. Kardzhaubayev *Reports of the Academy of Sciences of the Higher School of the Russian Federation*, **№ 4**. - P. 94–104. (2015).
3. *Methodological guidelines on the stability of power systems.* (Approved by the Order of the Ministry of Energy of Russia dated June 30, 2003 № 277. M.: Publishing house "ENAS SC", 2004).
4. P.I. Bartolemey, T.Y. Panikovskaya, D.A. Chechushkov *Joint symposium as part of the APEC project "Energy links between Russia and East Asia: development strategies in the XXI century"*. Irkutsk, P. 4–5. (2010).
5. A. Fishov, M. Shiller, A. Dekhterev, V. Fishov *Journal of Energy and Power Engineering*, **№9** (1). P.59-67. (2015).
6. J.A. Pecas Lopes, N. Hatziargyriou, J. Mutale, P. Djapic, N. Jenkins. *Electric Power Systems Research*, **№77**. P. 1189-1203. (2007).
7. P.V. Ilyushin *The possibility of ensuring reliable power supply of consumers from distributed generation facilities* http://www.cigre.ru /research_commitets/ik_rus/c6_rus.
8. *The potential benefits of distributed generation and rate-related issues that may impede their expansion* (A study pursuant to section 1817, U.S. Department of Energy, 2007. https:// www.ferc.gov/legal/fed-sta/exp-study.pdf).
9. P.V. Ilyushin *Energy Expert*, **№1**. P 58-62. (2015)
10. Y.N Kucherov, P.V. Ilyushin *Electro*, **№5**. - P.2-7. (2014).
11. A.V. Pazderin, S.A. Yeroshenko, A.A. Karpenko, S.Y. Kokin, S.A. Dmitriyev, A.I. Haljasmaa *Scientific review*, **№ 6**. P. 49-56. (2013).
12. G.S. Nudelman *Modern trends in development of the systems of relay protection and automation of power systems: papers of the 4th International Research and Practice Conference.* Yekaterinburg: Publishing House "Russian National Committee SIGRE", P. 1-8. (2013).
13. B.B. Mukatov, A.G. Fishov, I.A. Yefremov *Reports of the Academy of Sciences of the Higher School of the Russian Federation.* **№ 4** P. 90–103. (2014).
14. D.A. Cohen *Power and Energy Society General Meeting – Conversion and Delivery of Electrical Energy in the 21st Century*, IEEE. P. 1–5. (2008). DOI: 10.1109/PES.2008.4596818.
15. A.G. Fishov, S.T. Ismoilov, S.S. Trufakin *Modern trends in development of the systems of relay protection and automation of power systems: papers of the 4th International Research and Practice Conference.* Yekaterinburg: Publishing House "Russian National Committee SIGRE", P. 99-100. (2013).
16. J.M. Solanki. *Integrated Computer-Aided Engineering*, **17 (4)**. P. 331-346. (2010).
17. Chong Shao,Chen Xu, Shan He, Xiangning Lin *TENCON 2013 – 2013 IEEE Region 10 Conference (31194)*, P. 1–4. (2013).
18. S. D. J. McArthur, E. M. Davidson, V. M. Catterson, A. L. Dimeas, N. D. Hatziargyriou, F. Ponci, T. *Multi-agent systems for power engineering applications - part 2: technologies, standards and tools for building multi-agent systems.* IEEE Transactions on Power Systems, **22 (4)**. P. 1753-1759. (2007).
19. Arshad Saleem, Morten Lind, Manuela M. Veloso. *9th International Conference on Autonomous Agents and Multiagent Systems AAMAS.* (2010).
20. B.B. Mukatov *Managing division and restoration of network using expert technologies: Abstract of a Ph.D. Thesis: 05.14.02* (Novosibirsk. - 2016. – 24 p).

On power quality and reliability of supply

Lidiia Kovernikova[1*], *Roman* Shamonov[2]

[1]Melentiev Energy Systems Institute SB RAS, Irkutsk, Russia
[2]Public Joint Stock Company "Federal Grid Company of Unified Energy System", Moscow, Russia

Abstract. Power quality determines reliability of electrical equipment operation, performance of the assigned functions by it, service life. In the last years problem of power quality attracted increased attention because of power quality degradation and a great number of consumer complaints. Many problems of power quality control, such as determination of liabilities and responsibility of power suppliers and consumers for its quality, call for solution. The paper presents the types of damages caused by harmonics and the results of an analysis of harmonic active power at the node connecting an aluminum smelter block to a supply network. The analysis is performed on the basis of measurements. Measured current and voltage were used to calculate harmonic active power. It was found out that the aluminum smelter block generates the harmonic active power to the supply network only during part of the measurement time. For the other part harmonic active power flows from the supply network to the aluminum smelter block. The paper presents the proposals on the use of harmonic active energy for determination of the power quality distortion source and its contribution.

1 Introduction

Power quality and supply reliability are closely interrelated. Power quality determines reliability of electrical equipment operation, performance of the assigned functions by it, service life. At the present voltage and its quality are the main parameter and index of power quality. In [1] voltage quality and supply reliability were combined into term – quality of supply, which is confirmed by practice of power system operation. In Russia before the adoption of the federal laws "On technical regulation" [2] and "On electric power industry" [3] the problem of power quality was treated as a problem of power system condition control. The power quality had to correspond to the requirements of the state standard [4]. The only barrier to meet its requirements consisted in absence of proper instruments for measuring the power quality indices. However, despite the lack of measuring instruments significant results were achieved in the area of power quality control. The top achievement is the adoption of two regulations, which formulated the economic mechanism of power quality control: "The rules of consumer connection to the public network in terms of the effect on power quality" [5], "The rules of applying the rate discounts and surcharges for power quality" [6]. Adoption of the law "On technical regulation" [2] made it possible not to fulfill requirements of the state standard on power quality, since the state standards were granted the status of documents of optional application. In the law "On electric power industry" [3] the necessity of the quality of power supplied to consumers, which complies with the technical regulations and other mandatory requirements, is indicated in more than ten clauses. However, as fairly noted in [7], it is done in the form of "general terms". Adoption of the law "On electric power industry" gave rise to emergence of many players of the electric power industry. Therewith, each of them solved the problems of satisfying own interests and the problem of power quality evolved into a secondary one. In the last few years this problem attracted increased attention because of power quality degradation and a great number of consumer complaints. Many problems of power quality control, such as determination of liabilities and responsibility of power suppliers and consumers for its quality, call for solution.

In the Customs Union countries the voltage quality requirements are specified in the GOST 32144-2013 [8]. Deviations in voltage characteristics from the rated ones lead to different adverse consequences for power suppliers and consumers.

2 Examples of the adverse impact of poor quality voltage

Interruptions of the supply voltage and dips. Voltage interruptions and dips arise as a result of short currents, connection of high power loads. The voltage dips can lead to spontaneous disconnection of contactors and magnetic starters [9] and, as a result, to interruption of power supply to consumers causing a technological damage. In [10] the continuous productions and sectors based on digital technologies are described as the most

* Corresponding author: kovernikova@isem.irk.ru

 DOI: 10.1051/e3sconf/20172504001

sensitive to voltage dips. In these branches the price of one dip is 2120–4682 EUR. On the average 13 voltage dips and 6 momentary voltage interruptions are observed annually at the node of load connection. Table 1 presents typical costs of voltage interruptions for different types of enterprises in one of the US states [11]. The cost rises by 2–4 times with the increasing duration of voltage interruption up to 1 hour.

Table 1. Typical costs of momentary interruptions.

	Type of enterprise	Cost, USD/kW
Industrial	Paper	1.5–2.5
	Textile, metal fabrication, mining	2.0–4.0
	Rubber and plastics	3.0–4.5
	Petrochemical, food processing	3.0–5.0
	Pharmaceutical	5.0
	Automobile manufacturing	5.0–7.5
	Electronics	8.0–12.0
	Semiconductor manufacturing	20.0–60.0
Services	Communication	1.0–10.0
	Banks, hospitals, civil services	2.0–3.0
	Restaurants, hotels, commercial shops	0.1–1.0

Voltage unbalance. In 2016 at one of the substations of the Volga-Don enterprise of the backbone networks the operation of gas protection resulted in tripping the 63000 kVA transformer. The cause of the transformer tripping was a turn-to-turn fault in the primary winding that occurred because of quick insulation ageing because of the long-term current unbalance.

Harmonic voltage. The paper [12] presents the fact of damage of household appliances of consumers that was caused by tripping the 220 kV transmission line "Pokosnoe – Tulun" in the Irkutsk region. The power was supplied to consumers over the transmission lines of 10 and 35 kV from the 35 kV buses of the substation "Pokosnoe". After tripping the 220 kV transmission line at the substation "Pokosnoe" the values of the 11-th harmonic voltage factor ($K_{U(11)}$) sharply increased. Their maximum values at phases A, B and C reached 14.5 %, 17.2 % and14.4 %, respectively compared to the admissible value of 4 %. In the settlements situated at distances of 56, 87 and 101 kilometers from the substation "Pokosnoe" the maximum values of $K_{U(11)}$ proved to be 35.7 %, 46.0 % and 47.9 %, which caused the damage of household appliances.

Because of the harmonic problems the DC link at the substation "Mogocha" operated unsteadily. As indicated in [13], "due to availability of harmonic voltage of a high amplitude at a frequency of about 8 kHz in the 35 kV network of converters the 35 kV cable terminals were heated and damaged during commissioning works in the ring condition of power transmission between IPS of Siberia and IPS of East, which required additional elaboration of design options".

The harmonic currents caused the overheating of both phase and neutral wires of cable transmission lines. This problem is particularly vital in low voltage systems, where the single-phase loads are the harmonic sources. Even if the nonlinear loads turn out to be symmetric, considerable current can flow in the neutral wire primarily with the harmonics multiple of 3. Figure 1 presents the curve of current variation in the neutral wire of the network for external lighting of the trading and entertainment center in the city of Krasnoyarsk [14].

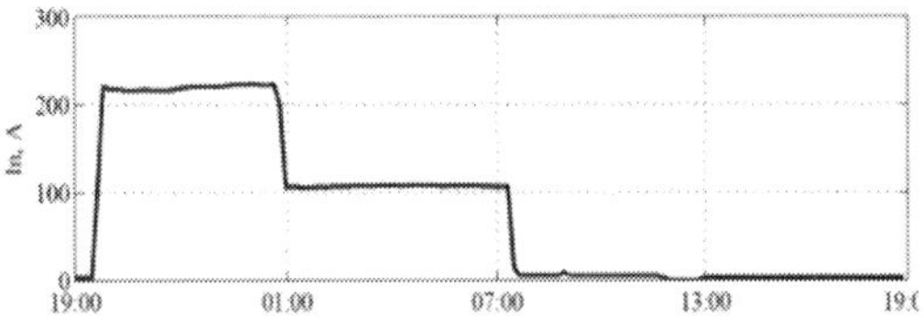

Fig. 1. Daily current variation in the neutral wire.

The harmonics cause malfunction of circuit breakers because of a high value of the cross ratio, transformer overheating because of additional eddy current losses, capacitor damage because of high values of harmonic currents, increase losses in the asynchronous motors, create transient overvoltages in the cosine capacitors, lead to malfunction of protection and automation systems and many other adverse effects [15–17]. The costs induced by harmonics amount to 5.4 % of all the costs due to poor power quality [10].

3 Damages from harmonics

All adverse effects, i.e. damages, can be classified in the form presented in Fig. 2.

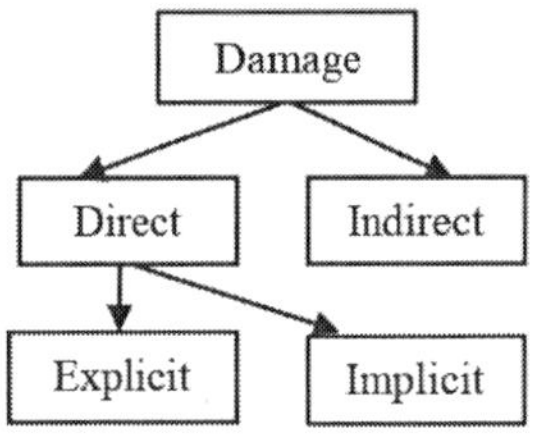

Fig. 2. Types of damages from harmonics.

The direct damage includes damages caused directly by equipment failures, malfunction of protection and automation systems, increase of power losses during its production, transmission, distribution and consumption. The direct damage can be divided into explicit and implicit damages. The explicit damage is linked to the effects, which can be seen and measured. The implicit damage is accumulated during a long time and can be estimated by the indirect indicators, special experiments and special calculations. The indirect damage comprises damages leading to violation of different production processes which cause quality degradation and decrease in output product quantity, drop in productivity and idle time of manpower. In addition, this damage includes an intangible damage that implies discomforts, for example, impossibility to switch a computer and enter an Internet because of poor voltage quality.

The effects of damages require additional financial expenses. According to the expert estimates the

 DOI: 10.1051/e3sconf/20172504001

economic loss because of poor power quality amounts to billions dollars and euros annually [10, 11, 18, 19]:

• in the US in the process industry, financial services and food industry – 60-80 billion USD,

• in the European Unity only in the trade and process industry – 10 billion EUR, and the total loss is estimated at 100 billion EUR,

• in Russia "by the minimal estimate they are about 25 billion USD".

The financial losses can be reduced by power quality improvement, which is possible by detecting a degradation source first of all. The problem of determining the distorting load contribution to power quality degradation remains unsolved for many decades and is topical.

4 Determination of the power quality distortion source and its contribution based on the harmonic active energies

The nonlinear load receives active energy at the fundamental frequency (at the first harmonic). Part of this energy is consumed by load and part is converted into active energies at harmonics for orders h>1. The harmonic active energies of nonlinear load flow to the supply network, where they dissipate in equipment resistances, heat it and shorten service life. In the case of numerous nonlinear loads the harmonic active energies come from the supply network to the loads together with the active energy of the first harmonic and have adverse impacts. If the values of harmonic active energies flowing from the load to the supply network are determined, then it becomes possible to determine the nonlinear load contribution to power quality distortion. The paper [20] proposed that the asymmetrical consumer contribution to power quality distortion should be determined by the negative phase-sequence active energy.

We use an equivalent circuit to analyze harmonic active energies at the node of nonlinear load connection to the supply network that is presented in Figure 3 [21].

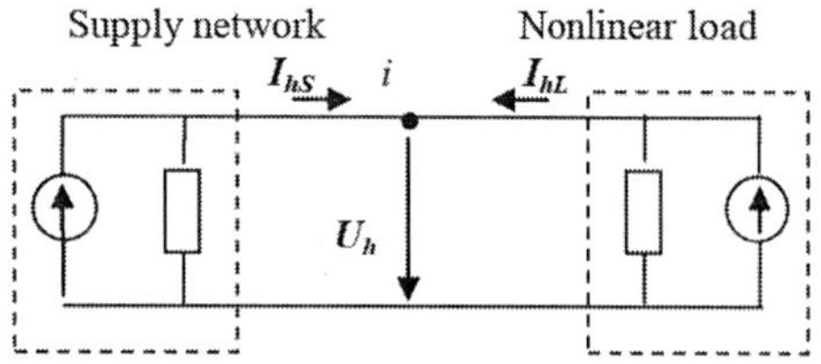

Fig. 3. An equivalent circuit of the supply network and nonlinear load for the h-th harmonic.

The notations used in the scheme are:

i – the number of the node;

I_{hS} – the vector of the h-th harmonic current of the supply network that is the resultant vector of the h-th harmonic current of all nonlinear loads available in the supply network except for current I_{hL};

I_{hL} – the vector of the h-th harmonic current of the nonlinear load connected to the i-th node;

U_h – the vector of the h-th harmonic voltage at the i-th node.

The resultant vector of the h-th harmonic current I_h at the i-th node is determined by the vector sum of currents

$$I_h = I_{hS} + I_{hL}. \tag{1}$$

The parameters of harmonic conditions depend on numerous factors: the wave effect appearing in transmission lines at the harmonic frequencies [22], the resonances emerging in the networks at harmonics [23], the values of phase angles of harmonic currents of nonlinear loads distributed in the network [24], the changes in network load and scheme, and the others, i.e. they are random values. Therefore, the values of current (I_h), voltage (U_h) and their phase angles (φ_{Uh}, φ_{Ih}) should be obtained by measurements. The harmonic active powers are calculated based on the measured parameters

$$P_h = U_h I_h cos\varphi_h, \tag{2}$$

where $\varphi_h = \varphi_{Uh} - \varphi_{Ih}$ – phase displacement between the h–th harmonic voltage and current. The power flow direction is determined by the angle value φ_h [25]. If φ_h lies in the range 0 to $\pi/2$ or $3\pi/2$ to 2π, the active power of the h-th harmonic flows from the network to the node connected load. If φ_h lies in the range $\pi/2$ to $3\pi/2$, then the active power of the h-th harmonic flows from the load to the network.

In the case of power supply from the source with sinusoidal voltage, the power flow of the first harmonic ($P_{1(+)}$) is directed to the nonlinear load and most of this power is consumed by load to perform effective work (P_w). Part of the active power of the first harmonic is converted by the nonlinear load into harmonic active powers that are directed from the load to the network ($P_{h(-)}$). Then the active power of the first harmonic from the network to the load can be represented as

$$P_{1(+)} = P_w + P_{h(-)} \tag{3}$$

In the network with other nonlinear loads the harmonic active powers can be directed from the network to the load ($P_{h(+)}$). The harmonic numbers of active powers $P_{h(-)}$ and $P_{h(+)}$ can be different. Thus, the harmonic active powers flowing through the node at any time interval are

$$P = P_w + P_{h(+)} + P_{h(-)}. \tag{4}$$

The parameters of harmonic conditions were measured during 24 hours with the time interval of one minute, which was taken into consideration in calculation of harmonic active powers. At the interval of one minute for one measurement the value of active power of any harmonic was assumed to remain invariable. Therefore, the active energy of any harmonic for the time interval of one minute is calculated as a product of harmonic active power by the time equal to one minute.

The presented theoretical principles were applied to the analysis of harmonic active powers flowing through the connection node of the aluminum smelter block to the 220 kV network. Figure 4 demonstrates the total time of harmonic active power flow for 24 hours through the node from the aluminum smelter block to the supply network (*Time* $P_{h(-)}$), and from the supply network to the aluminum smelter block (*Time* $P_{h(+)}$).

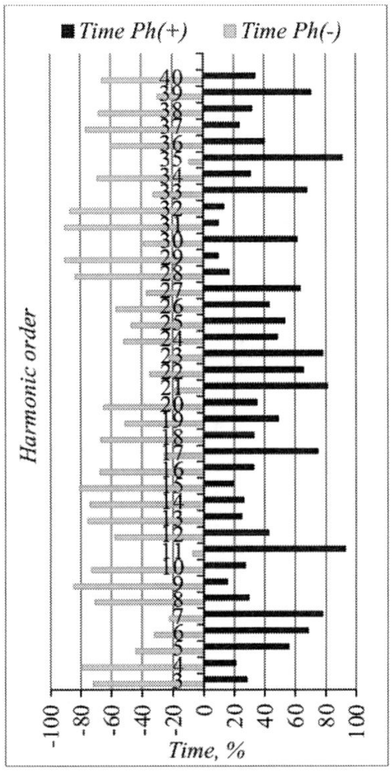

Fig. 4. Time of harmonic active power flow through the connection node of the aluminum smelter block to the supply network in two directions.

Table 2 presents statistical estimates of harmonic active powers, i.e. detrimental powers $P_{d(-)}$ and $P_{d(+)}$ [26] flowing through the connection node in both directions. The harmonic active powers are defined as

$$P_{d(-)} = \sum_{h=2}^{40} P_{h(-)}, \quad (5)$$

$$P_{d(+)} = \sum_{h=2}^{40} P_{h(+)}. \quad (6)$$

Values $P_{d(-)}$ and $P_{d(+)}$ were calculated for 1440 measurements.

Table 2. Statistical estimates of harmonic active powers, kW, and time of their flows, hours.

Parameter	Phase		
	A	B	C
$P_{dmax(-)}$	55.0	71.4	61.2
$P_{dmin(-)}$	0.1	0.4	0.6
$P_{daver(-)}$	11.4	19.5	9.0
$t_{h(-)}$	528.6	499.9	549.2
$P_{dmax(+)}$	59.4	22.4	72.9
$P_{dmin(+)}$	1.5	1.1	1.1
$P_{daver(+)}$	21.2	22.4	25.9
$t_{h(+)}$	407.4	436.1	386.8

The Table shows that the values of average harmonic active powers at three phases, which flow from the load to the supply network, are lower than the similar values of powers flowing from the supply network to the load. On the other hand, the time of harmonic active power flows to the supply network is longer than the time of harmonic active power flows to the load.

Table 3 presents the calculated active energies in both directions for some characteristic harmonics and the total energies for 39 harmonics at three phases for 24 hours of measurements. In the Table $W_{d(-)}$, $W_{d(+)}$ are harmonic active energies flowing to the supply network and load, respectively. From the Table it follows that during 24 hours the aluminum smelter block generates 686.4 kW·h of active energy of the 13-th harmonic to the supply network, but receives 1185.0 kW·h of active energy of the 11-th harmonic from the supply network. Taking into account the energy of 39 harmonics, the aluminum smelter block generates 957.0 kW·h of harmonic active energy to the supply network, and receives 1669.7 kW·h.

Table 3. Harmonic active energies, kW·h.

h	$W_{d(-)}$	$W_{d(+)}$
11	44.0	1185.0
13	686.4	157.0
23	11.3	46.6
25	19.4	31.8
35	7.1	173.6
37	67.2	10.4
2–40	957.0	1669.7

In addition to the active energy of positive sequence of the first harmonic, some up-to-date measuring instruments of power quality indices and smart meters fix the active energy of all harmonics, which makes it possible to calculate the total distorting active energy. This energy can be used to determine overpayment or underpayment for electricity by consumers. If instead of the total harmonic active energy the instruments and smart meters measure harmonic active energies flowing in different directions separately, this will allow determination of the value of consumer contribution to power quality distortion. In such a case the harmonic active energies could be applied to assess the nonlinear load contribution to power quality distortion and taken into consideration by electricity rate charge. If the harmonic active energies are fed from the network to the load, the enterprise could be granted a rate discount, for example, as in [6]. This issue should be elaborated thoroughly and tested at the first stage on the basis of the information obtained during continuous monitoring of harmonic condition parameters and power quality indices during a long period of time.

5 Conclusion

In Russia the power quality is not given due attention because of imperfection of the legislative and legal documents. Violations of the requirements of GOST 32144-2013 are of large-scale and systematic

 DOI: 10.1051/e3sconf/20172504001

character in all national power companies. Practically the tasks on provision of the levels of harmonic voltages and negative sequence of the first harmonic are not stated and solved.

The poor power quality causes damages to electrical equipment of electricity consumers and suppliers, interruptions of power supply to consumers, and as a consequence, financial losses to them.

The need to immediately tackle the power quality as the problem of the national importance became urgent, since the number of distorting electrical equipment of consumers and complaints about poor power quality increased.

Under market relations in electric power industry, it is necessary to develop the economic mechanism of maintaining the power quality, which will comply with the requirements of GOST 32144-2013.

References

1. CIGRE. Working Group 37.28, *Quality of supply – customer requirements*, (June, 2001)

2. Federal law of RF of 27.12.2002 №184–ФЗ "On technical regulation"

3. Federal law of RF of 26.03.2003 № 35–ФЗ "On electric power industry"

4. GOST 13109–97. Electric energy. Electromagnetic compatibility of technical equipment. Power quality limits in public power supply systems.

5. The rules of consumer connection to the public network with respect to the effect on power quality, *Promyshlennaya energetika* **8**, 45-48 (1991).

6. The rules of applying rate discounts and surcharges for power quality, *Promyshlennaya energetika* **8**, 49-51 (1991).

7. V.A. Ovseichuk, Novosti elektroenergetiki **1(91)**, (2015), http://news.elteh.ru/arh/2015/91/03.php

8. GOST 32144–2013. Electric energy. Electromagnetic compatibility of technical equipment. Power quality limits in public power supply systems, (2014).

9. V. Fishman, Novosti elektrotekhniki **5(29)**, (2004), http://news.elteh.ru/arh/2004/29/06.php

10. R. Targosz, D. Chapman, (2015), http://www.leonardo-energy.org/resources/297/the-cost-of-poor-power-quality-5800e490f1e14

11. A. de Almeida, L. Moreira, J. Delgado, *Power quality problems and new solutions, Proceedings of ICREPQ*, (2013).

12. S.S. Smirnov, L.I. Kovernikova, *Promyshlennaya energetika* **8**, 45-48 (2000).

13. Presentation of Public Joint Stock Company "Rosseti – Ministry of Energy", https://minenergo.gov.ru /system/download-pdf.

14. M.O. Chernyshov, V.H. Dovgun, V.V. Novikov, S.V. Brutsky, *Energy of Russia in XXI century, Innovative development and control, Proceedings of Russian Conference*, 225-228 (2015), http://isem.irk.ru/energy21/papers/Чернышов%20МО%20и%20др%20Проблемы%20обеспечения%20качества%20электроэнергии.pdf.

15. R. Hartungi, L. Jiang, *Investigation of power quality in health care facility. Proceedings of ICREPQ*, (2010).

16. L.I. Kovernikova, *Estimation of damage from higher harmonic, Proccedings of the 3rd International Workshop Liberalization and Modernization of Power Systems*, 201-204 (2006).

17. V.N. Gromov, *Power quality control: Proceedings of the International Scientific and Practical Conference*, 85-92 (Moscow, 2014).

18. L.N. Dobrusin, *Report at the VI All-Russian Energy Forum "Energy of Russia in XXI century*, Moscow, State Kremlin Palace, (2008).

19. Economic evaluation of power quality. *IEEE Power engineering. Review*, **22**, (2002).

20. V.S. Sokolov , M.A. Ermilov, A.V. Serkov, A.V. Gromov, N.V. Chernyshova, *Promyshlennaya energetika* **8**, 52-55 (2000).

21. L.I. Kovernikova, *Electrichestvo* **3**, 12-20 (2017).

22. A.E. Emanuel, *Power definitions and physical mechanism of power flow*, (John Wiley &Sons, 2010).

23. J. Arrilaga et al. *Harmonics in electrical systems: Translation from English*, (Energoatomizdat, 1990).

24. W.E. Kazibwe, T.H. Ortmeyer, M.S.A.A. Hammam, IEEE Transactions on PD **4**, 621-628 (1989).

25. R.H. Stevens, IEEE Transactions on PAS **102**, 3018-3021 (1983).

26. L.S. Czarnecki, *IET Generation, Transmission & Distribution* **1**, 1-7 (2012).

Software for optimal selection of places for installation of balancing devices in 0,4 kV electric power systems loaded with electric motors

Victoria Romanova[1]*, *Sergey* Khromov[1], *Ivan* Suvorov[1] and *Kristina* Balbekina[1]

[1]Trans-Baikal State University, Russia, Chita, 30, Aleksandro-Zavodskaya Str.

Abstract. This publication considers the issues of development of the software program for designing of 0,4 kV power supply systems with motor-actuated load under voltage unsymmetry conditions (using the example of the Trans-Baikal Territory). Voltage unsymmetry is practically constant phenomenon in the electric power networks of different voltage types. Voltage unsymmetry effects significantly the electric power consumers, including the supply mains itself. It has especially negative impact on the electrical equipment operation process and its lifetime. The urgency of the problem is confirmed by multiple research on the same topic and by significant number of damages suffered by the electric power consumers staying in service (especially in the Trans-Baikal Territory and in the Far-East regions). Voltage unsymmetry causes economic loss and reduction of the electromagnetic interference value by the voltage unsymmetry coefficient in negative-phase sequence (K_{2U}) gives inevitable economic effect accordingly. However, the payback period for the activities aimed at reduction of electromagnetic interference, will vary from some months to several years. The more accurate value of the payback period may be obtained using the developed software program. The developed software design program is implemented by means of the programming language C# in Microsoft Visual Studio environment, using the built-in cross-platform database SQLite. The software program shall allow making quick and accurate calculation of the power losses, to determine the economic feasibility of provision special measures for removal of the voltage unsymmetry, for determination of optimal application and location of the balancing devices. The software implementation in power systems loaded with electric motors will improve reliability and efficiency of asynchronous motors. The software is of interest for developers of projects on power supply systems for regions with non-linear loads.

1 Introduction

Currently, a special attention is given to the problems of electric energy quality. In modern conditions quality parameters of electric power such as voltage unbalance, unsinusoidality, become common factors which significantly reduce an efficiency of the power supply systems themselves and, accordingly, an efficiency of consumers connected to them.

A study of the main electric power quality parameters for the network equipment of 0.4, 6 and 35 kV power system in Zabaykalsky Krai showed that deviation of main quality parameters of electric power, such as voltage unbalance, voltage unsinusoidality is a common factor in Zabaikalsky power system. In some cases, unbalance level significantly (2-3 times) exceeds specified in GOST 32144 – 2013[1] value [2].

Due to unsatisfactory condition of the electric energy quality in distribution networks of Zabaikalsky power system, there is a large number of electrical failures at the consumer's facilities. Experience of electrical equipment operation shows that one of the main causes of electrical equipment failure is voltage unbalance.

A reliability and efficiency of electrical equipment depends directly on electric energy quality in distribution networks. For stable operation of electrical equipment, number of measures aimed at improving of electric energy quality, in particular, voltage unbalance elimination should be taken.

In this paper, one of possible methods for electric energy quality improvement which uses balancing devices in the electric power system both at the stage of electric network design, and in conditions of real operation is proposed.

The aim of this work is to develop an application software for optimal selection of places for installation of balancing devices in 0,4 kV electric power systems loaded with electric motors under conditions of voltage unbalance focused on engineering practice

* Corresponding author:romanova181@mail.ru

2 Materials and methods of research

In the research of the asynchronous motors, the problem of assessing the impact of unbalanced phase voltages on the motor operation is highly important.

To get a more complete picture of voltage unbalance impact on asynchronous motors it is necessary to define a dependence of asynchronous motors phase currents versus voltage unbalance coefficient for negative sequence K_{2U}.

The research of asynchronous motors operation modes can be efficiently carried out with by means of computer simulation using virtual model created in clear and effective simulation tool Simulink in interactive programming environment of MatLab, using a T-shaped equivalent circuit of induction motor [3-5].

The studied objects were asynchronous squirrel-cage motors of 4A series: 4A132S4Y3 with Pn = 7,5 kW and AI series: AI132M4 with Pn = 11 kW; the motors have the following rated parameters (Table 1) [6,7].

Authors developed a simulation model (Fig.1), which allows performing analysis of asynchronous motors operation modes, moreover, using this model, it was possible to study transient processes in the induction motor, measurements of operation and artificial mechanical characteristics, voltage, current and frequency in the power supply network [3].

Computer simulation of direct start of asynchronous motors was performed for motors of 4A, AI type at three-phase supply voltage 380 V with frequency 50 Hz.

Table 1. Technical data on three-phase asynchronous motors

Power of motor, P, kW	Type of the motor	cos φ	Starting torque ratio, Kmp	Starting current ratio, Kp
7,5	4A132S4Y3	0,86	2,2	7,5
11	AI132M4	0,87	2	7,5

Using the developed model, the asynchronous squirrel-cage motors were studied with regard to the following parameters:

-for different motor shaft loads (idle mode - 120%),

-for change of value of voltage unbalance coefficient for negative sequence (K_{2U}) from 0 to 5%.

The simulation scheme for study of operation modes of asynchronous squirrel cage motor created by means of simulation in Matlab Simulink is shown in (Fig.1).

In the course of process simulation, an analytical curves for the phase currents versus K_{2U} for series 4A, AI asynchronous motors were obtained.

For each phase of asynchronous motors under consideration, curves for the phase currents versus K_{2U} were obtained for different values of the motor shaft load.

The general form of the current change in the most loaded phase (phase "A") for series 4A asynchronous motors at Pn = 7,5 kW, AI series asynchronous motors at Pn=11 kW, for different K_{2U} values and different values of the motor shaft load are shown in (Fig.2,3).

Analysis of the dependences obtained in the course of study of the asynchronous motors modes showed the following:

1) when K_{2U} increases, current in one of the motor phases decreases and in the other two phases the current increase is observed,

2) it can be seen from the general view of the dependencies that: the more the motor shaft load, the more the voltage unbalance impact on the operating modes of electric motors,

3) the curves general view and directions are the same, that shows that impact of voltage unbalance on different series asynchronous motors is similar.

But currents at full load of the motor are of the greatest interest. When the load is 100%, current in phase "A" of 4A series asynchronous motors with Pn = 7,5 kW exceeds permissible overload current equal to 1,1 of rated current for K_{2U} = 1,4% [8,9]. For AI series motor with Pn = 11 kW, the permissible current overload is exceeded at K_{2U} = 2 %.

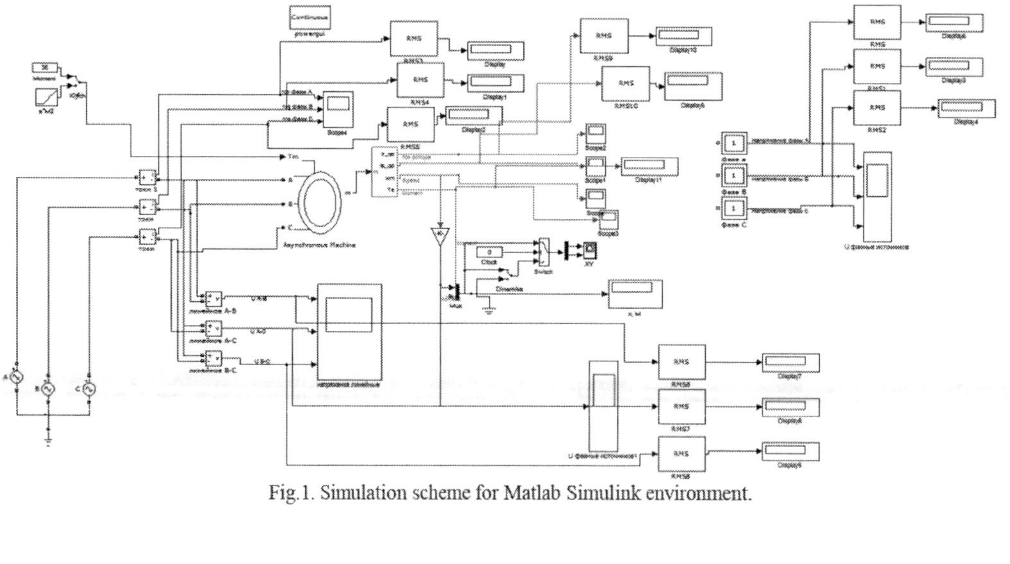

Fig.1. Simulation scheme for Matlab Simulink environment.

DOI: 10.1051/e3sconf/20172504002

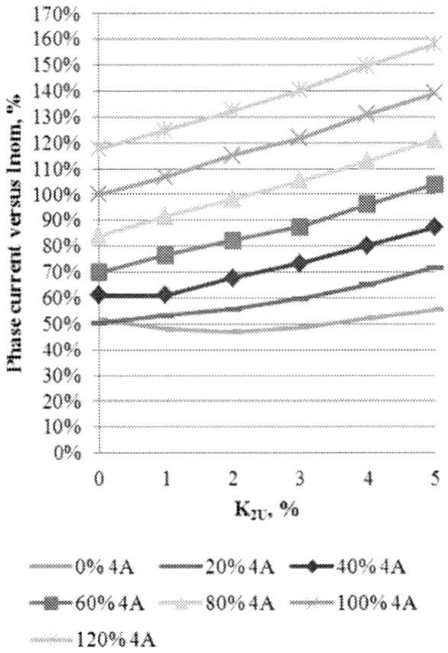

Fig.2. Curve of current for phase A for 4A series asynchronous motors with Pn = 7,5 kW versus K2U.

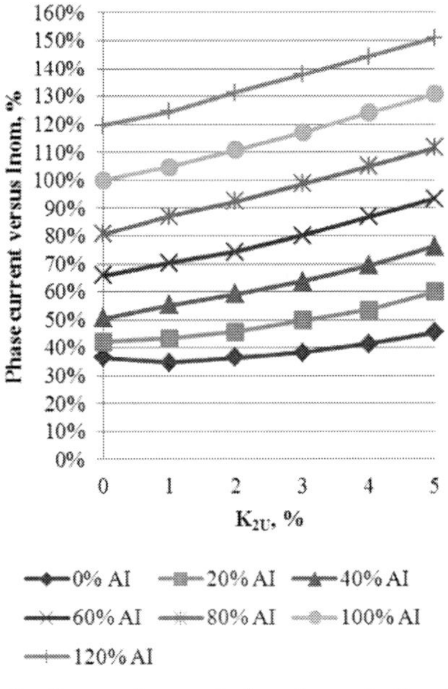

Fig.3. Curve of current for phase A for AI series asynchronous motors with Pn = 11 kW versus K_{2U}.

This shows that the motors not only have different range of acceptable values for K_{2U}, but they not always meet GOST requirements. Accordingly, if the motor is operated in areas with poor electric energy quality, it is necessary to regulate the supply voltage being guided not by GOST on power quality, but by motor operation mode.

It is obvious that the obtained values of permissible K_{2U} values for asynchronous motors will have a key role in the selection and implementation of balancing devices in the power system loaded with electric motors.

3 Formulation of the problem

The work of calculation algorithm development for power supply system taking into account balancing devices and their location, which will be a base of the software for optimal selection of places for installation of balancing devices in 0,4 kV electric power systems loaded with electric motors was performed in accordance with the intended purpose. Block diagram of the calculation algorithm for power system taking into account balancing devices and their location is shown in (Fig.4).

Based on data of the experimental study for main basic quality parameters of electric power, data on power losses in the electric power system components, and based on the allowable modes of consumers operation it is possible to select balancing devices parameters and to define optimal location of these devices in the power system.

Thus, calculation of additional power losses due to voltage unbalance will make it possible to define an economic feasibility of the special measures on unbalance elimination, and in addition to determine the optimal application and location of these devices.

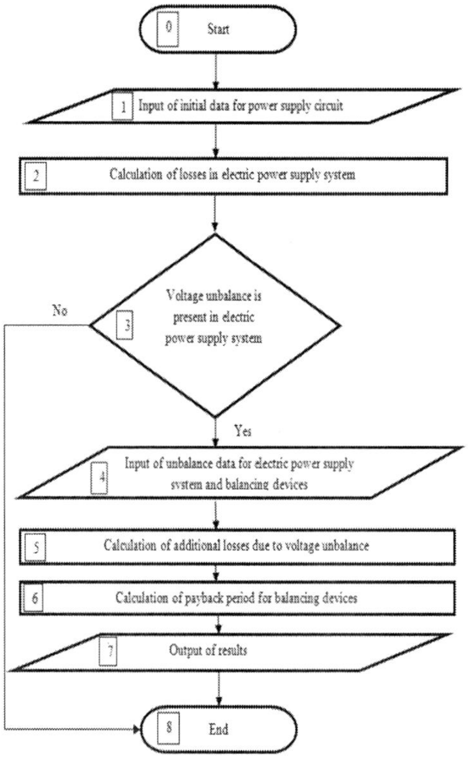

Fig. 4. Block-diagram of calculation algorithm for power supply system taking into account balancing devices and their location.

4 Calculation algorithm for power supply system taking into account balancing devices and their location

Main criterion for the software application is excess of the actual value of voltage unbalance coefficient for negative sequence (K_{2U}) in the particular electric power system over the permissible value K_{2U} specified for asynchronous motors (Table 2) installed in the same power supply system. Required K_{2U} value should be selected on the base of the permitted consumer operation mode. Permissible K_{2U} values for asynchronous motors are taken from results of the study [3].

Table 2. Table of allowable values K_{2U}.

Type of motor	The motor power, Pn, kW	Permissible value of K_{2U}, %
4A132S4Y3	7,5	1,8
AI132M4	11	1,26

The calculation algorithm for the power supply system taking into account balancing devices and their location can be implemented in the following sequence:

1) block 1 – entering of the initial data into the software.

The initial data are:

- characteristic of the circuit segment (general information about the site);
- parameters of the transmission lines (quantity of lines, lines numbers, length of lines; grade of wire and cable used in the line; the line loading);
- parameters of electric motors connected to the power line (the line number to which the motor is connected; quantity of motors in the line, the motor type);
- parameters of the transformer (quantity and type of the transformer);
- value of the voltage unbalance coefficients for negative and zero sequence in the design power supply system;
- cost of electric power.

2) block 2 – performing losses calculation for balanced operation mode in the basic elements of power supply system.

Balanced mode current (current for positive sequence) is defined from the expression:

$$I_1 = S/U, \tag{1}$$

where S – full load of the substation.

Losses in power lines are defined from the expression:

$$\Delta P_{eltl} = 3I_1^2 R_{PH} \tag{2}$$

where I_1 – balanced mode current (current for positive sequence) [10];

R_{PH} – resistance of the phase wire.

Full loading of the substation (sum of the loads due to individual lines) is defined from the expression:

$$S = \sum_{i=1}^{n} P_i , \tag{3}$$

where P_i – loading of the line.

Losses of active $\Delta P'$ and $\Delta Q'$ reactive powers for double-winding transformers are defined from the expression:

$$\Delta P' = \frac{\Delta P_k}{n} \left(\frac{S}{S_{nom}}\right)^2 \tag{4}$$

$$\Delta Q' = \frac{U_k}{100 \cdot n} \cdot \frac{S^2}{S_{nom}} \tag{5}$$

where $\Delta P_k, U_k, S_{nom}$ – passport data for particular transformer [11];

S – full load of the substation;

n – quantity of the same type transformers at the substation connected in parallel.

3) block 3 – transfer to condition: voltage unbalance is present in the power supply system.

At condition "NO" – calculation is stopped.

At condition "YES" – transfer to block 4.

4) block 4 – transfer to special measures for unbalance elimination, definition of optimal application and location of these devices. Unbalance voltage coefficients are input for negative and zero sequences for the two variants of calculation, and also selection from the database of suitable devices for the balancing.

a) variant 1 – considers installation of balancing device into common equipment of the lines comprising electric motors;

b) variant 2 – considers installation of balancing device into each line of the circuit segment comprising asynchronous electric motors.

5) block 5 – calculation of additional losses in key elements of the supply system for unbalanced operation mode.

In the course of calculation of increase in additional losses compared to the symmetric mode, presence in the network of amplitude and angular asymmetry can be accounted by using the following formula:

$$K_{unsym} = 1 + K_{2I}^2 + K_{0I}^2 \cdot \left(1 + 3 \cdot \frac{R_N}{R_{PH}}\right), \tag{6}$$

where K_{2I}, K_{0I} – current unbalance coefficients for negative and zero sequences [10].

Based on the results of data processing for the main quality parameters of electric power in the equipment of 0,4 kV network located in Mogocha district, Zabaikalsky Krai, we take the following values of the coefficients: $K_{2I} = 5\%$, $K_{0I} = 4{,}6\%$, $K_{2U} = 4{,}95\%$;

R_N, R_{PH} – resistance of neutral and phase wires.

Accordingly, taking into account (6), the expression (2) can be rewritten as follows:

$$\Delta P_{suppl.eltl} = \Delta P_{eltl} K_{unsym} \tag{7}$$

Additional loss of active power in asynchronous motor is defined from the formula:

$$\Delta P_{suppl.am} = 2{,}41 K_{AM} K_{2U}^2 P_n , \tag{8}$$

where K_{AM} – coefficient which accounts for the specific parameters of the motor (rated power, copper loss of the stator, starting current ratio);

K_{2U} – voltage unbalance coefficient for negative sequence;

P_n – the motor rated active power.

For industrial loads, it is generally recommended to use coefficient with value of 1,85.

When the system is operated in power transformers in the unbalanced mode for a long time, additional

DOI: 10.1051/e3sconf/20172504002

power losses due to negative sequence currents appear that can be defined by the following formula

$$\Delta P_{suppl.tf} = K_{2U}^2 \quad \Delta P_{IM} + \frac{\Delta P_{SC}}{u_{SC}^2}\ , \qquad (9)$$

where K_{2U} – the voltage unsymmetry coefficient in negative-phase sequence;

ΔP_{IM} – loss during idle mode;

ΔP_{SC} – losses during short-circuit mode;

U_{SC} – short-circuit voltage.

Then total losses are calculated.

For balanced mode, losses are defined from the following formula:

$$\Delta P_{sym} = \sum_{i=1}^{n\_line} \Delta P_{eltl} + \Delta P' \qquad (10)$$

where n_line – quantity of the lines;

i – the line number.

For unbalanced mode, losses are defined from the following formula:

$$\Delta P_{unsym} = \Delta P_{sym} + \sum_{i=1}^{n_{line}} \Delta P_{suppl.eltl} + \sum_{j=1}^{n_{dv}} \Delta P_{suppl.am} + \Delta P_{suppl.tf} \qquad (11)$$

where j – motor number;

n_{dv} – quantity of the motors in the transition line.

6) block 6 – calculation of the payback period for each installation of balancing devices and the least one is taken.

Additional losses without balancing devices are defined from the expression:

$$S_{suppl\frac{w}{o}} = \Delta P_{suppl\frac{w}{o}} + \Delta Q_{suppl\frac{w}{o}} = \sum_{i=1}^{n_{line}} \Delta P_{suppl.eltl} + \sum_{i=1}^{n_{line}} \sum_{j=1}^{n_{dv}} \Delta P_{suppl.am} + \Delta P_{suppl.tf}\ , \qquad (12)$$

at the initial values of the voltage unbalance coefficients for negative and zero sequences.

If balancing device is installed additional losses can be defined from the expression:

$$S_{suppl} = \Delta P_{suppl} + \Delta Q_{suppl} = \sum_{i=1}^{n\_line} \Delta P_{suppl.eltL} + \sum_{j=1}^{n\_dv} \Delta P_{suppl.am} + \Delta P_{suppl.tf}, \qquad (13)$$

at the initial values of the voltage unbalance coefficients for negative and zero sequences.

Selection of the balancing device parameters is performed depending of conditions:

a) variant 1 – power of the balancing device should be more than or equal to the sum of loads in lines outgoing from the central equipment:

$$P_{sym} \geq S \qquad (14)$$

b) variant 2 – power of the balancing device should be more than or equal to the sum of loads in each line comprising a motor:

$$P_{sym} \geq P_i \qquad (15)$$

Calculation of the total cost of balancing devices is performed by the expression:

$$S = \sum_{k=1}^{k\_d} C\_d, \qquad (16)$$

where S – total costs of installed balancing devices;

C_d – costs of particular balancing device;

k_d – quantity of installed balancing devices;

k – balancing device number.

Calculation of payback period for installation of the balancing device is performed by the formula:

$$PP = \frac{S}{\Delta Psuppl\frac{w}{o} + \Delta Qsuppl\frac{w}{o} - \Delta Psuppl + \Delta Qsuppl \quad * C} \qquad (17)$$

where PP – payback period;

S – total costs of installed balancing devices [12,13];

$\Delta P_{suppl\frac{w}{o}} + \Delta Q_{suppl\frac{w}{o}}$ – without the balancing devices;

$\Delta P_{suppl} + \Delta Q_{suppl}$ – supplementary losses with the balancing devices;

C – cost of the electric power.

7) block 7 – output of the resulting data.

The resulting data are:

- losses for balanced operation mode;
- additional losses for unbalanced operation mode (without balancing devices; variant 1 - for case when balancing device is installed into common equipment of the lines comprising electric motors; variant 2 - for case when balancing device is installed into all lines comprising asynchronous electric motors);
- costs of balancing devices (costs of balancing device; total costs of balancing device for variant 1; total costs of balancing device for variant 2; payback period for variant 1; payback period for variant 2).

Calculation algorithm above for electric power supply system will be a base for implementation of software for optimal selection of places for installation of balancing devices in 0,4 kV electric power systems loaded with electric motors operating under conditions of voltage unbalance.

For the software implementation and testing, let us consider segment of power supply system in village Taptugary, Mogocha district, Zabaikalsky Krai.

For this work performance, following initial data were used:

1) results of data processing for main quality parameters of electric power in the equipment of the 0,4 kV network in Mogocha district, Zabaikalsky Krai,
2) configuration of the power supply system in village Taptugary, consumers characteristics.

Power supply system, which was used for the software testing is shown in (Fig.5). The site power system consists of:

1) line 1 – wired with AC-25 (steel-aluminium conductor), the line length is 350 metres, feeds 7 private residence houses with furnace heating,
2) line 2 – cabled with GRSh 3*50+1*35 (cable with flexible core, rubber cover and poly amide silk isolation), the line length is 50 metres, feeds the two-storeyed residence house, the primary school, the kindergarten, the administration building and the library,
3) line3 – cabled with AVVshv 4*50 (cable with aluminium core, PVC jacket, with protective layer in the form of pressed out hose), the line length is 70 metres, feeds the boiler house with two boilers and

 DOI: 10.1051/e3sconf/20172504002

electric motors: models 4A132S4Y3 with Pn =7,5 kW in amount of 2 pieces, smoke exhaust fan Pn=3 kW, draught fan Pn=1,4 kW,

4) line 4 - wired with SIP 4*25 (self-supporting insulated conductor), the line length is 150 metres, feeds two ban mills with motors model AI132M4 Pn=11kW.

Table 3. Initial data for segment of power network in village Taptugary.

Parameter	Line 1	Line 2	Line 3	Line 4	TS
Power, kVA	30,4	49,4	20,9	25	100
Current, A	80	130	55	65	330

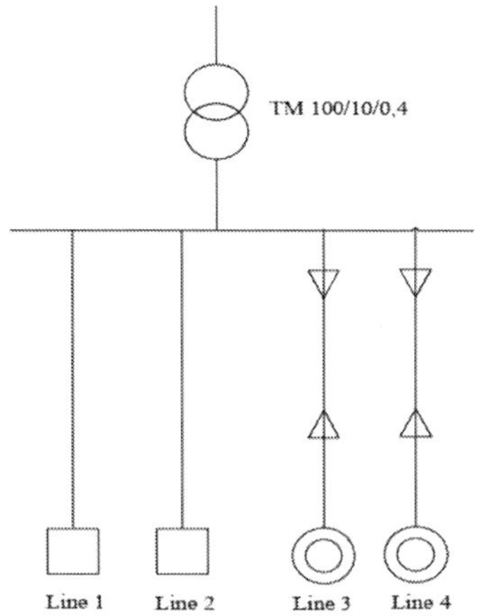

Fig.5. Diagram of segment of power supply system in village Taptugary.

5 Software implementation

The developed software is intended for reduction of voltage unbalance in segments of 0,4 kV power supply systems for the highest economic efficiency. The software can be used for calculation of economic efficiency of existing power systems modernization [14,15].

The software application will allow performing fast and accurate calculation of power losses to determine the economic feasibility of special measures for elimination of voltage unbalance, determining optimal use and location of balancing devices, with significant increase in productivity of the designer [16,17].

The developed software is implemented using C# programming language in Microsoft Visual Studio environment with the embedded cross-platform SQLite database [18-21].

The database contains tables for such areas: reference characteristics of the power system elements, table of the power supply system input data, table of output calculation data for power supply system. Each database table has its unique attributes and its purpose in the system.

The software consists of a database that stores all information necessary for calculation and design of 0,4 kV power supply systems loaded with electric motor; calculations module, which implements the algorithm of power losses calculation in the elements of power supply system; module for entering of input data and output of the calculation results for 0,4 kV power supply systems loaded with electric motor and operated under voltage unbalance conditions.

The initial data are data on the site of 0,4 kV power supply system loaded with electric motor.

The data is developed into database (DB) with its structure enabling easy manipulation of the data, and is more flexible, removing redundancy and non-conformity relations.

The database contains the tables of the following areas: reference parameters of the power supply system elements, tables of the power supply system input data, tables of the output data for calculation of the power supply system designing. Each BD Table has its unique attributes and its own designation in the system. Relationships of BD tables (logic model) are given in (Fig. 6).

The logic model represents relation between BD table.

The software interface is clear and simple (Fig. 7,8). The software interface includes means for entering of reference data for the individual elements of the power supply system, entering and display of the initial data for section of the power supply system, output of the calculation information for the power supply system, history of calculations.

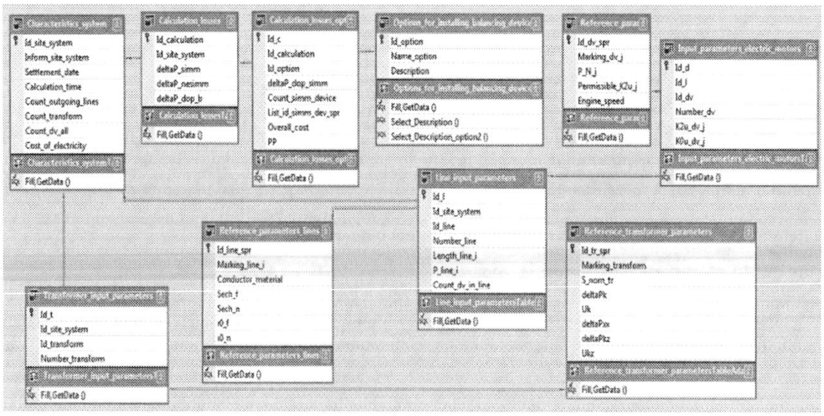

Fig.6. Dependencies of the database tables (logical model).

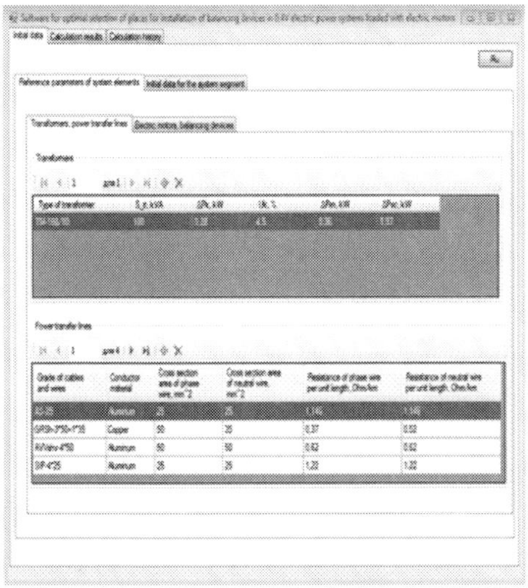

Fig.7. The software interface: input data window - reference parameters of the power supply system (transformers, power lines).

The software interface is presented in two languages (Russian and English). There is a button for language selection in the upper right corner of the software window.

Thus, using the software we defined value of power losses in main elements of power supply system for balanced operation mode, for unbalanced operation mode without installed balancing devices, for unbalanced operation mode with installed balancing devices; we defined quantity, power and installation place for the balancing device.

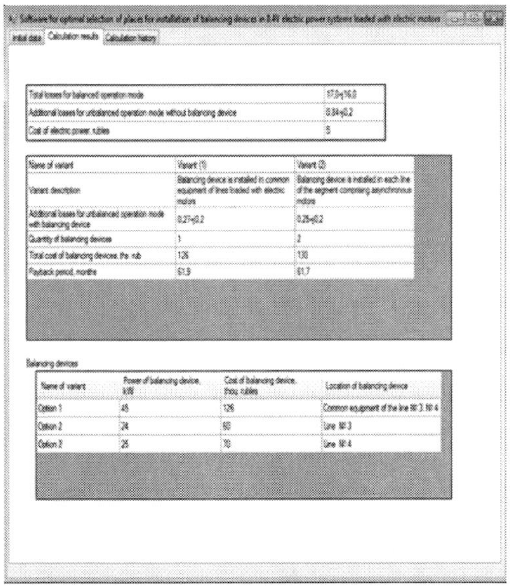

Fig.8. The software interface: window for output of calculation results.

It can be concluded from the results of the software use for the scheme under consideration (Fig.5) that it is economically advisable to install one balancing device with a capacity of 45 kW in the common equipment of the line 3,4 loaded with electric motor, but not on each motor individually.

As a result of the software implementation, necessity of use of balancing devices for industrial consumers in the considered scheme was confirmed. The most cost-effective location of the balancing device in electric power supply system, based on the payback period of measures aimed at reducing of electromagnetic interference was defined.

6 Conclusion

The software is designed for specialists involved in design of 0,4 kV power supply networks and systems operating in presence of voltage unbalance. The proposed software simplifies process of designing of power supply systems; in addition, it can be widely used by project organizations for real conditions of power supply systems operation.

Use of balancing devices reduces current and voltage unbalance in 0,4 kV distribution networks significantly improving the voltage quality at the terminals of electrical consumers, so solutions for these cases are implemented in the software. This increases the operational reliability of electric motors.

Software for optimal selection of places for installation of balancing devices in 0,4 kV electric power systems loaded with electric motors operated under voltage unbalance conditions is the first developed version. Further improvement of the software is possible by expanding the functionality so that it would make it possible to display graphic images of schemes and output a graphical image of calculation diagrams showing location of all balancing devices in electric power system, rendering of the graphs of various parameters dependencies, and also import and export of input and output data of the software.

Support of research

The work was executed within the framework of FSBEI HPE Trans-Baikal State University Council for Research and Innovative Activities Scientific Grant No. 223-GR of 09/02/2017 implementation.

References

1. *GOST 32144 – 2013. Electric energy. Electromagnetic compatibility of technical equipment. Power quality limits in the public power supply systems*. – M.: STANDARTINFORM, **19** (2014).

2. V. Romanova, S. Khromov, *Ulan-Ude: Buryat Scientific Centre, Siberian Branch of the Russian Academy of Sciences*, 20 – 25 (2016).

3. V. Romanova, S. Khromov, I. Suvorov, *Up-to-date technology. Systemic analysis. Modeling. Scientific journal*. Irkutsk: Publishing house of the

DOI: 10.1051/e3sconf/20172504002

Irkutsk state University of transportation, №**1** (53), 209 – 214 (2017).

4. I. Chernykh. *Modeling of electrical devices in MATLAB, SimPowerSystems and Simulink.* (SPb.: Peter M.: DMK Press, 288, 2008).

5. S. German-Galkin, G. Karganov, *Electrical machines: Laboratory works on PC* (SPb.: Korona-print, **256,** 2003).

6. http://principact.ru/content/view/166/72/

7. http://www.elektrikii.ru/publ/6-1-0-55

8. http://www.bmrzzakharov.narod.ru/raschet/overload.html

9. V. Kogorodsky, S.Kurjenkov, L. Paperno. *Relay protection of electric motors with voltages more than 1 kV.* (M.: Energoatomizdat, 1987).

10. A.Ded, A. Parshukova, N. Khalitov, *International journal of applied and fundamental research. Scientific journal.* Moscow: Publishing house "Russian Academy of Natural History", № **10**, P. 3, 421 – 425 (2015).

11. O. Grunin, L. Sawicki, *Power systems and networks in examples and tasks.* (Chita: Publishing house ChitGU, 209, 2010).

12. http://voltmarket.ru/catalog/trekhfaznye_stabilizatory.

13. https://energo-24.ru/authors/energo-24/11017.html.

14. D. Aleksandrov, *Reliability and quality of electric power supply to enterprises.* (Ulyanovsk: UlSTU, 155, 2010).

15. N. Kuznetsov, *Reliability of electrical machines: textbook. the manual for high schools.* (M.: Publishing house of MPEI, 432, 2006).

16. V. Radkevich, *Design of power supply systems: tutorial.* (Minsk: NPOOO "PION", 288, 2001).

17. V. Mankov, *Principles of design of power supply systems* (St. Petersburg: NOU DPO "UMITTS "Electro Service", 640, 2010).

18. http://www.w3ii.com/ru/sqlite/default.html.

19. http://er.nau.edu.ua/bitstream/NAU/10100/1/SQLite.Allow.pdf.

20. R. Frost, D. Dei, C. van Slyke, *Database Design and Development.* (Moscow: Publishing house "NT Press", 570, 2007).

21. V. Medvedev *Features of object-oriented programming in C++/CLI, C# and Java.* (Kazan: Editorial and Publishing center "Shkola", 444, 2010).

 DOI: 10.1051/e3sconf/20172504003

Determination of rational transformation coefficients of transformers distribution networks

Dauren Akhmetbayev[1]*, *Arman* Akhmetbayev[2], *Aigerim* Aidarova[3]

[1] ENU named by L.N. Gumilev, Astana, Kazakhstan
[2] Directorate of Information Systems of JSC "Kazakhtelecom", Almaty, Kazakhstan
[3] OJSC «Severelectro», Bishkek, Kyrgyzstan

Abstract. The thesis proposed a systematic approach to the determination of rational coefficients transformation of transformers distribution networks. Analytic expressions are obtained for transformation coefficient of transformer from the point of view of synthesis of distribution networks in the field of permissible voltage regimes. The Newton-Raphson method was used to solve nonlinear equations for transformation coefficients. This eliminates the need for calculation of steady distribution network modes.

1 Synthesis formalization for the simplest networks

For the sake of clarity and simplicity of presentation, we consider the stationary mode of the distribution network shown in Pic. 1, a.

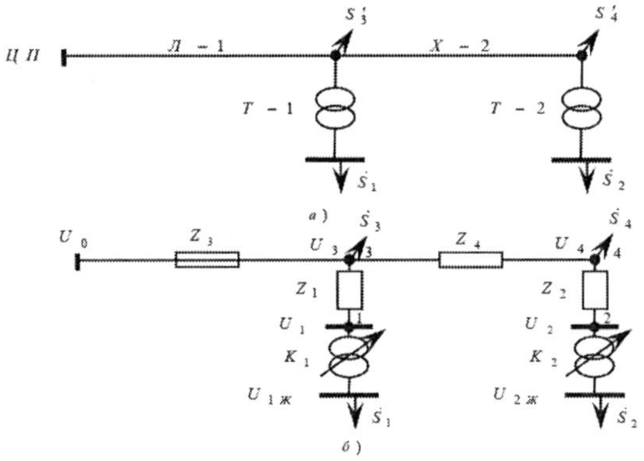

a – principle scheme; б - replacement scheme
Fig. 1. Electrical network scheme

- Z_1, Z_2 - complex resistance of transformers
- Z_3, Z_4 - line resistance of power transmission;
- $\dot{S}_1, \dot{S}_2, \dot{S}_3, \dot{S}_4$ - calculating power of nodal loads;
- K_1, K_2 - transformation coefficients of ideal transformers

Assume that providing the required voltage quality changes achieved transformation ratio transformers with tap changer on the basis of a counter voltage regulation. Adjusting range taps of transformers is considered sufficient to provide the required voltage levels. Consequently, the desired value of the stress on the tires reducing substation considered as given quantities.

The resulted voltages of the secondary windings of transformers with adjustable transformation coefficients, in accordance with the scheme of Fig. 6.3b, are determined by the expressions:

$$\left.\begin{aligned}
\dot{U}_1 = U_0 &- \frac{\sum_{j=1}^{4} \underline{C}^t_{1j} \underline{Z}_j \underline{C}_{j1} \hat{S}_1}{\hat{U}_1} - \frac{\sum_{j=1}^{4} \underline{C}^t_{1j} \underline{Z}_j \underline{C}_{j2} \hat{S}_2}{\hat{U}_2} - \\
&- \frac{\sum_{j=1}^{4} \underline{C}^t_{1j} \underline{Z}_j \underline{C}_{j3} \hat{S}_3}{\hat{U}_3} - \frac{\sum_{j=1}^{4} \underline{C}^t_{1j} \underline{Z}_j \underline{C}_{j4} \hat{S}_4}{\hat{U}_4} \\
\dot{U}_2 = U_0 &- \frac{\sum_{j=1}^{4} \underline{C}^t_{2j} \underline{Z}_j \underline{C}_{j1} \hat{S}_1}{\hat{U}_1} - \frac{\sum_{j=1}^{4} \underline{C}^t_{2j} \underline{Z}_j \underline{C}_{j2} \hat{S}_2}{\hat{U}_2} - \\
&- \frac{\sum_{j=1}^{4} \underline{C}^t_{2j} \underline{Z}_j \underline{C}_{j3} \hat{S}_3}{\hat{U}_3} - \frac{\sum_{j=1}^{4} \underline{C}^t_{2j} \underline{Z}_j \underline{C}_{j4} \hat{S}_4}{\hat{U}_4}
\end{aligned}\right\} \quad (1)$$

where, $\sum_{j=1}^{4} \underline{C}^t_{1j} \underline{Z}_j \underline{C}_{j1}, \sum_{j=1}^{4} \underline{C}^t_{1j} \underline{Z}_j \underline{C}_{j2}$, $\dots \sum_{j=1}^{4} \underline{C}^t_{2j} \underline{Z}_j \underline{C}_{j1}, \sum_{j=1}^{4} \underline{C}^t_{1j} \underline{Z}_j \underline{C}_{j2}$ - partial systemic functions of resistance relatively to the first and second nodes.

Without taking into account the influence of the phases of the node voltages on the magnitude of the driving currents, the system (1) is written in the form:

* Corresponding author: axmetbaev46@mail.ru

$$\left.\begin{aligned}\dot U_1 &= U_0 - \frac{\sum_{j=1}^{4}\underline{C}'_{1j}\underline{Z}_j\underline{C}_{j1}\hat S_1}{U_1} - \frac{\sum_{j=1}^{4}\underline{C}'_{1j}\underline{Z}_j\underline{C}_{j2}\hat S_2}{U_2} - \\ &- \frac{\sum_{j=1}^{4}\underline{C}'_{1j}\underline{Z}_j\underline{C}_{j3}\hat S_3}{U_3} - \frac{\sum_{j=1}^{4}\underline{C}'_{1j}\underline{Z}_j\underline{C}_{j4}\hat S_4}{U_4} \\ \dot U_2 &= U_0 - \frac{\sum_{j=1}^{4}\underline{C}'_{2j}\underline{Z}_j\underline{C}_{j1}\hat S_1}{U_1} - \frac{\sum_{j=1}^{4}\underline{C}'_{2j}\underline{Z}_j\underline{C}_{j2}\hat S_2}{U_2} - \\ &- \frac{\sum_{j=1}^{4}\underline{C}'_{2j}\underline{Z}_j\underline{C}_{j3}\hat S_3}{U_3} - \frac{\sum_{j=1}^{4}\underline{C}'_{2j}\underline{Z}_j\underline{C}_{j4}\hat S_4}{U_4}\end{aligned}\right\} \quad (2)$$

Known values of node voltages are such that the system (2) describes the stationary condition of the electrical network within the allowable regimes by voltage.

Desired voltages on tires 6-10 kV of reducing substations are provided by voltage regulation by changing the coefficient transformation of transformers. It is known that the selection of the regulating branches of the transformers is carried out individually for each substation according to the results of calculation of the stationary regime. In this case, the mutual influence of the regulating branches can not be taken into account when choosing the most suitable modes within the range of admissible regimes. The most suitable mode can be found in the system approach to the determination of the regulating branches of all transformers of the lowering substations.

Let the necessary expedient regime of the voltage of the network in question be ensured by changing the transformation ratios of the transformers.

Then the system of equations (2), taking into account the transformation coefficients of ideal transformers, can be written as follows:

$$\left.\begin{aligned}\dot U_1 &= U_0 - \frac{\sum_{j=1}^{4}\underline{C}'_{1j}\underline{Z}_j\underline{C}_{j1}\hat S_1}{K_1U_{1\text{Ж}}} - \frac{\sum_{j=1}^{4}\underline{C}'_{1j}\underline{Z}_j\underline{C}_{j2}\hat S_2}{K_2U_{2\text{Ж}}} - \\ &- \frac{\sum_{j=1}^{4}\underline{C}'_{1j}\underline{Z}_j\underline{C}_{j3}\hat S_3}{U_3} - \frac{\sum_{j=1}^{4}\underline{C}'_{1j}\underline{Z}_j\underline{C}_{j4}\hat S_4}{U_4} \\ \dot U_2 &= U_0 - \frac{\sum_{j=1}^{4}\underline{C}'_{2j}\underline{Z}_j\underline{C}_{j1}\hat S_1}{K_1U_{1\text{Ж}}} - \frac{\sum_{j=1}^{4}\underline{C}'_{2j}\underline{Z}_j\underline{C}_{j2}\hat S_2}{K_2U_{2\text{Ж}}} - \\ &- \frac{\sum_{j=1}^{4}\underline{C}'_{2j}\underline{Z}_j\underline{C}_{j3}\hat S_3}{U_3} - \frac{\sum_{j=1}^{4}\underline{C}'_{2j}\underline{Z}_j\underline{C}_{j4}\hat S_4}{U_4}\end{aligned}\right\} \quad (3)$$

Taking into account $U_1 = U_{1\text{Ж}}/K_1; U_2 = U_{2\text{Ж}}/K_2$ and the values of the distribution coefficients, the system (3) can be written in the form:

$$\left.\begin{aligned}&U_{\Delta 11}K_1^2 + U_{\Delta 12}K_2K_1 - \left(U_0 - U_{\Delta 13} - U_{\Delta 14}\right)K_1 + U_{1\text{Ж}} = 0 \\ &U_{\Delta 21}K_1K_2 + U_{\Delta 12}K_2^2 - \left(U_0 - U_{\Delta 23} - U_{\Delta 24}\right)K_2 + U_{2\text{Ж}} = 0\end{aligned}\right\} \quad (4)$$

where the coefficients of the unknowns for the scheme under consideration are equal to:

$$U_{\Delta 11} = \sqrt{\left(\frac{P_1(R_1+R_3)+Q_1(X_1+X_3)}{U_{1\text{Ж}}}\right)^2 + \left(\frac{P_1(X_1+X_3)-Q_1(R_1+R_3)}{U_{1\text{Ж}}}\right)^2}$$

$$U_{\Delta 12} = \sqrt{\left(\frac{P_2R_3+Q_2X_3}{U_{2\text{Ж}}}\right)^2 + \left(\frac{P_2X_3-Q_2R_3}{U_{2\text{Ж}}}\right)^2}$$

$$U_{\Delta 13} = \sqrt{\left(\frac{P_3R_3+Q_3X_3}{U_3}\right)^2 + \left(\frac{P_3X_3-Q_3R_3}{U_3}\right)^2}$$

$$U_{\Delta 14} = \sqrt{\left(\frac{P_4R_3+Q_4X_3}{U_4}\right)^2 + \left(\frac{P_4X_3-Q_4R_3}{U_4}\right)^2}$$

$$U_{\Delta 21} = \sqrt{\left(\frac{P_1R_3+Q_1X_3}{U_{1\text{Ж}}}\right)^2 + \left(\frac{P_1X_3-Q_1R_3}{U_{1\text{Ж}}}\right)^2}$$

$$U_{\Delta 22} = \sqrt{\left(\frac{P_2(R_2+R_3+R_4)+Q_2(X_2+X_3+X_4)}{U_{2\text{Ж}}}\right)^2 + \left(\frac{P_2(X_2+X_3+X_4)-Q_2(R_2+R_3+R_4)}{U_{2\text{Ж}}}\right)^2};$$

$$U_{\Delta 23} = \sqrt{\left(\frac{P_3R_3+Q_3X_3}{U_3}\right)^2 + \left(\frac{P_3X_3-Q_3R_3}{U_3}\right)^2}$$

$$U_{\Delta 24} = \sqrt{\left(\frac{P_4(R_3+R_4)+Q_4(X_3+X_4)}{U_4}\right)^2 + \left(\frac{P_4(X_3+X_4)-Q_4(R_3+R_4)}{U_4}\right)^2}$$

As an illustration, calculations are made by the proposed and conventional methods for the circuit in Fig. 1 with the specified parameters of the transformers, line and load:

$\square P_\kappa = 45{,}5\kappa BA, \square P_x = 9{,}4\kappa Bm, U_\kappa = 7{,}5\%, I_x = 0{,}9\%,$
$\pm 6\times 1{,}5\%, U_H = 11\kappa B$

Л1-2ХАС-70, Л2-2ХАС-35, $L_1=L_2=18$км., $S_1= S_2= 3{,}5+$ j 2,5 MBA.

Design parameters: transformers $Z_1= Z_2=1{,}6+j16{,}1$ Ом,

line -$Z_3=4{,}05+j3{,}78$ Ом, $Z_4=8{,}19+j4$ Ом.

Calculations of non-linear equations were carried out by the Newton-Raphson method in the MATLAB environment, and the working mode of the network by the two-stage method according to the data of the beginning. The results are shown in Table 1.

Table 1. Results of calculations

Direct method	Classical method
Mode of maximum loads	
K_2=0,3217 K_1=0,3123 U_2=34,19 kV U_1=35,22 kV	$U_{жел}$=10,5 kV $U_{отв}$= $U_н$-nx1,5% $U_{отв2}$=34,47 kV $U_{отв1}$=35,52 kV
Minimum mode	
K_2=0,279 K_1=0,277 U_2=39,42 kV U_1=39,71 kV	$U_{жел}$=10,0 kV $U_{отв2}$=39,4 kV $U_{отв1}$=39,78 kV

As can be seen from the table, the results obtained by different methods give good coincidences, which allows us to conclude that the proposed method can be used to solve both operational and design tasks.

2 Synthesis of distribution networks of complex structure

Assume that the desired voltage busbar transformer substations included first n complex electrical circuit nodes network provided by the transformation ratio transformers changes. Then a nonlinear matrix equation (8.8), after simple transformation can be written in the form of voltage balance, for example i-th node:

$$\dot{\underline{U}}_{i\text{ж}} = \dot{\underline{U}}_{\Delta i1} K_i K_1 + \dot{\underline{U}}_{\Delta i2} K_i K_2 + \ldots + \dot{\underline{U}}_{\Delta in} K_i K_n + \Big(\sum_{j=n+1}^{N} \dot{\underline{U}}_{\Delta ij} + \dot{\underline{U}}_0\Big) K_i \qquad (5)$$

where $\dot{\underline{\mathbf{U}}}_{\Delta i1} = \sum_{j=1}^{m} \underline{\mathbf{C}}_{ij}^{t} \underline{\mathbf{Z}}_j \underline{\mathbf{C}}_{j1} \dot{\underline{\mathbf{U}}}_{1\text{ж}}^{-1} \dot{\underline{\mathbf{S}}}_1$, $\dot{\underline{\mathbf{U}}}_{\Delta in} = \sum_{j=1}^{m} \underline{\mathbf{C}}_{ij}^{t} \underline{\mathbf{Z}}_j \underline{\mathbf{C}}_{jn} \dot{\underline{\mathbf{U}}}_{n\text{ж}}^{-1} \dot{\underline{\mathbf{S}}}_n$ - partial nodal voltage reduces of the i-th node.

$\dot{\underline{U}}_{i\text{ж}}$ - complex desired voltage of the i-th node;

$\dot{\underline{U}}_0$ - the voltage of the basic node;

$K_1, K_2, \ldots, K_n$ -transformation coefficients of transformers;

N - number of independent nodes.

The dependent variables in equation (4) are the node voltage phases (δ) and the transformation coefficients transformer (K).

The system of nonlinear equations for the balance of the stress of the node under consideration concerning the variables (δ, K) can be obtained from (5) by selecting the real and imaginary parts in the form:

$$\left.\begin{aligned}
&\omega_{U_i'}(\delta,K) = -U_{i\text{ж}}\cos\delta_i + \sum_{j=1}^{n} U_{\Delta ij} K_i K_j \cos(\delta_j - \varphi_j + \psi_{ij}) + \\
&+\Big(\sum_{j=n+1}^{N} U_{\Delta ij}\cos(\delta_j - \varphi_j + \psi_{ij}) + U_0\Big) K_i \\
&\omega_{U_i''}(\delta,K) = -U_{i\text{ж}}\sin\delta_i + \sum_{j=1}^{n} U_{\Delta ij} K_i K_j \sin(\delta_j - \varphi_j + \psi_{ij}) + \\
&+\Big(\sum_{j=n+1}^{N} U_{\Delta ij}\sin(\delta_j - \varphi_j + \psi_{ij})\Big) K_i
\end{aligned}\right\} \qquad (6)$$

Where:

U_i', U_i'' are the real and imaginary parts of the voltage;

$U_{\Delta ij}$ - module of partial nodal voltage drop;

φ_j - phase shift of the load power of the j-th node;

ψ_{ij} - angle of mutual complex nodal resistance.

It is seen from (6) that a system of 2n equations with 2n unknowns is compiled, the solution of which can be obtained by the Newton method. The Jacobi matrix is not degenerate and can be written in block form in the form:

$$\frac{\partial \mathbf{W}_U(x)}{\partial x} = \begin{Vmatrix} \dfrac{\partial \omega_{U'}(\delta,K)}{\partial \delta} & \dfrac{\partial \omega_{U'}(\delta,K)}{\partial K} \\ \dfrac{\partial \omega_{U''}(\delta,K)}{\partial \delta} & \dfrac{\partial \omega_{U''}(\delta,K)}{\partial K} \end{Vmatrix}$$

The partial derivatives of the i-th row of each block matrix can be calculated as follows:

$$\frac{\partial \omega_{U'_i}(\delta,K)}{\partial \delta} = U_{i\text{ж}}\sin\delta_i - U_{\Delta ii} K_i^2 \sin(\delta_i - \varphi_i + \psi_{ii}) - \sum_{\substack{j=1 \\ j\neq i}}^{n} U_{\Delta ij} K_i K_j \sin(\delta_j - \varphi_j + \psi_{ij});$$

$$\frac{\partial \omega_{U'_i}(\delta,K)}{\partial K} = 2U_{\Delta ii} K_i \cos(\delta_i - \varphi_i + \psi_{ii}) + \sum_{j=n+1}^{n} U_{\Delta ij}\cos(\delta_j - \varphi_j + \psi_{ij}) + U_0 +$$

$$+\sum_{\substack{j=1 \\ j\neq i}}^{N} U_{\Delta ij} K_i \cos(\delta_j - \varphi_j + \psi_{ij});$$

$$\frac{\partial \omega_{U''_i}(\delta,K)}{\partial \delta} = -U_{i\text{ж}}\cos\delta_i - U_{\Delta ii} K_i^2 \cos(\delta_i - \varphi_i + \psi_{ii}) + \sum_{\substack{j=1 \\ j\neq i}}^{n} U_{\Delta ij} K_i K_j \cos(\delta_j - \varphi_j + \psi_{ij});$$

$$\frac{\partial \omega_{U''_i}(\delta,K)}{\partial K} = 2U_{\Delta ii} K_i + \sum_{j=n+1}^{N} U_{\Delta ij}\sin(\delta_j - \varphi_j + \psi_{ij}) + \sum_{\substack{j=1 \\ j\neq i}}^{n} U_{\Delta ij} K_i \sin(\delta_j - \varphi_j + \psi_{ij}).$$

If we neglect the influence of the phases of the node voltages on the values of the driving currents, then the system (6.32) can be written in the form:

$$\left.\begin{aligned}
&\omega_{U_i'}(\delta,K) = -U_{i\text{ж}}\cos\delta_i + \sum_{j=1}^{n} U_{\Delta ij} K_i K_j \cos(-\varphi_j + \psi_{ij}) + \\
&+\Big(\sum_{j=n+1}^{N} U_{\Delta ij}\cos(-\varphi_j + \psi_{ij}) + U_0\Big) K_i \\
&\omega_{U_i''}(\delta,K) = -U_{i\text{ж}}\sin\delta_i + \sum_{j=1}^{n} U_{\Delta ij} K_i K_j \sin(-\varphi_j + \psi_{ij}) + \\
&+\Big(\sum_{j=n+1}^{N} U_{\Delta ij}\sin(-\varphi_j + \psi_{ij})\Big) K_i
\end{aligned}\right\} \qquad (7)$$

Then the partial derivatives of the matrix are defined as:

$$\frac{\partial \omega_{U'_i}(\delta,K)}{\partial \delta} = U_{i\text{ж}}\sin\delta_i;$$

$$\frac{\partial \omega_{U'_i}(\delta,K)}{\partial K} = 2U_{\Delta ii} K_i \cos(-\varphi_i + \psi_{ii}) + \sum_{j=n+1}^{N} U_{\Delta ij}\cos(-\varphi_j + \psi_{ij}) + U_0 +$$

$$+\sum_{\substack{j=1 \\ j\neq i}}^{n} U_{\Delta ij} K_i \cos(-\varphi_j + \psi_{ij});$$

$$\frac{\partial \omega_{U''_i}(\delta,K)}{\partial \delta} = -U_{i\text{ж}}\cos\delta_i;$$

$$\frac{\partial \omega_{U'_i}(\delta, K)}{\partial K} = 2U_{\Delta ii} K_i \sin(-\varphi_i + \psi_{ii}) + \sum_{j=n+1}^{N} U_{\Delta ij} \sin(-\varphi_j + \psi_{ij}) +$$

$$+ \sum_{\substack{j=1 \\ j \neq i}}^{n} U_{\Delta ij} K_i \sin(-\varphi_j + \psi_{ij}).$$

In distribution networks calculation formula for determining the elements of the Jacobian matrix, greatly simplified and reduced amount of calculations performed since the voltage balance equation I - th node can be reduced to the form:

$$\omega_{\underline{U}_i}(K) = -U_{i\text{ж}} + \sum_{j=1}^{n} U_{\Delta ij} K_i K_j \cos(-\varphi_j + \psi_{ij}) + \left(\sum_{j=n+1}^{N} U_{\Delta ij} \cos(-\varphi_j + \psi_{ij}) + U_0\right) K_i \tag{8}$$

which can easily be written in the matrix form as follows:

$$\mathbf{K}_{Д} \mathbf{U}_{\Delta} \mathbf{K} + (\mathbf{U}_{\Delta\Sigma} + \mathbf{U}_0 \mathbf{E}) \mathbf{K} - \mathbf{U}_{\text{ж}} = 0 \tag{9}$$

$\mathbf{U}_{\Delta}$ - square matrix of nodal voltage drops from load currents, powered from transformer buses;

$\mathbf{U}_{\Delta\Sigma}$ - column matrix of nodal voltage drops from load currents without transformer links;

$\mathbf{U}_{\text{ж}}$ - column matrix of the desired stresses;

$\mathbf{K}$ - column matrix of transformation coefficients of transformers;

$\mathbf{E}$ - the identity matrix.

Given the initial approximations of the transformation coefficients of transformers and expanding in a Taylor series, we can obtain a linear approximation of each nonlinear equation of the system (6.34), for example, the nonlinear approximation of the i-th node has the following form:

$$-U_{i\text{ж}} + U_{\Delta i1} \cos(-\varphi_1 + \psi_{i1}) K_i^0 K_1^0 + \ldots + U_{\Delta ii} \cos(-\varphi_i + \psi_{ii}) (K_i^0)^2 + \ldots +$$

$$+ U_{\Delta in} \cos(-\varphi_n + \psi_{in}) K_i^0 K_n^0 + \left(\sum_{j=n+1}^{N} U_{\Delta ij} \cos(-\varphi_j + \psi_{ij}) + U_0\right) K_i^0 +$$

$$+ U_{\Delta i1} \cos(-\varphi_1 + \psi_{i1}) K_i^0 K_1^{(1)} + \ldots + (2U_{\Delta ii} \cos(-\varphi_i + \psi_{ii}) K_i^0 +$$

$$+ \sum_{j=n+1}^{N} U_{\Delta ij} \cos(-\varphi_j + \psi_{ij}) + U_0) \Delta K_i^{(1)} + \ldots + U_{\Delta in} \cos(-\varphi_n + \psi_{in}) K_i^0 \Delta K_n^{(1)} = 0 \tag{10}$$

We write the system of linearized equations in the matrix form as follows:

$$W(K^0) + \frac{\partial W(K^0)}{\partial K} \Delta K^{(1)} = 0 \tag{11}$$

Where $\Delta K^{(1)}$ is the matrix of corrections of the first iteration of Newton. Each step of the iterative process involves solving the linear system (11) with the subsequent determination of the approximation:

$$K^{(i+1)} = K^i + \Delta K^{(i+1)} \tag{12}$$

The Newtonian iteration process, in general, can be written in matrix form:

$$K^{(i+1)} = K^{(i)} - \left[\frac{\partial W(K^i)}{\partial K}\right]^{-1} W(K^i) \tag{13}$$

Control of convergence is carried out on the vector of residuals:

$$W_U(K) \leq \varepsilon$$

Where ε - is a predetermined small value.

From the calculated values of the transformation coefficients, it is easy to find the tap voltage and the regulating stage of each transformer.

3 Conclusions

1. A systematic method for calculating the transformer ratios of distribution network transformers has been developed.
2. Nonlinear equations are obtained with respect to transformer transformation coefficients.
3. To solve the problem, the Newton-Raphson method is proposed.

References

1. V.A. Venikov, *Optimization of the regimes of power plants and power systems* (M.:Energoatomizdat, 1990).
2. V.I. Idelchik, *Calculations and optimization of modes of systems electrical networks* (M.: Energoatomizdat, 1988).
3. A.G. Rusina, T.A. Filippova. *Modes of power plants and electric power systems*. (Novosibirsk: SGTU, 2016).
4. D.S. Ahmetbayev and oth. *Development of topological method for calculating current distribution coefficients in complex power networks*, Results in Physics, (2017).
5. D.S. Akhmetbayev *Topological method of calculating current distribution indices* (Worid of Science. Praha:IADE, 2010).
6. D.S. Akhmetbayev *Electricity*, **№ 11**, (2010).

E3S Web of Conferences 25, 04004 (2017) DOI: 10.1051/e3sconf/20172504004
RSES 2017

Evaluation of electrical equipment service life on the basis of its status index trend

Alexander Nazarychev[1]*, *Alexey* Tadzhibaev[1], *Dmitry* Andreyev[2]

[1] PPEIPD, Russia
[2] JSC «Zarubezhenergoproyekt», Russia

Abstract. This paper presents an approach designed to evaluate electrical equipment service life indicators, taking into account the changes of its technical status integral characteristic, i.e., of status index. This approach combines some previously developed methods of evaluating EE technical service life and its status index, and it allows to carry out a quantitative evaluation of electrical equipment service life depending on the actual accrued operating time of equipment, considering actual operating conditions and modes. This report also presents an example of application of this approach to evaluate the reliability of different types of equipment.

1 Introduction

In the capacity of the main index of electrical equipment (EE) service life Paper [1] proposes to use technical service life, which can most completely reflect the property of reliability.

EE service life is the total accrued operating time of EE following its operation start or its latest major repairs, till it reaches a limit state. Accrued operating time (a calendar-based operating time) can be determined by either the time length or the volume of energy facility operation [1]. A limit state is a condition, when any continued equipment operation must be terminated due to unavoidable violations of safety regulations or because of an unavoidable operation capacity reduction, or any impermissible operation level drop. Accrued operating time can be measured by both continuous units and/or discrete units of measurement. Applying the notion of accrued operating time, we shall undertake to solve the problem of evaluating EE service life, taking into account its technical status variations under different operating conditions.

2 Evaluation of electrical equipment service life indicators

Any actual service life of EE depends from its technical status (TS). The evolution of equipment TS variations is determined by influencing factors, operating conditions and modes [1]. In terms of quantity, the effect of EE operating conditions and modes on EE TS, for example, in the group of companies belonging to the PJSC "Rosseti", is determined by the status index (SI) value, which is defined on the basis of textual algorithms (TSI), applying major equipment operation parameters [1, 2, 3]. The most significant operation conditions are stipulated in normative-technical documentation (NTD). Fig.1 displays a diagram showing service life drawdown and the corresponding SI changes when EE is run under normative (standard) conditions (n.c.).

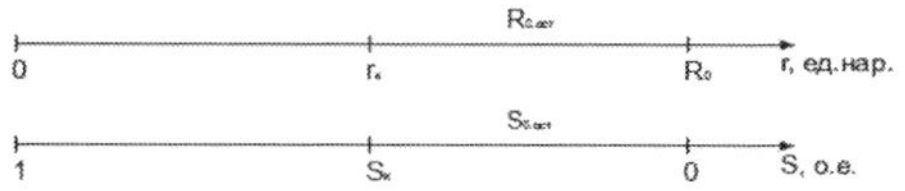

Fig. 1. A diagram showing EE service life drawdown and SI changes during operation under normative conditions. The units of measurement of used service life are given along Axis r, accrued operation time units (u.aot).

The R_0 moment on Axis r, which shows accrued operation time, corresponds to EE transition to a limit state (EE SI, designated by "S", is equal to 0). Then the accrued operation time interval in terms of its volume will be equal to EE normative service life R_0 , provided that EE was operated on the whole under n.c. stipulated in normative-technical documentation. Let us assume that EE was running under n.c. during an interval starting from 0 (the EE SI "S" is 1) up to a certain checking point r_{ch}. (the SI of EE is equal here to an intermediary value S_{ch}.) and therefore EE has used its service life in the volume of r_{ch}. In that case the remaining normative service life can be defined as $R_{0.rem} = R_0 - r_{ch}$., on the condition that, following the moment r_{ch} on the accrued operation time Axis r (Fig. 1), EE will also be running under n.c. In that case EE will have a spare SI $S_{0.rem}$ prior to its transition into a limit state, when the SI value is S = 0.

The wear and tear of EE is determined by its operation modes, by some external operation conditions and by influencing factors. Thus, if operation conditions differ from n.c., the service life value r_{ch} has to be

* Corresponding author: rector@peipk.spb.ru

corrected, taking into consideration EE operation modes and conditions. If it is known that within the accrued service period from r_{ch} till R_0 EE will also be running under conditions which differ from n.c., then the $R_{0.rem}$ value will have to be corrected, too, taking into account EE actual operating conditions. As a result of these corrections certain values of actually used service life R and of actually remaining service life R_{rem}. will be obtained. Figure 2 shows a diagram displaying the used service life and the corresponding status index changes (SI), if EE was run under different conditions which differ from standard ones.

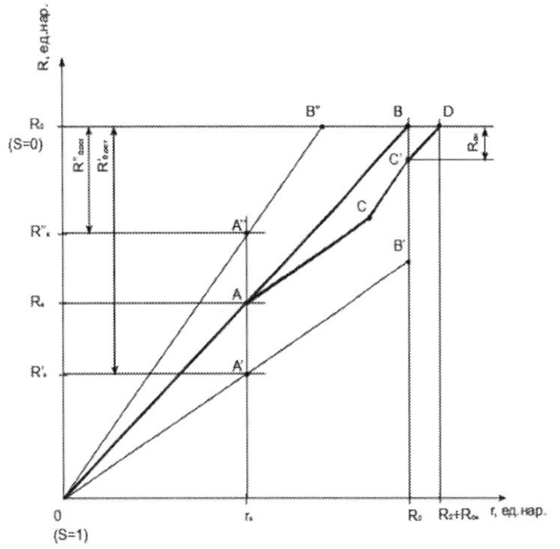

Fig. 2. A diagram showing EE used service life and the corresponding EE SI changes in cases, when equipment is operated under non-normative conditions.

The abscissa and the ordinate axes (ref. Fig. 2) are axes which present the service life drawdown. On the x-axis the accrued operation time of EE is marked in such a way, as if equipment were running under n.c., while along the axis of ordinates a corrected accrued operation time is marked, considering EE actual operation conditions. Therefore, the abscissa axis is an axis which shows a calendar-based operating time, while the axis of ordinates shows an actually accrued operation time. The origin of coordinates corresponds to the beginning of operation of a new piece of equipment, whose SI S=1 r.u.(relative unit). Let us then mark on the x-axis a certain calendar-based accrued operation time r_{ch}. This r_{ch} point is a point where the SI of EE is to be checked. If during the interval of accrued calendar-based operation time between 0 and r_{ch}. EE was on the whole running under n.c., then its TS was worsening, as this is stipulated in NTD, and that corresponds to the 0AB direct line. If equipment running conditions were of a lighter character (direct line 0A'B'), then equipment TS was worsening at a slower rate, if compared with its worsening rate under normative conditions (its SI will be higher, than it would be under n.c.). If operating conditions were on the whole of a heavier character (direct line 0A''B''), then, conversely, the said TS was worsening faster (the SI will be lower, than it would be under n.c). Consequently, along the y-axis the value of accrued operating time can be marked, which corresponds to the actually consumed service life R during the accrued operating time r_{ch}. under certain possible operation conditions. Besides, the $R \leq R_0$ relationship will always be true. Point A, with its coordinates being (r_{ch}., R_{ch}.), characterizes the accrued operation time r_{ch}. of EE in such a way, as if EE were running under n.c., and the corresponding actually used service life R_{ch}. In this instance the equality $R_{ch} = r_{ch}$ will be true. Point B corresponds to the beginning of EE limit state (the SI of TS is S = 0), provided that within the accrued operation time length $r = R_0$ EE was running under normative operation conditions only. Point A', with its coordinates being (r_{ch}, R'_{ch}), characterizes the accrued operation time r_{ch} of EE in such a way, as if it were on the whole running under lighter conditions. In this case, the following relation $R'_{ch} < r_{ch} < R_{ch}$ will be true. However, point 'B' does not correspond to the onset of any EE limiting state, because the actually used service life R'_{ch} has not reached yet the value R_0. Then, following the expiry of the accrued operation time $r = R_0$, the service life of EE can be extended and its above-standard operation can be maintained till the moment, when SI becomes equal to S = 0. Point A'', with its coordinates being (r_{ch}, R''_{ch}), characterizes the accrued operation time r_{ch} of EE in such a way, as if it were running under heavier conditions only, and also its corresponding actually used service life R''_{ch}. In this case the following relation $R''_{ch} > r_{ch} > R_{ch}$ will be true. Besides, Point B'' corresponds to the onset of EE limit state (SI TS S = 0), however, the actually spent service life R'_{ch} had reached the R_0 value faster, than that would have been under n.c.. Therefore, EE has to be decommissioned at an earlier time, than that would take place under n.c., i.e., when SI is S = 0, but that will happen before r reaches the R_0 value. If subsequently, after Point A, EE is going to be operated under normative conditions and operation modes, then its accrued operating time will afterwards correspond to normative values. In that case the remaining normative service life for Point A will be defined as follows: $R_{0.rem} = R_0 - R_{ch}$, while for Points A' and A'' it will be, accordingly, expressed as: $R'_{0.rem} = R_0 - R'_{ch}$ and $R''_{0.rem} = R_0 - R''_{ch}$. If though the conditions of further operation are going to differ from standard values, then the value $R_{0.rem}$ must be corrected, too, considering probable operation conditions, in order to determine the value of the actually remaining service life R_{rem}. When the actual accrued operating time reaches the value of the normative service life R_0, EE will pass on to its limit state and it has to be decommissioned.

Let us now examine the 0ACC'D broken line which characterizes the EE operation under different operating conditions. Prior to Point A, as it has already been mentioned above, EE had been running under n.c. From Point A till Point C equipment operation on the whole was under lighter conditions, while from Point C till point C' it was on the whole being carried out under heavier conditions. At the moment of time when accrued operating time r reached a normative service life value, EE had not really used it up completely, i.e., the actually used service life proved to be shorter, than the normative one. Therefore, EE operation can be kept on. In this case the remaining normative service life will be equal to EE service life in the conditions of above-standard operation R_{as}. If the said above-standard operation is going to be carried out under conditions, which differ from normative ones, then R_{as} should be corrected, taking into considering any planned out operation conditions.

For the sake of convenience, while making calculations and in order to compare the results of calculations concerning different EE types and voltage classes, and also so as to be able to consider the effect of operation conditions on the process of EE service life drawing down, it is expedient to pass on to relative units for measuring accrued operation time. In such a case the normative service life R_0 should be taken from NTD as the basic service life value for normative conditions. By relating all the other values of accrued operating time and service life values to R_0 it is then possible to obtain relative units for measuring EE accrued operating time.

The SI indicator "S" allows to make a quantitative evaluation of equipment TS and it is determined by means of calculations, considering the expert values of weight factors [3]. Besides, the SI value will be changing, depending upon the calendar-based accrued operating time, this why in a common case the said SI is a function of accrued operating time. The EE status will become a limiting one, when its SI S = 0, which also corresponds to the moment, when EE service life had been completely drawn down. The status of EE, when the SI value is S = 1, corresponds to the TS of a new piece of EE, whose operation has not yet started, while its actually spent service life is R = 0.

Evidently, the operating life of EE and, consequently, its technical service life, will proportionally depend upon the dynamic character of EE TS variations. That is to say, any changes of the actual EE technical service life over a certain calendar interval of accrued operating time are determined by the changes of EE SI.

In a common case, the value of SI can be presented in the form of the following function:

$$S = f(r, \{X_i(r)\}), \tag{1}$$

where r – is accrued operating time, $\{X_i(r)\}$ – is a multitude of parameters upon which SI depends.

Let us now examine an infinitely small interval of the dr calendar-based operating time, and to which the actually drawn down service life dR corresponds. If during its accrued operating time dr EE was running under n.c., then dR = dr. If, however, its working conditions differed from its normative conditions, then, taking into account the fact that dr tends towards zero, it is reasonable to assume that the relationship between the actually spent service life will be proportional to the relationship of EE SI over the dr interval. The SI function S is monotonically diminishing within the interval of time between repairs, and, accordingly, the relationship of EE SI variations during an infinitely small interval dr in case of normative and non-normative operation will tend towards the relationship of the SI value at the beginning of the dr operating time interval. Furthermore, during the said infinitely small interval of accrued operating time dr any $\{X_i\}$ parameter changes can be neglected. Then it is logically correct to write down the following expression:

$$\frac{dR}{dr} = \frac{1 - S(r)}{1 - S_0(r)}, \tag{2}$$

$$dR = dr\frac{1 - S(r)}{1 - S_0(r)}, \tag{3}$$

where $S_0(r)$ – is the function of EE SI variations under normative conditions; S(r) – is the function of EE SI changes under actual operation conditions.

Expression (3) is a differential equation of the first order. Let us integrate both its parts over the interval of accrued calendar-based operating time from r_1 till r_2:

$$R_{12} = \int_{r_1}^{r_2} \frac{1 - S(r)}{1 - S_0(r)} dr, \tag{4}$$

where R_{12} – is the actually drawn out service life over the interval of accrued operating time $[r_1; r_2]$, depending on the value of SI S(r).

Expression (4) is most commonly used to determine the actually consumed service life of EE over a prescribed accrued operation time interval, considering the EE TS evaluation made on the basis of the TS value. The formula used to calculate a specific accrued operating time can change depending on the way Function S(r)-SI is presented, and depending on the accrued operating time r.

For the moment of time, when EE operation is started, $r = r_1 = 0$. Let us assume that $r = r_2 = r_{ch.i}$ is a certain point of checking, for which it is necessary to determine the actually drawn down service life $R_{ch.i}$. If Functions S(r), $S_0(r)$ are known for the whole interval of calendar-based operating time $r \in [0; r_{ch.i}]$, i.e., it is known in what way EE TS was changing, then Expression (4) will acquire the following form:

$$R_{K.i} = \int_{0}^{r_{K.i}} \frac{1 - S(r)}{1 - S_0(r)} dr, \tag{5}$$

where $R_{ch.i}$ – is the actually drawn down service life of EE during its operating time from the beginning of equipment operation in the scope of r_{ch}.

If we assume that $r_{ch.i} = R_0$, then Expression (5) will acquire the following form:

$$R = \int_{0}^{R_0} \frac{1 - S(r)}{1 - S_0(r)} dr, \tag{6}$$

where R – is the actually drawn down service life of EE during its operating time in a scope, which corresponds to its normative service life R_0.

The actually used service life of EE $R_{ch.i}$ at the moment of checking $r_{ch.i}$ must always be compared with the value R_0. If $R_{ch.i} \approx R_0$, then the operation of EE has to be terminated. If $R_{ch.i} < R_0$, then EE operation can be continued. In this case the remaining service life of EE is to be determined and it must be determined, how long more it can still keep running. The value of the actually spent service life $R_{ch.i}$ at the moment of checking must not exceed the R_0 value. If during K-intervals the intensity of factors affecting EE is going to differ from normative values, then the actual summed up service life of EE, which has been used during these intervals of EE operating time Δr_{Σ}, can be calculated by the following expression:

$$\Delta R_{\Sigma} = \sum_{j=1}^{K} \int_{0}^{r_j} \frac{1-S(r)}{1-S_0(r)} dr, \quad (7)$$

where j = 1…K – is the number of intervals, during which some abnormal operating conditions can be observed; r_j – is a calendar-based accrued operating time during the j-th interval. And then: $\sum_{j=1}^{K} r_j \le R_0$.

Expression (7) is true in case the spent service life period of EE and, consequently, any changes of its TS do not depend on EE accrued operating time before the r_j interval, which is being examined.

Thus, any service life value possess an additivity property [1], whose essence lies in the fact the total actually spent service life ΔR_{Σ} consists of the summed up values of EE actually used service life at any single j-th section of accrued operating time r_j. The actually spent service life $R_{ch.i}$ during the accrued operating time in the scope of $r_{ch.i}$ will be determined as the sum of EE accrued operating time under normative Δr_0 and under non-normative operation conditions Δr_{Σ} :

$$R_{\kappa.i} = \Delta r_0 + \Delta r_{\Sigma}. \quad (8)$$

The values of ΔR_0 are determined by the following expression:

$$\Delta r_0 = r_{k.i} - \sum_{i=1}^{K} r_j. \quad (9)$$

By introducing Expressions (7) and (8) into (9) a formula can be obtained to determine the actually spent service life of EE $R_{ch.i}$ during a calendar-based operating time in the scope of $r_{ch.i}$:

$$R_{\kappa.i} = r_{k.i} - \sum_{i=1}^{K} r_j + \sum_{j=1}^{K} \int_{0}^{r_j} \frac{1-S(r)}{1-S_0(r)} dr. \quad (10)$$

Let us transform Formula (10) and we shall get then:

$$R_{\kappa.i} = r_{k.i} + \sum_{j=1}^{K} \left(\int_{0}^{r_j} \frac{1-S(r)}{1-S_0(r)} dr - r_j \right). \quad (11)$$

If $r_{ch.i} = R_0$, then the actually spent service life of EE R during the calendar-based operating time of EE in the volume of normative service life can be determined as follows:

$$R = R_0 + \sum_{j=1}^{K} \left(\int_{0}^{r_j} \frac{1-S(r)}{1-S_0(r)} dr - r_j \right). \quad (12)$$

Formulae (6) and (12) allow us to determine any actually spent service life during the period of normative EE operation (during the accrued operating time in the scope of R_0), while Formulae (5) and (10) allow to determine any actually spent service life during the calendar-based accrued operating time $r_{ch.i}$ till the i-th moment of checking. If EE had already drawn down a certain service life span $R_{ch.i}$ by the moment of checking $r_{ch.i}$, then we can determine the value of its remaining normative service life $R_{0.rem}$, if any subsequent operation of EE is going to take place under n.c., on the basis of the following expression:

$$R_{0.ocт} = R_0 - R_{\kappa.i} = R_0 - r_{\kappa.i} + \int_{0}^{r_{\kappa.i}} \frac{1-S(r)}{1-S_0(r)} dr \quad (13)$$

or (for the K-intervals of non-normative operation till the point of checking):

$$R_{0.ocт} = R_0 - R_{\kappa.i} = R_0 - r_{\kappa.i} + \sum_{j=1}^{K} \left(r_j - \int_{0}^{r_j} \frac{1-S(r)}{1-S_0(r)} dr \right) \quad (14)$$

If $r_{ch.i} = R_0$, i.e., checking is undertaken, when the calendar-based accrued operating time r is equal to the normative service life R_0. In this case Expressions (13) and (14) will, accordingly, acquire the following form:

$$R_{0.ocт} = \int_{0}^{R_0} \frac{1-S(r)}{1-S_0(r)} dr \quad (15)$$

$$R_{0.ocт} = \sum_{j=1}^{K} \left(r_j - \int_{0}^{r_j} \frac{1-S(r)}{1-S_0(r)} dr \right) \quad (16)$$

Under certain known, non-standard conditions of further EE operation the value of $R_{0.rem}$ has to be corrected, considering the forecast TS value of EE according to Expression (4), by adopting $r_1 = R_{ch.i}$ (the current point of checking), and $r_2 = R_0$:

$$R_{ocт} = \int_{R_{\kappa.i}}^{R_0} \frac{1-S(r)}{1-S_0(r)} dr, \quad (17)$$

If necessary, it is also possible to apply Expression (11) by introducing $r_{ch.i} = R_{0.rem}$.

 DOI: 10.1051/e3sconf/20172504004

$$R_{ocr} = R_{0.ocr} + \sum_{j=1}^{K}\left(\int_{0}^{r_j}\frac{1-S(r)}{1-S_0(r)}dr - r_j\right). \qquad (18)$$

After making this calculation, it is necessary to evaluate the difference between R_{rem}. and $R_{0.rem}$ so as to be convinced that $R_{rem} \leq R_{0.rem}$. In the opposite case, EE operation, with due account taken of its TS forecast value, should not be continued during the entire interval of the remaining accrued calendar-based operating time $r = R_{0.rem}$. EE must then be operated either under lighter operation conditions, which will allow to reduce the tempo of its TS worsening and thus to extend the EE service life, or the period of equipment operation will have to be reduced by the value of (R_{rem}-$R_{0.rem}$). Since it is rather a complicated matter to forecast the operation conditions and modes of EE for any future periods, it is assumed that EE, after the point of checking, will be running in the same mode, as its mode had been before the said checking point, and in that case its TS and SI will be changing in a similar way.

For the sake of convenience, expressions used to determine the actually spent service life of EE can be described by means of relative units (r.u.), to measure EE accrued operating time; for example, Expression (3) will then have the following form:

$$R_{12}^{*} = \int_{r_1^*}^{r_2^*}\frac{1-S(r)}{1-S_0(r)}dr^{*}, \qquad (19)$$

where $R^*_{12} = R_{12}/R_0$ – is the relative, actually used service life of EE during its accrued operating time $[r^*_1;\ r^*_2]$, depending upon its SI $S(r^*)$; $r^*_1 = r_1/R_0$; $r^*_2 = r_2/R_0$.

Let us examine the possibility to determine EE service life, if all the initial data are available. The graphs shown in Figures 3 and 4 are not linked to any specific EE type, they just display the potential of the currently developed approach.

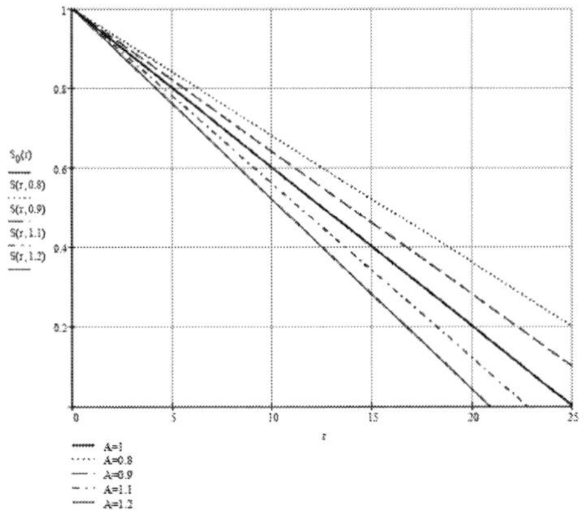

Fig. 3. Dependences of the values of SI S(r) of EE from its accrued calendar-based operating time r, for different Factor A values The value of A = 1 corresponds to EE normative operation conditions.

By applying a linear sort of approximation for EE SI [1, 2] the value of actually used service life, taking into account any changes of SI S(r), will be determined in the following way:

$$R_{к} = \int_{0}^{r_к}\frac{1-S(r)}{1-S_0(r)}dr = \int_{0}^{r_к}\frac{m}{m_0}dr = Ar_{к}, \qquad (20)$$

where $r_{ch.}$ – is the accrued calendar-based operating time of EE at the point of checking; $R_{ch.}$ – and it is the actually consumed service life period at the point of checking.

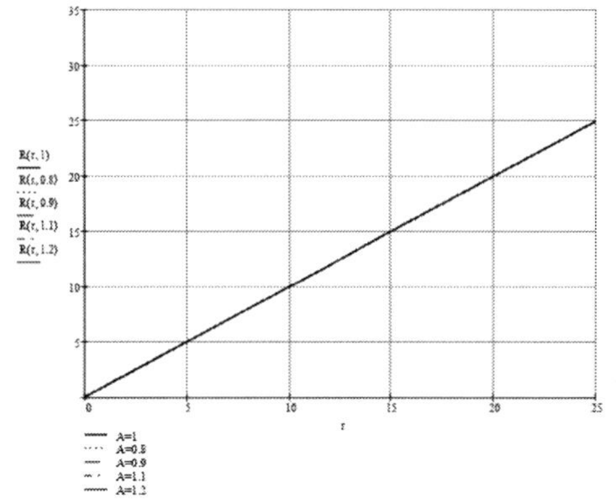

Fig. 4. Dependences of the actually spent service life R of EE from its accrued calendar-based operating time, for different values of Factor A. The value of A = 1 corresponds to n.c.

In Formula (20) Factor A expresses the relation of EE SI, when EE is operating under certain actual conditions, to the value of EE SI, under n.c., with the accrued operating time here having the volume r_{ch}. If A > 1, then the TS of EE will be worsening faster (under heavier conditions), if A < 1, then EE TS will be worsening at a slower rate (under lighter conditions). If A = 1, then EE is running under n.c.

Figure 3 presents the dependences of EE SI variations from the accrued calendar-based operating time r for different A values. Fig. 4 shows the dependences of the actually used service life R from the accrued calendar-based operating time r for different values of A. These dependences are formed on the basis of Expression (20). The A factor for certain concrete EE can be formed on the basis of its operation process data. The examples of initial data, received from one of the Inter-Regional Network Companies in order to obtain Factor A, are presented in Table 1.

Table 1. Approximation Factor Values of the Function of SI Variations.

EE Type	EE Model	m	m_0
Transformer	ТМН(TMN)-6300/110/10	–0.02670	–0.04
Station transformer	ТМГСУ(TMGSU)11-160/10	–0.02280	–0.04
Circuit breaker 10 kV	ВМП(VMP)-10П(P)-630-20	–0.01850	–0.04
Circuit breaker 10 kV	VD-4(VD-4)	–0.14570	–0.04
Circuit breaker 110 kV	ВМТ-110Б-25/1250 (VMT-110B-25/1250)	–0.02230	–0.04
Circuit breaker 110 kV	ВМТ-110Б-25/1250УХЛ1 (VMT-110B-25/125UKhL1)	–0.01170	–0.0357
Tripper 10 kV	РВЗ-10/400 I (RVZ-10/400 I)	–0.01350	–0.0333
Tripper 10 kV	РЛНД-10/630 (RLND-10/630)	–0.01010	–0.04
Tripper 35 kV	РНТА-35/125 (RNTA-35/125)	–0.0081	–0.04
Current transformer	ТФНД-110М-II (TFND-110M-II)	–0.02300	–0.0333
Voltage transformer	ЗНОЛП-10 У2 10000/100 (ZNOLP-10 U2 10000/10)	–0.16670	–0.04
Voltage transformer	НАМИТ-10-2УХЛ2 (NAMIT-10-2UKhL2)	–0.05880	–0.04

Let us consider some operation examples concerning the EE, which is presented in this report.

Example 1. Let us assume that some EE has been in operation during seven years under n.c. of operation. In this case its actually consumed service life length will also be equal to seven years. If that EE were running during seven years under heavier conditions, with A being equal to 1.1, then in reality it would draw down a period of 7.7 years. If EE during that period of seven years were running under lighter conditions, with A being equal to 0.8, then it would spend 5.6 years. If later on EE is going to be operated under n.c., then its remaining normative service life will be defined by Expression (13) and, accordingly:

$$R_{0\,rem.} = 25 - 7 = 18 \text{ years;}$$

$$R_{0\,rem.} = 25 - 7.7 = 17.3 \text{ years;}$$

$$R_{0\,rem.} = 25 - 5.6 = 19.4 \text{ years;}$$

If after that it is scheduled to operate that EE under certain conditions which will differ from its normative ones, then the value $R_{0\,rem.}$ will have to be corrected.

Example 2. Let us imagine that some EE during two years, following the start up of its operation, was running under normative conditions, and after that during the next three years it was running under lighter conditions, with A being equal to 0.9, and then during the following five years it was running under heavier conditions, with Factor A = 1.2. In that case the actually consumed service life of that EE, bearing in mind the additivity property referred to earlier (ref. to Expressions (7)), will be determined as the sum of actually spent service life during any single interval of its accrued calendar-based operating time:

$$R_к = 2 \cdot 1 + 3 \cdot 0.9 + 5 \cdot 1.2 = 10.7 \text{ years.}$$

Therefore, during the ten years of (2 + 3 + 5) of its accrued calendar-based operating time, that EE actually had used up 10.7 years of its service life. If later EE is going to be operated under normative conditions, then its remaining normative service life can be determined by the following expression (13):

Normative conditions:

$$R_{0.ост} = 25 - 10.7 = 14.3 \text{ years.}$$

If further on it is scheduled to operate EE under those conditions which will differ from its normative ones, then the value $R_{0.rem.}$ has to be corrected by applying Expression (17). If, for instance, it is scheduled to run EE under lighter operating conditions (with A = 0.8), then:

$$R_{ост} = \int_{R_{к.i}}^{R_0} \frac{1-S(r)}{1-S_0(r)} dr = \int_{10.7}^{25} A dr = \int_{10.7}^{25} 0.8 dr = 11.44 \text{ years.}$$

In that case we shall have the following relationship: $R < R_{0.rem.}$ (11.44 < 14.3), and, consequently, equipment operation can be prolonged by 14.3 – 11.44 = 2.86 years. If after that period EE is going to be run under service conditions which will differ from its normative ones, then the value of 2.86 years will have to be corrected too, in order to find out how long more it will be capable of running. If its operation conditions are going to be heavier, then EE will be able to run less time, not during 2.86 years. If, for example, A = 1.1, then EE will have used up the period of 2.86 after 2.6 years of its accrued calendar-based operating time:

$$2.86 = \int_{10.7+11.44}^{r_{кн}+10.7+11.44} 1.1 dr = 1.1 r_{кн} \Rightarrow r_{кн} = \frac{2.86}{1.1} = 2.6 \text{ years.}$$

Summing it up, EE will be able to run during 10.7 + 11.44 + 2.6 = 24.74 years, before it completely exhausts its service life and can pass into a limit state.

The examples presented herein prove the possibility to calculate EE technical service life, taking into account the well-known function of EE SI variations, depending on the accrued calendar-based operating time of electrical equipment.

DOI: 10.1051/e3sconf/20172504004

3 Conclusions

An approach has been worked out to evaluate the values of EE spent service life, with account taken of the changing nature of its integrated characteristic, i.e, that of its status index. This approach combines also some previously developed methods of evaluating equipment service life and status index, and it allows to evaluate the numerical values of EE used up service life as a dependence of its actual accrued operating time, taking into account its real operation conditions and modes, and this is supported by the examples given herein.

Therefore, during EE operation in practical applications it is necessary to correct EE normative service life, considering concrete operation conditions and modes, and in such a way to obtain actual data concerning both the EE accrued and remaining operating time. The said calculation of service life length on the basis of the trend of status index variations allows us to assess how EE had actually been running under concrete operation conditions and how long it will still be able to work under predicted conditions.

References

1. D.A. Andreyev, A.N. Nazarychev, A.I. Tadzhibaev. *Determination of the Probability of Equipment Failures in Power Network Companies on the Basis of Technical Status Evaluation* (FSAEE «PPEIPD». – St.Pb., 2017. – 194 p).

2. R.N. Berdnikov, D.B. Gvozdev, I.A. Kuzmin, A.N. Nazarychev, D.A. Andreyev, A.I. Tadzhibaev. *Collected Science-Technical Reports by a Team of Employees of the "ROSSETI" Group of Companies»*. (M.: "Electric Energy. Transmission and Distribution", p.p. 151 – 163. 2017).

3. A.N. Nazarychev, E.V. Novomlinskiy, D.A. Andreyev. *The Methodological Aspects of Investigations Concerning Large Energy Sector Systems. Issue 67. "The Problems of Reliability of Energy Sector Systems"*. (Syktyvkar: The Komi Republican Printing Press Publishing House", 2016. – p.p. 171 – 179).

Residual life assessment of overhead transmission lines 110 kV and above and determination of their reconstruction terms

Bauyrzhan Uteuliyev [1]*

[1]KEGOC JSC, Astana, Republic of Kazakhstan

Abstract. In this article is given a method for assessment the residual life of overhead transmission lines on reinforced concrete supports with centrifuged poles. Reinforced concrete poles of supports, wires and lightning protection cables are adopted as the main elements. The intensity of change in the actual state parameters of these elements is determined by the laws of random variables distribution that allow predicting the residual life and the timing of repairs and reconstruction of overhead transmission lines. The parameters of natural climatic conditions and other external factors are considered by including of coefficients into the formula for changing the actual state parameters.

1 Introduction

Practically, the reinforced concrete supports with centrifuged poles have a long service life of more than 40 years at all electric grid organizations who have an OHTL of 110 kV and higher.

With such an amount of outdated OHTL, electric grid organizations are trying to plan reconstruction considering their technical condition, but limited financial resources and network mode do not allow complete replacement of worn out components of OHTL. It should be noted that at first glance the deteriorated technical condition of the OHTL not always requires reconstruction, many defects can be eliminated by repair and OHTL can serve for several tens of years. In such cases, works on the OHTL reconstruction must be planned with consideration of the limiting state prediction, i.e. assessment of its residual life.

2 Life assessment of OHTL

The possibility of predicting the life assessment of objects is provided with simultaneous presence of the following conditions [8]:

1) the parameters determining the actual condition of the structure are known;

2) the criteria for the limiting state of the structure are known;

3) it is possible to periodically (or continuously) monitor the values of the technical state of the structure.

For the parameter of the actual state takes the characteristic of damage, defect or malfunction, changing during operation, whose numerical values exceed the threshold established by normative-technical documentation. The parameter of the actual state is monitored during the operation of the OHTL by diagnosing its elements.

For OHTL we take three levels of categories of technical condition [2]:

1) the first category is a workable state in which the change of the parameter of the actual state characterizes the breaking of the OHTL operational capability or a single failure case that slightly reduces reliability;

2) the second category is a critical state, in which frequent failures occur with high probability, which significantly reduce reliability;

3) the third category is the limiting state, in which the change of the parameter of the actual state leads to numerous failures, which restoration is costly.

The concept of the limiting state corresponding to the exhaustion of the resource allows for a different interpretation. In some cases, the reason for the termination of operation is the moral wear, in others - an excessive decrease in efficiency, which makes the further operation of the economic inexpedient, and thirdly - the reduction of safety indicators below the maximum permissible level. [1]. For OHTL with consideration of ongoing repairs and replacement of worn-out elements, it is advisable to adopt a state corresponding to the wear limit when the further operation of the OHTL is economically inexpedient.

As the actual state parameter of reinforced concrete centrifuged poles, we take the coefficient of concrete's state, derived by calculation and determined in the diagnosis of poles. For the actual state parameter of the wires and lightning protection cables, we take the fractional loss of strength, calculated from the results of diagnosing the conductor structure [3, 4].

For each category of the actual state parameter, determine the admit boundary values according to Table 1.

* Corresponding author: Uteuliyev@kegoc.kz

Table 1. Characteristics of the actual state parameter of the OHTL

Category of states	Level of functional ability	Reinforced concrete supports (Coefficient of concrete's state)	Wires and lightning protection cables (Fractional loss of strength, %)
First category	Workable state	$0 \le x_{\phi 1} \le 0{,}3$	$0 \le x_{\phi 1} \le 5$
Second category	Critical state	$0{,}3 \le x_{\phi 2} \le 0{,}6$	$5 \le x_{\phi 2} \le 10$
Third category	Limiting state	$x_{\phi 3} > 0{,}6$	$x_{\phi 3} > 10$

The experience of OHTL maintenance shows that in the process of their operation there is a diverse nature of the changing process of the actual state parameter in real conditions during the time that characterizes a workable or inoperative state.

During the operation in the survey of the OHTL is measure the value of the actual state parameter at a fixed moment of time. The results of the OHTL survey show that the change of the actual state parameter tends to increase during operation time. Naturally, the highest intensity of change of the actual state parameter is possessed by OHTL with a service life more than 30-35 year.

For the mathematical model of the change of the actual state parameter during operation, it is necessary to consider the natural and climatic conditions and other external factors. The impact of natural and climatic conditions and other external factors is considered by the introduction of the climatic coefficient k based on data [5, 7, 9, 10]. The value of this coefficient is determined by multiplying the coefficients from the following tables 2-6, depending on the region in question.

Table 2. Values of the coefficient for ice

Area on ice	Normative ice wall thickness, mm	Coefficient k_z
I	10	1
II	15	1,04
III	20	1,07
IV	25	1,1
V	30	1,15
VI	35	1,17
VII	40	1,19
Special	above 40	1,2

The coefficient of ice k_z is corrected with multiplying by the coefficient k_i, considering the change in height over the earth's surface, and by a coefficient k_d that considering the change in the thickness of the ice wall, depending on the diameter of the wire [11].

Table 3. Values of the coefficient for wind

Area by wind	Normative wind speed at a height of 10 m above the ground, m/s	Coefficient $k_в$
I	25	1
II	29	1,04
III	32	1,07
IV	36	1,1
V	40	1,15
VI	45	1,17
VII	49	1,19
Special	above 49	1,2

The coefficient of wind $k_в$ is corrected with multiplying by the coefficient k_w that considering the height of the location of the gravity reduced center of the wires (cables) and the middle points of the structures zones of the OHTL towers above the ground [11].

Table 4. Values of the coefficient for absolute minimum air temperature

Climatic region	The average of the absolute annual minimum of air temperature, ° C	Coefficient k_m
Very cold	below -60	1,2
Cold with a very humid climate	from -60 to -45	1,19
Cold with a cold summer	from -60 to -45	1,15
Cold with a warm summer	from -60 to -45	1,1
Moderately cold	from -45 to -40	1,07
Moderate	from -40 to -25	1,03
Moderately warm	above -25	1

Table 5. Values of the coefficient of relative humidity

Average monthly relative air humidity, ° C	Coefficient $k_{ов}$
30	1
40	1,03
50	1,07
60	1,1
70	1,13
80	1,17
90	1,2

Table 6. Values of the coefficient of soil salinity

Types of soil	The content of water-soluble salts in soils, %	Coefficient k_n
Non-saline	up to 0,5 inclusive	1
Weakly saline	from 0,5 to 1,5 inclusive	1,1
Salted	over 1,5	1,2

2.1 The intensity of the change of the actual state parameter

To select the distribution law, a variation of the developments values [11, 17] was taken on the failure of reinforced concrete supports, wires and lightning protection cables. Empirical characteristics have shown that the failures distribution of reinforced concrete supports obeys the normal distribution law of random variables, and the wires and lightning protection cables – to the Weibull distribution.

In actual conditions, the change of the actual state parameter is difficult to observe continuously, and the control is performed periodically. At each moment of the next control, one of two solutions are possible: to continue operation or perform a repair. This decision is made based on an analysis of the intensity of the change of the actual state parameter.

The intensity of the change of the actual state parameter is determined by the formula

$$\mu = \frac{f(t)}{F(t)} \tag{1}$$

where $f(t)$ - the probability density of the change in the value of the actual state parameter in time t;

$F(t)$ - the probability of failure-free operation in time t.

In the normal distribution law, these two quantities are determined by the formulas

$$f(t) = \frac{1}{\sigma\sqrt{2\pi}} exp\left(-\frac{(t_i - T)^2}{2\sigma^2}\right) \tag{2}$$

$$F(t) = \frac{1}{\sigma\sqrt{2\pi}} \int_{-\infty}^{t} exp\left(-\frac{(t_i - T)^2}{2\sigma^2}\right) dt \tag{3}$$

where t_i - is the time of trouble-free operation (independent variable);

T - average predicted failure time (mathematical expectation), years;

σ - standard deviation.

In the Weibull distribution law, $f(t)$ and $F(t)$ are determined by formulas

$$f(t) = \frac{b}{T}\left(\frac{t_i}{T}\right)^{b-1} exp\left(-\frac{(t_i)^b}{T}\right) \tag{4}$$

$$F(t) = 1 - exp\left(-\frac{(t_i)^b}{T}\right) \tag{5}$$

where b - is the form parameter.

Given the permissible values of the actual state parameter, it is possible to determine with a high probability the transition time of state from one category to another. In this case, the stochastic process for OHTL with a service life of more than 30 years will look as follows [2]

$$X(t) = x_{\phi} + k \cdot t_i \cdot \mu \tag{6}$$

where x_{ϕ} - the average value of the actual state parameter recorded during the survey.

k - climatic coefficient, considering the features of changing of the actual state parameter in the region under consideration;

t_i - number of years under consideration, $t_i = 1,2,3...$

μ - the intensity of the change of the actual state parameter.

2.2 Predicting residual resource of OHTL

When predicting the residual resource of OHTL structural components, with consideration of the change of the actual state parameter of its elements, the following calculations are made for plotting the graph:

1) determine the average value of the actual state parameter of the elements;

The average value of the actual state parameter of reinforced concrete supports is determined by the expression [16]

$$x_{\phi.on} = \frac{\sum_{i=1} n_{oni} \cdot K_{ci}}{N_{on}} \tag{7}$$

where n_{oni} - the number of poles with the same coefficient of concrete's state;

K_{ci} - coefficients of concrete's state;

N_{on} - total number of surveyed poles.

Instead of the average value of the actual wire condition parameter (lightning protection cables), we take the maximum value of the section loss in the surveyed section

$$x_{\phi.n.m} = max\,(p_{ci}) \tag{8}$$

where p_{ci} - the loss of the cross-section of wires (lightning protection cables).

2) determine the average operating time of the OHTL for failure [6, 11, 17];

3) determine the intensity of the change of the actual state parameter (for reinforced concrete supports according to the normal distribution law, for the conductor structure according to the Weibull distribution law);

4) determine the trajectory of the change of the actual state parameter;

5) determine the average value of the actual state parameter of the elements for subsequent stages, consider-

 DOI: 10.1051/e3sconf/20172504005

ing the residual wear, which is determined by the following formula

$$x_{ocmi} = x_{\phi 1} \cdot \frac{\sum n_0}{N} \cdot \frac{t_n - t_0}{t_0} \quad (9)$$

where $x_{\phi 1}$ - the actual initial mean value of the actual state parameter at the time of the survey, determined by formula (3);

n_0 - the number of defective items at the time of survey;

N - the total number of elements on the OHTL;

t_n - predicted time when the actual state parameter reaches the boundary value;

t_0 - time of the survey.

6) determine the distribution densities of each actual state parameter according to the normal distribution law;

7) determine the period of carrying out repairs or reconstruction at the highest peak of the distribution densities for each drift of the actual state parameter.

Figures 1 and 2 show the results of calculating the residual life of the OHTL 220 kV, examined at a service life of 41 years.

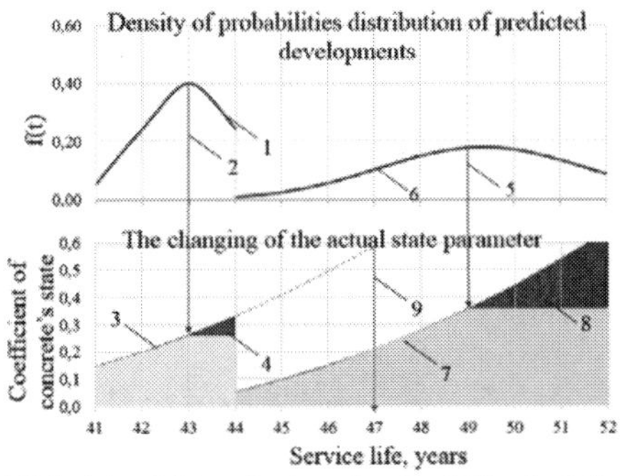

Fig. 1. Schedule of determination of residual life of reinforced concrete supports.

In Fig. 1 the following designations are accepted:

1 - density of probabilities distribution of predicted developments for repair;

2 - the projection of the distribution density peak on the trajectory of the drift of the actual state parameter, which determines the beginning of the repair;

3 - drift trajectory of the actual state parameter for repair;

4 - the period of repair;

5 - density of probabilities distribution of predicted developments for reconstruction;

6 - the projection of the distribution density peak on the trajectory of the drift of the actual state parameter, which determines the beginning of the reconstruction;

7 - drift trajectory of the actual state parameter for reconstruction;

8 - the period of reconstruction.

9 - the projection of the distribution density peak on the axis, which determines the guaranteed life of reinforced concrete supports.

By increasing the number of repairs, it is possible to extend the residual life of reinforced concrete supports.

If we compare the peaks of the distribution densities for each drift of the actual state parameter in Fig. 1, then it is possible to predict the average time for the onset of a OHTL failure. The predicted failure will occur in the region of decline in the distribution density, which is colored in a dark color on the graph. Accordingly, it is necessary to carry out the first repair to replace the supports with a coefficient of concrete's state of 0.6 during the service life of 43-44 years.

The predicted residual life of reinforced concrete supports is 52 years. Therefore, with a service life of 49 years corresponding to the peak of the probability distribution density, it is necessary to begin reconstruction of the OHTL 220 kV with the replacement of reinforced concrete supports.

According to [8], the guaranteed life of reinforced concrete supports was 47 years, which is marked by a green arrow in Fig. 1, i.e. the trajectory of the first predicted drift of the actual state parameter when it reaches the limit state is a guaranteed resource.

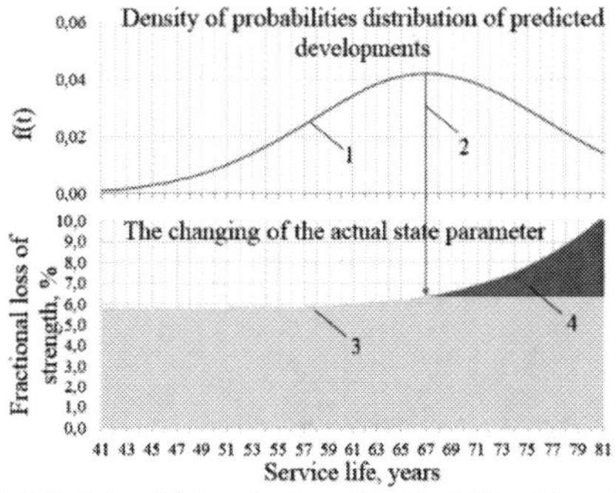

Fig. 2. Schedule of determination of residual life of the wire.

In Fig. 2 the following designations are accepted:

1 - density of probabilities distribution of predicted developments for replacement of a wire;

2 - the projection of the distribution density peak on the trajectory of the drift of the actual state parameter, which determines the beginning of the replacement of the wire;

3 - drift trajectory of the actual state parameter to replace the wire;

4 - the period of the replacement of the wire.

According to the calculated data, shown in Fig. 2, the residual life of the wire was 81 years. Therefore, with a service life of 67 years corresponding to the peak of the probability distribution density, it is necessary to begin replacement of the wire throughout the entire OHTL 220 kV.

Similarly, the calculation was made for a lightning protection cable, the residual life of which was 80 years, and with a service life of 66 years, it is required to begin the replacement of the lightning protection cable throughout the entire OHTL 220 kV.

3 Conclusion

When predicting the residual resource and determining the reconstruction period of the OHTL, choose the worst-case version of the limiting state of the OHTL elements. The above calculations showed that the worst option for achieving the start of reconstruction are reinforced concrete supports. Therefore, the reconstruction of the OHTL 220 kV must begin with a service life of 49 years, subject to repairs with the replacement of reinforced concrete supports. The number of necessary repairs before reconstruction should be determined by the decision of the operating organization.

Thus, it is possible to determine the planned timeframes for the reconstruction of all surveyed OHTL, which will make it possible to draw up a perspective schedule for the OHTL reconstruction of electric grid organizations.

References

1. WD 26.260.004-91, *Methodical instructions. Predicting the residual life of the equipment by changing the parameters of its technical state during operation*, (M.: NIIChimmash, 1991).
2. G.M. Borovik, *Modeling and predicting of operational reliability indicators of artificial structures in conditions of severe climate*, (Khabarovsk: Publishing house of DVGUPS, 2005).
3. V.V. Bolotin, *Predicting the resource of machines and structures*, (M: Mechanical engineering, 1984).
4. V.Yu. Volokhovski, A.N. Vorontsov, V.V. Sukhorukov, V.V. Cukanov, V.V. Chernetsov. *Energoexpert*. **3**, p. 62-67. (2014).
5. V.Yu. Volokhovski, A.N. Vorontsov, V.V. Sukhorukov, V.V. Cukanov, V.V. Chernetsov. *Energy of a single network*, **1**, p. 42-51. (2016).
6. State St 25870-93, *Macroclimate regions of the globe with a cold and temperate climate. Zoning and statistical parameters of climatic factors for technical purposes*, (M.: Publishing house of standards, 1983).
7. *Rules for the installation of electrical equipment*, (7th Edition. – OAO «VNIIE», 2003).
8. WD 34.51.101-90, *Instructions for choosing insulation of electrical equipment*, (NIIPT, Energosetproekt, 1990).
9. Building Regulations of KR 2.04-01-2010, *Building Climatology*, (Almaty: KazNIISSA, 2011).
10. E.V. Sugak, *Theory of random processes: a tutorial*, (Krasnoyarsk: KFAGA, 2004).
11. V.N. Shklyar, *Reliability of management systems: a tutorial*, (Tomsk: Publishing house of TPTU, 2009).
12. B.A. Uteuliyev *Energetic*, **3**, p. 13-17. (2017).
13. S.A. Egorov, N.E. Egorova, *Laws for the distribution of time between failures: Guidelines for practical exercises*, (Ivanovo.: Ivanovo State Textile Academy, 2011).

AUTHOR INDEX